JASON WILKES

BURN MATH CLASS

And Reinvent Mathematics for Yourself

BASIC·BOOKS

A Member of the Perseus Books Group
New York

Library of Congress Control Number: 2015956108
ISBN: 978-0-465-05373-5 (hardcover)
ISBN: 978-0-465-07381-8 (e-book)

10 9 8 7 6 5 4 3 2 1

Contents

Act II

Act III

Dedication

This is harder than I expected. Let's see...

Let $J \equiv$ "always being himself."
Let $W \equiv$ "the same reason."
Let $R \equiv$ "all the help. Hopefully that will make sense eventually."

Okay...

$\frac{\alpha}{2}$(To E. T. Jaynes, for J)
$\frac{\alpha}{2}$(To David Foster Wallace, for W)
$(1 - \alpha)$(To Reader, for R)
where $\alpha \in [0, 1]$ is determined by the final dedicatee.[1]

$\dots ($

---

1. I promise the book isn't this confusing.

Preface

Good fiction's job [is] to comfort the disturbed and disturb the comfortable.
—David Foster Wallace, interview with Larry McCaffery

Fiction and nonfiction are not so easily divided.
—Yann Martel, *Beatrice and Virgil*

Burn Math Class

Okay, don't really burn math class. Or anything. Arson is mean, and extremely illegal. I...No, nevermind. I don't want to start the book like that.

(*Author thinks for a moment.*)

Alright, I think I've got it. Sorry about that.

(*Ahem.*)

We should all be very angry. Something beautiful has been stolen from us, but we've never felt its absence because the theft happened long before we were born. Imagine that by some massive historical accident, we had all been convinced that music was a dull, tedious, rigid enterprise to be avoided except when absolutely necessary. Suppose that we had all attended music classes for more than a decade during our youth, and that through some feat of brilliant sadism among the instructors, we all left these classes with a firm belief that music is at most a means to an end. We might agree that everyone should have a basic level of familiarity with the subject, but only for reasons of practicality: you need music because it might — rarely — help you with *other things*. But the consensus view would hold that music more closely resembles plumbing than an art form.

The world would still be full of artists, of course. Just as it's full of them now. By artists, I don't necessarily mean art-school students, or professional artists, or the guy who wrote some stuff on a toilet and put it in a museum. I mean the people who pull things into existence out of nothingness; who refuse not to be themselves; who have their own way of fracturing reality that's so authentic you can feel it in your nerve endings; who enter the world on fire and too often die young. "Music is not for them," we'd agree. "Music is for the accountants among us, and it's best if we leave it to them." As far-fetched as such a situation might seem, this is exactly what has happened to mathematics. Mathematics has been stolen from us, and it is time we take it back.

With this book, I am advocating a process of conceptual arson. The state of mathematics education all over the world has degenerated to a point where it no longer makes sense to do anything but burn it all down and start over. We begin by doing just that. In this book, mathematics is not approached as a preexisting subject that was created without you and must now be explained to you. Beginning on the first page, mathematics does not exist. We invent the subject for ourselves, from the ground up, free from the historical baggage of arcane notation and pretentious terminology that haunts every mathematics textbook. The orthodox terminology is mentioned throughout, and used when it makes sense to do so, but the mathematical universe we create is entirely our own, and existing conventions are not allowed in unless we explicitly choose to invite them.

The result is an approach that requires zero memorization, encourages experimentation and failure, never asks the reader to accept anything we have not created ourselves, avoids fancy names that hide the simplicity of the ideas, and presents mathematics like the adventure it is, in a conversational form that could easily be read as if it were a novel. While the primary goal of our journey is hedonism rather than practicality, we are lucky to find no conflict between the two. You will actually learn the subject — a lot of it — and learn it well.

When one attempts to construct a narrative through mathematics that does not ask the reader to accept facts established elsewhere, it is impossible not to notice what I think is the fundamental tragedy of existing mathematical pedagogy — a tragedy that is never mentioned, even in the harshest criticisms of orthodox educational practice:

> We have been teaching the subject *backwards*.

Let me explain what I mean with a story. I got a C in basic algebra. All I learned was to hate the word "polynomial." I got a C in trigonometry. All I learned was to hate the words "sine," "cosine," and "hypotenuse." Mathematics had never been anything but memorization, boredom, and arbitrary authority — these are a few of my least favorite things. By my senior year of high school, I had completed all of the required mathematics courses, and I can't describe how happy I was that I could now die without ever setting foot in a mathematics classroom again. Free at last.

One night during my senior year, I was hanging around in a bookstore, as I often did, and I saw a book on calculus. I had always heard that calculus was pretty difficult, but I had never taken a course in it, and I would never have to... a relaxing thought. This lack of any obligation to learn the subject somehow made the book seem more appealing, so I thought I would flip through it for a few seconds. I expected to see some scary symbols, think "Yep, that sure looks difficult," and then put the book down and be done with it forever.

But when I opened the book, it wasn't just the usual garbage. In completely honest and unpretentious language, the author was saying something like this: Straight things are easier to deal with than curvy things, but if you zoom in far enough, then each tiny piece of a curvy thing almost looks like a straight thing. So whenever you have a curvy problem, just imagine zooming in until things look straight, solve the problem down there at the microscopic level where it's easy, and then zoom out. You've solved the problem. You're done.

Anyone can understand this idea, and it has nothing to do with mathematics. If you've got a hard problem, then break it up into a bunch of easy problems, solve those, and put them back together. The idea had a feel of elegance and necessity that I had never experienced in a math class. I flipped through the book some more, and when I saw that it had a section where the author complained about the way mathematics is usually taught, I knew this guy was my kind of person.

So I bought the book, and I started reading it whenever I didn't have anything else to do. I liked the way the author wrote. It gave me an odd sense of justification for having always disliked mathematics in school, while at the same time convincing me that I had been entirely wrong about the subject. I wasn't planning to learn calculus, and I didn't remember any of the prerequisites from high school, so I didn't even know how to solve the "easy problems" down at the microscopic level. But that didn't matter, because I was free from all the restrictions of formal education, and there was no one to punish me for doing things wrong.

Thus began my strange journey of learning calculus before I knew algebra, trigonometry, what a "logarithm" is, or any of the other stuff they say you have to learn before calculus. I bought a notebook and started playing around. Whenever I didn't understand something, I'd draw a picture and try to convince myself that it was true. I usually didn't succeed in doing this.

Weirdly, the calculus concepts were by far the simplest parts of the book. Much harder were the so-called "prerequisites" to calculus: the algebra, trigonometry, and other conceptual packing peanuts with which modern high school courses are filled. All the stuff about zooming in made sense to me: derivatives and integrals were not only simple computationally but easy to understand from first principles. From their pre-mathematical motivation to their definitions to the methods of computing them, there was a coherent and well-motivated narrative tying everything together. But once in a while the author would make use of things that were supposed to be more "basic" — things I couldn't understand at all, though I vaguely remembered hearing them from some teacher in a quiet, boring classroom. I couldn't for the life of me figure out where all those supposedly simple things came from, like the area of a circle, or the pile of unexplained "trig identities."

Fortunately, there was no one to force me to memorize any of it, so I kept learning the calculus bits without learning the algebra and trigonometry bits. I'd be reading something about calculus in the book, understanding it just fine, and then get lost because I didn't remember how to add fractions. Occasionally, in cases like this, staring at the confusing step for a while was enough to eventually realize, "Oh, they're just multiplying by 1 twice. It's like they're *lying* to make the problem easier, and then *correcting for the lie* so that they don't get the wrong answer. Interesting ..." Other things weren't so easy to figure out by staring at them, and they continued to stump me. Logarithms, sine and cosine, the "quadratic formula," and "completing the square" were not part of my vocabulary, and the terms themselves felt like a hangover from all the negative experiences I'd had with the subject in school.

After learning a bit more calculus, I still didn't understand the "prerequisites," but I started noticing some interesting things. I noticed that the derivative of a sphere's volume was its surface area, and the derivative of a circle's area was the distance around it. I still had no idea where the area and volume formulas themselves came from, but this weird "zooming in" operation suggested that they were all related somehow. This was my first exposure to a strange fact about mathematics: we may be completely defeated by two different questions, unable to make progress on either in isolation, and yet nevertheless manage to demonstrate with certainty that they have the *same* answer, all while remaining completely ignorant of what that answer is. This fact, which appears at first to be some sort of black magic, turns out to be a fundamentally important feature of abstract mathematics at all levels. This was clearly not the dull, authoritarian field I had been exposed to in school.

When it came time to start my first year of college, I did the unthinkable: I decided to take a calculus class. Having always hated mathematics with every fiber of my being, because of a freak accident in a bookstore I found myself taking calculus 1 for fun. Then calculus 2. Then my calculus 2 professor suggested that I take a graduate-level math course during my second year. I reminded him that I didn't know anything and that he was insane. I took it anyway, and got the highest grade. By my senior year of college the department gave me one of those plaque thingies that said something like "Congrats on being the best math major we have." I want to stress that I have absolutely no innate mathematical talent whatsoever, and nothing in my thirteen years of pre-university mathematics education suggested I'd find any enjoyment in the subject either. In any education system where the above series of events is possible, something has gone horribly wrong.

In the end, the Mathematics Department, my sworn nemesis in high school, ended up being the department where I felt most at home.[1] After college I went on to enter a Ph.D. program in mathematical physics at the University of Alberta. In the summer after my first year, following a lifelong pattern of doing everything except what I'm supposed to be doing, I became obsessed with

---

1. I was lucky to have amazing mathematics teachers in college, and I should mention (R/y+V)icky Klima, Eric Marland, and Jeff Hirst. I had many other wonderful teachers, but these four deserve special mention for being unbelievably helpful, and always putting up with me storming into their offices with bizarre questions unrelated to any course.

psychology and neuroscience. I eventually applied to some Ph.D. programs in the field, somehow got accepted, left my mathematical physics program with a master's degree, and I'm now living in Santa Barbara, California, studying the brain and behavior using mathematics. During my first year of graduate school in the Department of Psychological and Brain Sciences, I met tons of extremely smart students who had the same unjustified fear of mathematics that I always had in high school. Every time I see the flash of fear in someone's eyes when advanced mathematics is mentioned, I want to tell them that their entire experience of the subject is a lie. The perceived difficulty of mathematics is entirely the fault of how we teach it, and I hold myself to that standard as well. If there is anything in this book that you've repeatedly attempted to understand, but failed to do so, that is *my* fault, not yours. The underlying ideas are extremely simple. All of them. I promise.

Throughout my first year in Santa Barbara, I couldn't help but think that research in every area of science could be accelerated significantly if only everyone in the various fields of science knew more mathematics. By "knew more mathematics," I don't mean "had more mathematical facts in their heads." I mean "had been explicitly trained in abstract reasoning." What's worse, I'm fairly certain that nine out of ten people have more "innate skill" at mathematics than I do (whatever that means). The only reason I happen to know more of it than my fellow graduate students is because of a random accident in a bookstore that led me toward a subject I never thought I'd love.

It's summer now, and I'm writing this book for everyone who ever hated mathematics. Not only for the young and disenchanted, but also for the many scientists who secretly regard mathematics as a distasteful but necessary professional requirement, and have dutifully endured it but have never felt the fire, the anarchy, and the hedonistic pleasure of the subject. Unless I have failed miserably, we'll have a lot of fun along the way.[2] However, I should stress that this is not another one of those tired attempts to "make math fun," which usually translates to spreading a thin layer of silly faces and bad puns on top of the same old approach. While for some this might be a minor improvement over the standard textbooks, such books never present the subject in the way I always wished it would be presented: clearly pointing out everything that is arbitrary, everything that only looks the way it looks because someone is trying to sound fancy (consciously or not), separating historical accidents from timeless processes of reasoning, acknowledging the well-justified contempt that most students in most math classes feel most of the time by poking fun at the way that the subject is typically taught, and most importantly: *backwards.*

The subject as it is usually presented in modern educational institutions is something that no creative, independent thinker should be able to stand,

---

2. The above sentence was (of course) written by the author of this book, a heavily biased source whose opinions of his own work should not be trusted. However, the same principle applies to that last sentence as well, suggesting that the aforementioned distrust should itself be distrusted. We appear to have reached an impasse. Think what you will.

and books that try to remedy this fundamental flaw with chapter titles like "Funky Functions and Their Groovy Graphs" are missing a large part of why so many students find the subject so alienating.[3] But mathematics itself, when stripped of everything unnecessary, everything pretentious, and presented in as honest and human a manner as possible, is clearly one of the most beautiful things our species has discovered. It is a scientific art form that doesn't need to justify itself by being "useful," though learning it is one of the most useful things you can do.

At each point of our journey, I will focus on the ideas that I consider to be of largest conceptual importance, whether or not they're typically presented together. Although we'll start at an extremely basic level, we'll eventually start learning some things that aren't usually taught until the latter years of a four-year degree in mathematics. If there has ever been a book that goes from addition and multiplication to calculus in infinite-dimensional spaces, I've never found it. If you continue reading, I hope to show that this approach is not nearly as delusional as it seems.

I try at every stage to run the ideas through a conceptual centrifuge before I present them. What is presented in courses on any subject is usually a cloudy, confusing mixture of the essential with the historically contingent, a mixture that hides the simplicity of the underlying ideas from even the most attentive students. I've always wished that academics would spend much more time attempting to separate this mixture into its component parts *before* writing their books or giving their lectures. I've attempted to do this throughout the book, but for an example of what I mean, see the first few pages of Chapter 4, "On Circles and Giving Up." Also notice that circles first enter the story long after we've invented calculus. And they should; they're extremely confusing before then.

Here's an example of how we do things differently. One of the few things I remember hearing about in high school mathematics was the "Pythagorean theorem," but I didn't know why it was true, I didn't know why we should care, and I didn't like the unnecessarily fancy name. We'll avoid all three of these problems like this: I'll use the term "shortcut distance" instead of "hypotenuse," I'll think of a more descriptive name than "Pythagorean theorem," I'll offer the simplest explanation of why it's true that I'm aware of (it takes about thirty seconds to explain), and once we've invented it for ourselves, I'll show you a simple derivation of the fact that time slows down when you move.[4] This fact comes from Einstein's theory of special relativity, but the explanation uses no mathematical ideas more complicated than the "Pythagorean

---

3. To be fair, this is a chapter title from a very well-written book. Mark Ryan, I'd love to meet you in person someday. You're an incredible teacher.
4. To be a bit more precise, whenever two objects are moving in different directions or at different speeds, their "clocks" start moving at different rates. But it's not just a fact about clocks. It's a physical property of time itself. The universe is crazy. More on that later!

ent that is usually hidden behind the wall of formal proofs and pre-polished derivations (in the "unfriendly" textbooks) or behind cartoons and largely unexplained statements of fact (in the "friendly" ones). But at no point along the continuum between the friendly introductory books on the one hand and the awe-inspiring Grothendieck-style monographs on the other do we accurately represent the creation process in a pedagogically useful way.

It is impossible for any book to explore all of the pre-mathematics of a given concept before proceeding to its mathematics, and I do not attempt this impossible task. Rather, I attempt to construct a pre-mathematical narrative that leads from one concept to another, starting from addition and multiplication, proceeding immediately on to single-variable calculus, then backward through the (more advanced!) topics that we commonly think of as its prerequisites, and finally on to calculus in spaces of finitely many or infinitely many dimensions. A large amount of mathematics is found in this narrative, which is why it would not be a mistake to use this book as a mathematics textbook. However, once the pre-mathematics of any given concept has been developed at length, the mathematics itself often turns out to be startlingly straightforward, so we prefer to focus primarily on the former. That is not to say that the book contains an exhaustive and complete discussion of each of the topics it covers. Far from it! Rather, it is my estimate of everything that is missing, in terms of information, motivation, and where we should really begin teaching the subject. The book is a dirty proof of concept, not a polished diamond. I hope that it will start a conversation, but it is by no means the final word on any topic.

Further, it is important to be clear about what I am *not* criticizing. The problem of pedagogical foundations is fundamentally different from the problem of logical foundations, though they are implicitly conflated in most textbooks. I do not intend to criticize the logical starting point of the field, by which I mean choosing our logic to be first-order predicate calculus and our theory to be ZFC, NBG, or your favorite axiomatization of set theory.[1] What I want to criticize is the *pedagogical* starting point of the field, which is all that the vast majority of members of our society ever come into contact with.

Why Has Pre-mathematics Been Neglected?

Given the ubiquity of pre-mathematical reasoning inside the minds of professional mathematicians, it is worth asking why it so rarely appears in textbooks and journal articles. There are surely multiple reasons, but I believe that the primary culprit is *professionalism*. Though pre-mathematics is fundamentally important to understanding our field, it is structurally banned from any discussion that demands professionalism, including (but by no means limited

---

1. Though for a brilliant critique of the standard set-theoretic approach to logical foundations, see Robert Goldblatt's spectacular book *Topoi: The Categorial Analysis of Logic.*

theorem," so at that point you'll be able to completely understand the argument. The conclusion will still seem surprising, though. It seems surprising to anyone with a human mind, no matter how long you've known it! Given the fact that this argument is perfectly comprehensible once we've invented the formula for shortcut distances (formerly known as the Pythagorean theorem), it's a tragedy that this short argument isn't a mandatory part of every high school geometry class. They should ring a bell, throw confetti, and start explaining it to you five seconds after they teach you the Pythagorean theorem. But they don't. We will.[5]

Burn Math Class breaks a lot of conventions and a lot of rules, probably too many for its own good. No method of learning works for everyone, and I certainly don't claim that this book is a universal cure for the ailments of mathematics education, nor do I claim that it is guaranteed to be suited to everyone's learning style. If this book's approach doesn't work for you, please stop reading it and find one that does. Your time is valuable, and you shouldn't waste it trying to trudge through a book that isn't to your taste. This book was written as a labor of love, entirely for fun, not as part of a job. Ideally, it should be read for the same reasons.

Whether or not this experiment contributes anything of lasting value, radical changes simply *are* needed in education. As it stands, our educational institutions at all levels — from grade school to grad school to the style requirements of academic journals — appear to have been optimally designed to induce a kind of reverse Stockholm syndrome, causing us to revile subjects we might have otherwise loved. Students are graduating from these institutions deeply bored by the most phenomenally mind-blowing things our species has discovered. If they think mathematics, physics, evolutionary biology, molecular biology, neuroscience, computer science, psychology, economics, and other such fields of inquiry are dull and uninteresting, it isn't their fault. It's the fault of an education system that is brilliantly engineered to punish creativity; a system in which they are taught the spellings of words, but not how to think without deceiving themselves, as we all inevitably do; a system in which the laws of nature and arbitrary fiats like the prohibition of prepositions at the end of sentences are presented on equal footing, as if they were both equally valid descriptions of The Way Things Are; a system in which they are legally obligated to spend the majority of their young lives. For them, and for any of you who have ever had a similar experience, this book is my note of apology.

---

5. You'll have to provide the bell and confetti. Not that I'm unwilling to provide them, but I'm probably not where you are at the moment.

theorem," so at that point you'll be able to completely understand the argument. The conclusion will still seem surprising, though. It seems surprising to anyone with a human mind, no matter how long you've known it! Given the fact that this argument is perfectly comprehensible once we've invented the formula for shortcut distances (formerly known as the Pythagorean theorem), it's a tragedy that this short argument isn't a mandatory part of every high school geometry class. They should ring a bell, throw confetti, and start explaining it to you five seconds after they teach you the Pythagorean theorem. But they don't. We will.[5]

Burn Math Class breaks a lot of conventions and a lot of rules, probably too many for its own good. No method of learning works for everyone, and I certainly don't claim that this book is a universal cure for the ailments of mathematics education, nor do I claim that it is guaranteed to be suited to everyone's learning style. If this book's approach doesn't work for you, please stop reading it and find one that does. Your time is valuable, and you shouldn't waste it trying to trudge through a book that isn't to your taste. This book was written as a labor of love, entirely for fun, not as part of a job. Ideally, it should be read for the same reasons.

Whether or not this experiment contributes anything of lasting value, radical changes simply *are* needed in education. As it stands, our educational institutions at all levels — from grade school to grad school to the style requirements of academic journals — appear to have been optimally designed to induce a kind of reverse Stockholm syndrome, causing us to revile subjects we might have otherwise loved. Students are graduating from these institutions deeply bored by the most phenomenally mind-blowing things our species has discovered. If they think mathematics, physics, evolutionary biology, molecular biology, neuroscience, computer science, psychology, economics, and other such fields of inquiry are dull and uninteresting, it isn't their fault. It's the fault of an education system that is brilliantly engineered to punish creativity; a system in which they are taught the spellings of words, but not how to think without deceiving themselves, as we all inevitably do; a system in which the laws of nature and arbitrary fiats like the prohibition of prepositions at the end of sentences are presented on equal footing, as if they were both equally valid descriptions of The Way Things Are; a system in which they are legally obligated to spend the majority of their young lives. For them, and for any of you who have ever had a similar experience, this book is my note of apology.

---

5. You'll have to provide the bell and confetti. Not that I'm unwilling to provide them, but I'm probably not where you are at the moment.

Prefacer

Pre·fac·er [pruh-**fes**-er] (*noun*)
A preface for professors. Or for professional mathematicians, or
students with enough mathematical background to understand the
rambling in this section, or curious students with no mathematical
background, or high school teachers, or anyone who finds themselves thinking about mathematics often... or not.

This is not your standard "introductory" mathematics text. It is simultaneously more introductory and more advanced than most individual books you
are likely to encounter, and as such, it is something of an experiment.

What Kind of Experiment?

This book will be extremely easy to misunderstand if one comes to it expecting
a mathematics textbook, although it shares many features in common with
mathematics textbooks, and could indeed be used as one. To understand the
goal and structure of this book, I must first coin a term that our lexicon is
presently lacking: *pre-mathematics*. By "pre-mathematics" I do not mean
those tiresome non-subjects such as "pre-algebra" and "pre-calculus" that we
inflict on unsuspecting students. Rather, I will use the term to refer to the
entire set of ideas, confusions, questions, and motivations that occupy the
minds of the inventors of mathematical concepts, and which drive them to
define and examine one species of mathematical object rather than another.

For instance, the definition of the derivative and the various theorems that
follow from it are part of mathematics proper, and they can be found in any
mathematics textbook that covers calculus. The *reasons why* the concept is
defined the way it is, rather than any of the infinitely many other ways it could
have been defined, as well as the processes of reasoning that would lead one
to choose the standard definition over all other candidate definitions (in the
absence of a preexisting mathematics textbook), are much less frequently given
sufficient attention. It is this set of possibilities and processes of reasoning to
which the term pre-mathematics refers. Pre-mathematics includes not only all
of the possible alternative definitions of mathematical concepts that would lead
to essentially identical formal theories, but also — perhaps more importantly
— all of the blind alleys down which one would be led in attempting to invent
the standard mathematical definitions and theorems from scratch. It is the
conceptual heavy lifting that must be done to pull a mathematical concept into

existence out of nothingness. Mathematics is the sausage; pre-mathematics is how the sausage is made.

That is the main topic of the book: the rarely discussed process of moving from the vague and qualitative to the precise and quantitative, or equivalently, how to invent mathematics for yourself. By "invent," I mean not only the creation of new mathematical concepts, but also the more relevant process of learning how to *reinvent* bits of mathematics that were originally invented by someone else, in order to gain a deeper and more visceral understanding of those concepts than could be gained simply by reading a standard textbook. This is a process that we virtually never teach explicitly, yet one would be hard pressed to think of a more valuable skill that one could learn from any mathematics course. Learning how to (re)invent mathematics for yourself is of critical and fundamental importance in both the pure and applied domains. It includes pure questions like "How did mathematicians figure out how to define curvature in a way that lets them talk about it in seventeen dimensions, where we can't picture anything?" and applied questions like "Given what I know, how should I build a model of the phenomenon I'm studying?" Such questions are often addressed in textbooks, briefly, as an afterthought, but it is dramatically less common to place these questions in the spotlight, on equal if not higher footing than the theorems and results themselves.

An honest description of the informal, messy creation process is the missing piece of the puzzle in our exposition of mathematics at all levels, from elementary school to the postdoctoral level, and its absence is one of the primary reasons that our subject so often bores even the most attentive students. The elegance and beauty of our subject cannot be fully appreciated without a visceral understanding of the pre-formal conceptual dance by which mathematical concepts are created. This process is not nearly as difficult to explain as it may seem, but doing so requires a radical shift in the way we teach our subject. It requires that we include in our textbooks and lectures at least some of the false starts, mistakes, and dead ends that a normal human mind would first need to experience before arriving at the modern definitions. It requires that we write our textbooks as narratives in which the characters often get stuck and don't know what to do next. This book is a quirky, flawed, deeply personal attempt to outline what I believe are some of the core concepts and explanatory strategies of pre-mathematics: strategies that professional mathematicians use every day but rarely discuss openly in their textbooks and courses.

This highlights an important point. While a proper emphasis on pre-mathematics requires a radical change in how we teach mathematics, it does *not* require a change in how professional mathematicians *think about* mathematics. Pre-mathematics is their bread and butter. It is the language in which they think, since they are by definition the ones who create — or if you prefer, discover — the subject. In that respect, the content of this book is not novel. It is only novel insofar as it places under the spotlight all the content that is usually hidden behind the wall of formal proofs and pre-polished derivations (in the "unfriendly" textbooks) or behind cartoons and largely unexplained statements of fact (in the "friendly" ones). But at no point along the continuum between the friendly introductory books on the one hand and the awe-inspiring Grothendieck-style monographs on the other do we accurately represent the creation process in a pedagogically useful way.

It is impossible for any book to explore all of the pre-mathematics of a given concept before proceeding to its mathematics, and I do not attempt this impossible task. Rather, I attempt to construct a pre-mathematical narrative that leads from one concept to another, starting from addition and multiplication, proceeding immediately on to single-variable calculus, then backward through the (more advanced!) topics that we commonly think of as its prerequisites, and finally on to calculus in spaces of finitely many or infinitely many dimensions. A large amount of mathematics is found in this narrative, which is why it would not be a mistake to use this book as a mathematics textbook. However, once the pre-mathematics of any given concept has been developed at length, the mathematics itself often turns out to be startlingly straightforward, so we prefer to focus primarily on the former. That is not to say that the book contains an exhaustive and complete discussion of each of the topics it covers. Far from it! Rather, it is my estimate of everything that is missing, in terms of information, motivation, and where we should really begin teaching the subject. The book is a dirty proof of concept, not a polished diamond. I hope that it will start a conversation, but it is by no means the final word on any topic.

Further, it is important to be clear about what I am *not* criticizing. The problem of pedagogical foundations is fundamentally different from the problem of logical foundations, though they are implicitly conflated in most textbooks. I do not intend to criticize the logical starting point of the field, by which I mean choosing our logic to be first-order predicate calculus and our theory to be ZFC, NBG, or your favorite axiomatization of set theory.[1] What I want to criticize is the *pedagogical* starting point of the field, which is all that the vast majority of members of our society ever come into contact with.

Why Has Pre-mathematics Been Neglected?

Given the ubiquity of pre-mathematical reasoning inside the minds of professional mathematicians, it is worth asking why it so rarely appears in textbooks and journal articles. There are surely multiple reasons, but I believe that the primary culprit is *professionalism*. Though pre-mathematics is fundamentally important to understanding our field, it is structurally banned from any discussion that demands professionalism, including (but by no means limited

---

1. Though for a brilliant critique of the standard set-theoretic approach to logical foundations, see Robert Goldblatt's spectacular book *Topoi: The Categorial Analysis of Logic.*

to) all mathematical work in academic journals. Why? Because *precise* pre-mathematics is *not* formal. It is (by definition) the set of hunches, guesses, and intuitions that lead to the development of a formal mathematical theory in the first place, and the only precise, honest way to explain imprecise thought processes is with informal arguments expressed in informal language: language that accurately conveys to the reader that we are not 100% sure our intuitions are on the right track, and that we are always (to some extent) exploring in the dark. Such informal language is *not* simply dumbing things down. It is a precise manner of describing the chains of reasoning by which new mathematical concepts are created. And without a firm understanding of how mathematics is created, one's understanding of the subject will be crippled in comparison to what it could have been otherwise.

To be clear, this is also not a criticism of formal expositions of mathematics or of the concept of the formal proof. But formal proofs do not spring into existence fully formed, nor (more importantly) do the formal *definitions* of the mathematical concepts on which they are based. An overly formal description of informal thought processes misleads the reader by providing *evidence of nonexistent principles*, and in doing so tricks the reader into believing that their failure to realize how A follows from B must be a deficit in their own knowledge, when in fact it is often a lack of perfect precision in the underlying pre-mathematical reasoning itself. Full disclosure requires that we offer informal descriptions of that which is informal. Professionalism has its place, but fundamentally its function is to censor honesty, and it has redacted pre-mathematics almost entirely out of existence.

What I Wanted the Book to Be, from the Start

This book grew out of an attempt to explain a subset of the universe of mathematics in as honest and unpretentious a way as possible, making sure at each stage to give away the secrets of our trade. At every step I attempt to separate necessary deductions from historically contingent conventions; I emphasize that the often intimidating words "equation" and "formula" are just code for "sentence"; I try to make it clear that all of the symbols in mathematics are just abbreviations for things that we could be saying verbally; I try to engage the reader in the process of inventing good abbreviations; I always attempt to make clear the distinction between what other textbooks actually do and how they could have done it; I attempt to present each derivation not in the standard post hoc, cleaned up form that reflects nothing about the thought process that led to it, but in a way that makes clear at least some of the blind alleys that most of us would be tempted to wander down before finally arriving at the answer; I try to explain everything as deeply as I can without sacrificing the coherence of the narrative; and I vowed that I would burn the book before I ever allowed myself to say "memorize this," even once. There are many things I wish I had done differently, but at very least, the book is full of all the things mentioned above.

I also try to explain the strange dance our field does on the border between structured necessity and unrestrained anarchy. This is something that we virtually never explain to students, so I emphasize it whenever possible. Here's what I mean. On the one hand, there's the anarchy. We are free to use whatever axioms we please, even an inconsistent set. Defining and playing with an inconsistent formal system is not *illegal*, it's *boring*. For instance, "dividing by zero" is not *illegal*, and every mathematics professor knows that. We are perfectly free to define a symbol $\star$ by the property $\star \equiv a/0$ for all a, and many analysis books do just that, in a section on what is usually called "the extended real number system."[2] But if you insist on defining the above symbol, then the algebraic structure you're examining cannot be a field. You want to insist on saying it's still a field? That's perfectly fine, but then you can only be talking about a "field with one element." You want to insist that there's still more than one element, or that fields, by your definition, have at least two elements? That's fine too, but then you're working in an inconsistent formal system. You want to do that? Fine. But now any sentence is provable, so there's not much to do.

It is important to emphasize that even when we hit rock bottom like this, we still have not done anything *illegal*. Rather, we've made the discussion *boring*. Every mathematician knows that, at least in the choice of what to study, there are no *laws* in mathematics. There are only more or less elegant and interesting mathematical structures. Who gets to decide what counts as elegant and interesting? Us. QED.

On the other hand, there's the structured part of mathematics. Once we finish with the "anything goes" stage in which we say exactly what our assumptions are and what we're talking about, *then* we find that we have conjured up a world of truth that is independent of us, about which we may know very little, and which it is our job to explore.

Needless to say, when we fail to inform students of this most fundamental point about anarchy and structure, we completely mislead them about the nature of mathematics. For whatever reason, we almost never tell them about this odd interplay between anarchic creation and structured deduction. I'm convinced that this is one of the things that make so many students feel as if mathematics is a kind of totalitarian wasteland full of undefined laws that no one tells you about, and in which you always have to be afraid of accidentally doing something wrong. That's certainly how I always felt in high school, before the story I told in the first preface. This is one of the things I try to remedy in this book.

---

2. Though they usually write ∞ instead of $\star$ for obvious reasons. I'm using $\star$ to remind us that the following argument is not a problem with "infinity," it's a problem with the boredom that starts to corrupt our mathematical universe when we assume that the additive identity has a multiplicative inverse.

The Book Decides It Wants to Be About Something Else, Too

As much as I wanted to explain general things like the big-picture structure of mathematics, I eventually wanted to get around to explaining the ideas that *are* taught in the standard textbooks, so that the book might actually be helpful to students on Earth. To do this, I needed to build a narrative that somehow had to arrive at many of the standard textbook definitions before I could explain the mathematical arguments that spring from them. However, because of the goal of the book, I promised myself that I would not introduce these definitions in the standard way, which is usually to say "such and such is defined this way," often out of the blue, or at best with a few pages of motivation, either conceptual or historical, followed by a huge conceptual leap into the mathematical definition itself. In swearing-off this practice, I found that I had placed myself under a rather large set of constraints. The problem can be summarized as follows:

> Assume you don't know anything about mathematics except the basics of addition and multiplication. Not necessarily the algorithms for performing them, but you know what phrases like "twice as big" mean, and you get the gist of both operations. You're living in a world before textbooks. How could you discover even the simplest parts of mathematics? As a specific case, how would you figure out that the area of a rectangle is "length times width"?

It would be a non sequitur to answer this question by talking about how area is defined in measure theory, or by talking about axioms, or Euclid's fifth postulate, or how the formula $A = \ell w$ doesn't hold in non-Euclidean geometry. It is not a question about rigor, and it is not a question about history. It is a question about *creating something*. The question is about how to move from a vague, qualitative, everyday concept to a precise, quantitative, mathematical one, when there's no one around to help you or do it for you.

I was originally asked the above question by one of my closest friends, Erin Horowitz. Around the time I started writing this book, we would occasionally have many-hours-long sessions in which we would talk about mathematics. She doesn't have a mathematical background, but she's extremely curious, and always interested in knowing the "whys" of things. We would talk about formal languages, Taylor series, the idea of a function space, or any other crazy things we felt like talking about. One day she asked me the above question about how mathematical ideas are created. It wasn't a hard question once it was phrased that way, using the area of a rectangle as a test case, and I basically just gave her the simplest argument I could think of, which is the argument about area that you'll find in Chapter 1, in the section called "How to Invent a Mathematical Concept." After I yammered for a bit, she asked why we're never taught things like this in school. She completely understood the short argument, and so could anyone. Here's the weird part: the argument involves solving a functional equation.

There are very few courses in mathematics departments that focus just on functional equations. I'm not certain that there should be more, but it is a rather confusing fact once one realizes it. After all, every mathematics undergraduate encounters plenty of differential equations, as they should, and they inevitably encounter integral equations as well, but the area of mathematics devoted to studying and solving general expressions involving unknown functions has been largely neglected by history. Despite the fact that it is one of the oldest topics in mathematics, we tend not to hear about it very often. In his monumental work *Lectures on Functional Equations and Their Applications*, J. Aczél laments that "through the years there has been no systematic presentation of this field, in spite of its age and its importance in application."

Surprisingly, I started to discover that functional equations are enormously helpful in explaining even the simplest of mathematical concepts, as long as the ideas are presented in the right way.[3] Here's how. You don't use the term "functional equation," and in fact, if at all possible, don't even use the word "function." Most people have had bad enough experiences in math classes that it's easy to scare them and shut off their natural creativity by using a lot of orthodox mathematical terminology. Instead, you say something like this:

> We've got a vague, everyday concept that we want to make into a precise mathematical concept. There's no wrong way to do this, because we're the ones who get to decide how successfully we performed the translation. However, we want to cram as much of our everyday concept into our mathematical concept as we possibly can. We start by saying a few sentences about our everyday concept. Then we come up with abbreviations for those sentences.[4] Then we mentally eliminate all the possibilities that don't do everything we asked them to. We can rinse and repeat if we want to, putting more and more vague, everyday information into abbreviated form, and then mentally throwing away everything that doesn't behave like that. Occasionally, just by writing down examples, we can slowly become convinced that the precise definition we're looking for has to look a certain way. We may not end up with just one possibility, and even if we do, we may not know when we've found the only one, but that doesn't matter. If there's more than one candidate definition that does everything we want it to, we can just do what mathematicians do all the time without telling you, and pick the one we think is prettiest. What counts as "prettiest"? That's up to us.

In short, as crazy as it might sound, I believe that informal mathematical arguments involving functional equations not only provide a way to better explain where our definitions come from at all levels of mathematics, but also that such arguments offer a kind of anti-authoritarian pedagogical style that engages the reader in the process of creating mathematical concepts in a way

---

3. Not functional equations in the full generality of Aczél's monograph, but in a somewhat informal guise analogous to the calculus we teach before we teach analysis.

4. At this point they're writing down a functional equation without knowing it.

that is unheard of in introductory courses and textbooks. Surprisingly often (though certainly not always) the rarely discussed pre-mathematical practice of passing from a vague qualitative concept to a quantitative mathematical one turns out to involve the use of functional equations. In Chapter 1, we use this idea to "invent" the concepts of area and slope, arriving at the standard definitions not by simply postulating them, but by deriving them from qualitative correspondence with our everyday concepts. This is a simple illustration of what pre-mathematical pedagogy might look like, but it is only an example, and there is certainly room for improvement. In the meat of the book, we proceed to "invent" a large amount of mathematics this way, sometimes by an informal use of functional equations, sometimes not, but always making clear what we're trying to do and how else it could be done.

How This Might Help

To see how an emphasis on pre-mathematics differs from the standard approach, let's look at a specific example of how current teaching practice backs itself into a corner. Consider the problem faced by a teacher or an introductory textbook in attempting to explain where the definition of slope comes from. On the one hand, you want to motivate the idea. On the other hand, you eventually want to arrive at the conventional definition, $\frac{y_2-y_1}{x_2-x_1}$, or as they say in introductory textbooks, "rise over run." All of differential calculus rests on this formula plus the idea of a limit, so there could hardly be a more important concept to convey to students. Teachers and introductory textbook writers face the following problem. They might be able to think of some set of postulates that would single out "rise over run" as the unique definition satisfying all the postulates, but the proof of this would surely be too complicated for an introductory class, and it would probably just confuse everyone ten times more, so they just introduce "rise over run" as *the* definition of slope, possibly with a bit of motivation beforehand. Given the situation, this seems like a completely reasonable thing to do.

However, I believe that in this case and others like it, we're confusing many more students than we realize, and turning them off of mathematics. When I first heard the definition of slope in high school, it did nothing but speed the process of demotivation for me. Introducing the concept like this (i) leaves open infinitely many questions, (ii) makes any reflective student feel as if they are missing something, and (iii) implicitly suggests that it is their own fault for not understanding it. The students are indeed missing something, but it is *not* their fault; they are missing something because it is being deliberately hidden from them, and it is being hidden by the best intentions of their teachers. In my own experience, I felt something like "I couldn't invent any of these definitions on my own, from first principles, so there's something I don't understand about all of them." I certainly didn't put the feeling into those words at the time. All I thought explicitly was "I don't understand this stuff."

Years later, when I found myself explaining mathematics to others, I would always try to make the point that we *could* define slope to be "3 times rise over run," or "rise over run to the fifth power," or even "run over rise," and we could go on to develop calculus using any of these definitions. All of our formulas would look slightly different (or possibly very different, depending on which definition we chose), and we might have to state some familiar theorems in a slightly different or even unrecognizable form, but the essential content of the theory would be identical, however ugly and unfamiliar-looking it might prove to be. An analogous story holds for any mathematical concept. I've yet to explain this to anyone without being asked why this isn't explained in courses and textbooks. I don't know. It should be.

Burn Math Class: A Mathematical Creation Story

> *What am I supposed to publish? L. J. Savage (1962) asked this question to express his bemusement at the fact that, no matter what topic he chose to discuss and no matter what style of writing he chose to adopt, he was sure to be criticized for not making a different choice. In this he was not alone. We would like to plead for a little more tolerance of our individual differences.*
> —E. T. Jaynes, *Probability Theory: The Logic of Science*

Writing a book is an emotional experience. In the course of preparing this book for publication, I was lucky to have two wonderful editors, T. J. Kelleher and Quynh Do, who were both extremely helpful throughout. I primarily dealt with Quynh for most of the publication process. She showed unending patience in helping me improve what I can only imagine was a very difficult book to edit, and though we did not always agree, her comments made the book *tremendously* better than it would have been otherwise. After mentioning one's editors, it is customary to say "any remaining flaws in the book are my own," but the customary phrase is far too mild.

Even in its final form, the book will inevitably contain numerous instances of the following sins: typos, hyperbole, poorly worded sentences, repeating myself, contradicting myself, sounding too arrogant, sounding too insecure, saying "I'll never do X!" and then promptly doing X, saying "I'll never do X!" and then later doing X (but not promptly), Easter eggs no one will find or understand, unintentionally alienating or offending innocent readers, experimenting with the medium in ways that some will find distracting, too many prefaces, too many digressions, too many dialogues, too few dialogues, too much meta-commentary, the use of arcane Greek and Latin words despite having made fun of them (and the people who use them) for being more pretentious than is necessary, and at least one unforgivably large error, most likely resulting from an accidental copy/paste of a random paragraph into a completely different part of the book...ad infinitum.

This is my first book. It was built on a scaffolding of my flaws. Writing a book is something I never thought I would do, and I was taken completely by surprise when it started happening. I wrote this book over a four-month period in the summer of 2012, in a euphoric flurry of coffee, eyestrain, sixteen-hour days, forgetful meal-skipping, and loving every minute. Writing had never been so fun. I was 25 at the time. Since then, I feel like I've become a different person. Parts of the book now hurt to read. When a book is created in the manner described above, it will inevitably be shot through with certain defects that no amount of editing or polish can hide.

Most of these flaws are accidental, but others are present by design. When an error is simply a misstep, there is no harm in hiding it. When we fix a typo in sentence N, sentence $N + 1$ is not harmed as a result. The same principle holds for sloppy wording or unnecessary repetition, and (although many such missteps surely remain in the book) this is the type of error that one should attempt to fix.

However, in some cases, an error is not a misstep but a stairstep. It is something without which we never could have arrived where we are. Removing the Nth stair from a staircase harms the steps after it, whether that staircase is a narrative or a mathematical argument. Certain rare ideas require flaws in order to be properly conveyed. My goal is to let the reader in on the secrets of the creation process, both of mathematics and of books themselves, and the process of creation cannot be accurately represented in a spotless manner. If there is a single unifying theme that ties together all the quirks of this book, it is *full disclosure*. Full disclosure in the sense of complete openness and honesty, not only about the process of mathematical creation, but also about the process of writing a book, as well as the emotional experience of returning to something one has written after a long absence and realizing that some of its flaws run too deep to ever be excised. The thought of a person taking time out of their lives to read this book makes me so happy that I have no desire to hide anything from them. I want to show them everything. All of it. Inevitably, this results in a rather unusual book.

I hope to convince you in what follows that the reason why so many members of our society never come to love or understand mathematics is that we have been communicating the subject entirely wrong. *That does not mean I know how to do it right!* This book may turn out to be a colossal failure, but if I'm certain of anything, it is that mathematics deserves better than the methods we currently use to teach it, at every level. This book is my personal attempt to right a few of these wrongs by writing the book I always wanted to read. Ready to have some fun? Me too. Let's begin.

Act I

1 Ex Nihilo

> *If you want to build a ship, don't drum up people together to collect wood and don't assign them tasks and work, but rather teach them to long for the endless immensity of the sea.*
> —Antoine de Saint-Exupéry, *Citadelle*

1.1 Forgetting Mathematics

1.1.1 Hello, World

Forget everything you've been told about math. Forget all those silly formulas you've ever been told to memorize. Make a little room in your head with clean white walls and no math. Without leaving that room, let's reinvent mathematics for ourselves. Without the burden of teachers, without the burden of a classroom, without paying any attention to the thing called "mathematics" that has been handed down the generations to us, free of that ridiculous lie that the worst thing you can do is to be wrong. Only by doing this will we be able to understand anything.

I'm calling this chapter Ex Nihilo for two reasons. The first is to poke fun at the unnecessarily fancy terminology that shows up in all subjects, including math. We humans love to sound smart, and saying stuff in a different language (especially a dead language) makes it seem more important. Having said that, we can dispense with the Latin. The term Ex Nihilo means "out of nothing," and I chose this as the name for the first chapter to emphasize that in this book, mathematics is *ours*. The term no longer refers to that thing you learned in school. We are pulling mathematics into existence, out of nothing.

I'll assume that the language of addition and multiplication is familiar enough that you can speak it fluently. I don't mean that it should be obvious how to calculate the square of 111111111 or how to find the square root of 12345678987654321 or anything crazy like that. In general, mathematicians don't like to deal with numbers. All I mean is that I'll assume you can convince yourself of basic things like the order of addition doesn't matter. Same deal for multiplication. To say the same thing in more abbreviated form:

$$(?) + (\#) = (\#) + (?) \qquad \text{and} \qquad (?)(\#) = (\#)(?)$$

no matter what numbers $(?)$ and $(\#)$ are.

To begin our journey into mathematics, we will *not* need to waste our time learning how to do boring things like calculating a specific number in decimal notation for $\frac{1}{7}$. All we need to know is that the funny symbol $\frac{1}{7}$ refers to whichever number turns into 1 when we multiply it by 7. If you see something like $\frac{15}{72}$, then don't be fooled into thinking that there's some mysterious thing called "division" that you have to learn random facts about. A symbol like $\frac{15}{72}$ is just an abbreviation for $(15)(\frac{1}{72})$, which is just multiplication. What number does $(15)(\frac{1}{72})$ refer to? I have no idea, and you certainly don't have to either. But we do know that it is whatever number turns into 15 when we multiply it by 72. That's all.

Assuming that you get the basics of addition and multiplication, we're going to take an utterly bizarre path through mathematics. After this first chapter, which largely consists of learning how to invent our own mathematical concepts, we're going to jump straight into inventing calculus, and then use it to reinvent for ourselves all of the things that are usually thought of as prerequisites to calculus. By turning the subject on its head, we'll discover that calculus — the art of the infinitely large and the infinitely small — can not only be invented before its so-called prerequisites, but that those "prerequisites" cannot be fully understood without calculus itself.

This approach also frees us from the need to memorize anything. Since we'll never (intentionally) accept anything we have not created for ourselves, and since we can always look back at what we've already done, we find that mathematics — a field so often associated with memorization — actually requires less memorization than any other subject. While in other fields memorization may be unavoidable, in mathematics it is poison, and any mathematics teacher who makes you memorize something without apologizing for it on bended knee should be immediately teleported to the unemployment office and made to memorize the phone book.[1] Mathematics is a beautiful discipline in which nothing *ever* needs to be memorized. It's about time we started teaching it that way.

Our adventure will eventually lead us to some fairly "advanced" topics that typically aren't taught until the latter half of a four-year bachelor's degree in mathematics. We'll see that this "advanced" stuff is really no different from the "basic" stuff, but at each stage the textbooks change the way they write things, just to confuse you.

We're about to set out on an adventure into a beautiful world of necessary

---

1. **Author:** Okay, that was too extreme. I didn't really mean that. I just meant that memorization isn't very helpful. But writing a book is an exciting experience, and I might occasionally get carried away. So try not to take my editorializing too seriously, okay? I mean, I've never written a book before, and I'm scared I might burn out along the way. I know I'll never finish unless I make sure to enjoy the writing process. So if it's not too much to ask, please try to tolerate my extraneous hyperbole. I promise it's all in good fun. Anyways, let's keep moving, dear Reader. Can I call you Reader?
Reader: Works for me.
Author: Great! You can call me Author. Or whatever you want. I'll answer to any loud noise, really, so pick your favorite and let's keep moving. I can't believe I'm actually writing a book!

truth in which nothing is accidental. You may occasionally get discouraged (and it'll probably be my fault). You may have to play around with ideas on your own to convince yourself that you understand them. You may have to think very hard, and you'll have to try even harder not to be intimidated when you see symbols (abbreviations). But you won't have to trust me. You won't have to wonder what's being hidden from you. And you won't have to memorize anything. . . unless you want to. Here we go.

1.1.2 "Function" is a Ridiculous Name

The term 'function' got into mathematics, I was told by Prof. K. O. May, due to a misinterpretation of a proper usage by Leibniz. Nevertheless, it has become a fundamental concept of mathematics and whatever it is called, it deserves better treatment. There is perhaps no better example in mathematical education of missed opportunities than in the treatment of functions.
—Preston C. Hammer, *Standards and Mathematical Terminology*

Machines do all sorts of things. A bread-maker is a machine that eats bread ingredients and spits out bread. An oven is a machine that will eat anything and spit out that same thing at a much hotter temperature. A computer program for adding one to a number can be thought of as a machine that eats a number and spits out one plus whatever number you put in. A baby is a machine that eats things and spits them out all covered in spit.

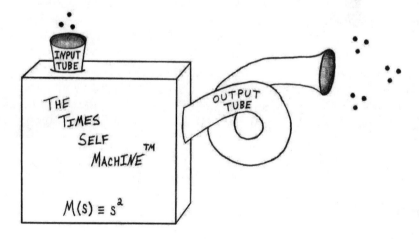

Figure 1.1: One of our machines.

For whatever reason, mathematicians have decided to use the strange word "function" to describe machines that eat numbers and spit out other numbers. A much better name would be. . . just about anything. We'll start by calling them "machines," and then once we're used to the idea, we'll occasionally start calling them "functions," but only occasionally.[2] Let's use the only tools we've got — addition and multiplication — to invent some machines that eat numbers and spit out numbers.

1. The Most Boring Machine: If we feed it a number, it hands the same number back.

2. Add One Machine: If we feed it a number, it adds one to what it ate and spits out the result.

3. Times Two Machine: If we feed it a number, it multiplies that number by two and hands back the result.

4. Times Self Machine: If we feed it a number, it multiplies that number by itself and hands back the result.

It takes a lot of words to talk about these machines, so let's invent some abbreviations. All of the symbols in every area of mathematics, however complicated they look, are just abbreviations for things we could be talking about in words, except we're too lazy. Because they usually don't tell you this, most people are really intimidated when they see a bunch of equations they don't understand, but less intimidated when they see an abbreviation like DARPA or UNICEF or SCUBA.

But math is just lots and lots of abbreviations plus reasoning. We'll be inventing a bunch of abbreviations on our journey, and it's important that we invent abbreviations that remind us of what we're talking about. For example, if you want to talk about a circle, two reasonable abbreviations would be C and $\bigcirc$. Some good abbreviations for a square would be S and $\square$. This is so obvious that you might wonder why I'm saying it, but when you look at a page full of equations and think "Ahhh! That's scary!", all you're really looking at is a bunch of simple ideas in a highly abbreviated form. This is true for every part of mathematics: deconstructing the abbreviations is more than half the battle.

We want to talk about our machines using fewer words, so we need to invent some good abbreviations. What makes an abbreviation good? That's for us to decide. Let's look at some of our options. We could describe the Times Two Machine by saying:

---

2. We'll use certain nonstandard terms throughout the book, but I should emphasize that I don't necessarily think my terminology is "better" than the standard terminology, and I'm certainly not arguing that other books should use it! The purpose of occasionally inventing our own terms is simply to remind ourselves that the mathematical universe we are creating is entirely *our own*. It's a universe we're building, from scratch, so we get to decide what to call things. But please don't think that my purpose is to convert everyone to a new set of terms. The word "function" may not be the best, but it's not all that bad once you get used to it.

If we feed it 3, it spits out 6.
If we feed it 50, it spits out 100.
If we feed it 1.001, it spits out 2.002.

And then we could just say what it does to every possible number. But that's a crazy waste of time, and we'd never finish. We could say that whole infinite bag of sentences at once, simply by saying "If we feed it (*stuff*), it spits out $2 \cdot$ (*stuff*)," where we're choosing to remain agnostic about which specific number (*stuff*) is. We could abbreviate this idea even further by writing *stuff* $\longmapsto 2 \cdot$ *stuff*.

So, just by being agnostic about which number we were putting into a machine, we managed to collapse an infinite list of sentences down to a single sentence. Can we always do that? Well, probably not. We don't know yet. But at this point we've decided to only think about machines that can be completely described in terms of addition and multiplication, and that's what let us summarize an infinite number of sentences with just one. We can describe the rest of our machines in this abbreviated way too:

1. The Most Boring Machine: *stuff* $\longmapsto$ *stuff*

2. Add One Machine: *stuff* $\longmapsto$ *stuff* $+ 1$

3. Times Two Machine: *stuff* $\longmapsto 2 \cdot$ *stuff*

4. Times Self Machine:[3] *stuff* $\longmapsto$ (*stuff*)^2

In case that doesn't make sense, here are some examples:

1. The Most Boring Machine:
   $3 \longmapsto 3$
   $1234 \longmapsto 1234$

2. Add One Machine:
   $3 \longmapsto 4$
   $1234 \longmapsto 1235$

3. Times Two Machine:
   $3 \longmapsto 6$
   $1000 \longmapsto 2000$

---

3. We're writing (*stuff*)^2 as an abbreviation for (*stuff*)$\cdot$(*stuff*). More generally, we'll use the abbreviation (*stuff*)^{number} to stand for "The thing you get when you multiply (*stuff*) by itself *number*-many times." You're not allowed to think "I don't understand powers," because at this point there's nothing to understand. It's just an abbreviation for multiplication.

4. Times Self Machine:
   $$2 \longmapsto 4$$
   $$3 \longmapsto 9$$
   $$10 \longmapsto 100$$

Let's try to abbreviate these machines as much as we possibly can without being ridiculous. By "being ridiculous," I mean "losing information." For example, we could abbreviate the entire collected works of Shakespeare by the symbol ♣, but that doesn't help very much, because we can't extract any of the information we're abbreviating from the abbreviation.

How many abbreviations do we need to *completely* describe our machines? Well, we need to come up with names for (i) the machine itself, (ii) what we're putting in, and (iii) what we're getting out. Then we need to do one more thing: (iv) we need to describe how the machine works.

Let's name the machines themselves with the letter M, so we don't forget what we're talking about. We might want to talk about more than one machine at a time, so let's use the letter M with different hats (M, $\hat{M}$, $\ddot{M}$, $\bar{M}$, etc.) to talk about different machines. We've been using the name *stuff* to refer to whatever we're putting into the machine, but let's abbreviate this even further and just write s (s stands for *stuff*). But now that we've got two abbreviations, we can *build the third abbreviation out of the first two*. This is a tricky idea, and I've never really heard anyone acknowledge that we're doing it, let alone what an odd process it is, but this is where most of the confusion about "functions" comes from.

What do I mean by saying that we can build the third abbreviation out of the first two? Well, what name should we invent to talk about "the thing that the machine M spits out when I feed it some stuff s"? If we build the name for this using just the abbreviations M and s, then we don't have to come up with any more abbreviations, and we're using as few symbols as possible. So let's call it $M(s)$. Again, $M(s)$ is the abbreviation we're using to refer to "the thing that the machine M spits out when I feed it some stuff s."

So we had to name *three* things, but instead we named *two* things, then we paused, looked around to see if anyone was watching, and sneakily used the two names we had already come up with as the "letters" to write down the third name. That's a really weird idea, but it helps tremendously once we get used to it. If you've been confused about "functions" before, don't worry. It's all simple stuff about machines and abbreviations. They just don't tell you that.

Okay, so we've got our three names, but we still haven't described any particular machines in this new abbreviated language we invented. Let's re-describe the four machines from earlier. I won't list them in the same order. See if you can figure out which machine is which (i.e., which is the Add One Machine, which is the Times Self Machine, etc.)

1. $M(s) = s^2$

2. $\hat{M}(s) = 2s$

3. $\ddot{M}(s) = s$

4. $\bar{M}(s) = s + 1$

The reason this intense abbreviation can be confusing is that, in one sense, we're only describing the output, or what the machines spit out. Both sides of ~~an equation~~[4] a sentence like $M(s) = s^2$ are talking about the thing that the machine M spits out. But in another sense, this sentence is talking about all three things at once: the machine itself, the stuff we put in, and the stuff we get out. Take another look at this crazy abbreviation:

$$M(s) = s^2$$

We're talking about the output on both sides, sure. But our abbreviation for the output — namely $M(s)$ — is a weird hybrid that we built out of the two other abbreviations: the abbreviation for the machine itself, which was M, and the abbreviation for the stuff we put in, which was s. So the sentence $M(s) = s^2$ has *three abbreviations* just on the left side. As if that weren't enough, we then go on to describe the operation of the machine. The right side of this sentence, s^2, is a description of the machine's output, written in terms of the input.

We've said the same thing in two ways: the $M(s)$ on the left side is our *name* for the output, and the s^2 on the right side is a *description* of the output. Since we've said the same thing in two ways, we throw an equals sign between them, and we've described this particular machine in a way that expresses infinitely many different sentences in a few symbols! It expresses infinitely many different sentences because it tells us: If you feed 2 to the machine M, it spits out 4. If you feed 3 to the machine M, it spits out 9. If you feed 4.976 to the machine M, it spits out whatever $(4.976) \cdot (4.976)$ is, and so on.

1.1.3 Things We Rarely Hear

The idea of these machines is very simple. Like I mentioned before, they're usually called "functions," which is an odd name, and it doesn't convey the

---

4. The term "equation" causes most people to experience a discomforting combination of fear and boredom, a mixture of emotions that anyone familiar with the sympathetic nervous system might have thought impossible. So what is an "equation"? We've already talked about how mathematical symbols are just abbreviations for things that we could be describing with words. Against this background, "equations" are just sentences. Abbreviated ones. Once we realize that, the term "equation" doesn't seem quite so bad. We'll use both terms throughout the book.

idea very well. Not only is the word "function" a bit confusing at first, but the common abbreviations used to talk about functions can be pretty counter-intuitive when we first encounter them. Here are some reasons such a simple idea can be so confusing:

1. They don't always explain that we're talking about machines.

2. They don't always explain that everything we're saying about these machines *could* be expressed in words, but we're lazy (the good kind of lazy!), so we're doing it in a highly abbreviated form.

3. They don't always explain that we're using the shortest abbreviations we can, or how we built a weird hybrid abbreviation out of two other abbreviations.

4. They don't always distinguish between the name of a machine, M, and the name for its output, $M(s)$. Sometimes books will talk about "the function $f(x)$," which isn't really what they mean. To be fair, sometimes it's useful to use our language incorrectly like this (it is *our* language after all, so we're allowed to), but we'll try not to do that until we're much more familiar with the idea.

We rarely hear all this. A sizable proportion of textbooks and lectures just say that a function is "a rule that assigns one number to another number," then they draw some graphs, pace back and forth a bit, and start writing stuff like $f(x) = x^2$ a lot. For some (including myself when I was first exposed to the idea), this is a rather confusing conceptual leap.

I want to draw your attention to something puzzling in the previous sentence. Why do they write x? We wrote s instead of *stuff* because we got tired of writing out the whole word. But then what on earth is x an abbreviation for? Maybe it's not an abbreviation for anything. There's no law that all of the names we give things have to be abbreviations. Maybe the x is like Harry S. Truman's middle name: it looks like an abbreviation, but it's really not! Maybe the letter is the name itself. As it turns out, the letter x *is* an abbreviation for something. What? Let's take a break and find out.

1.1.4 The Unbearable Inertia of Human Conventions

Why do textbooks always use x? The answer is pretty funny.[5] It's actually a bastardized translation from Arabic. See, back in the old days, some Arabic mathematicians went through a train of thought similar to the one we've gone through here, and they decided to use the word *"something"* for the same reason that we used *"stuff."* Perfectly reasonable. The idea is to always choose

---

5. This explanation comes from a guy named Terry Moore, in his wonderful short talk "Why is 'x' the Unknown?" So credit goes to him.

Figure 1.2: Generally speaking, humans are slow to change things.

abbreviations that remind you of the thing you're abbreviating, so that you don't have to memorize anything. Up to this point, everything made sense. Then came the problem. The first letter of the word "something" in Arabic makes a sound similar to the sound "sh" in English. It turns out that the Spanish language has no "sh" sound, so when all of this Arabic mathematics was translated into Spanish, the Spanish translators chose the closest thing they could think of. This was the Greek letter "chi," which makes a "ch" sound (as in Bach, not Cheerios). The letter chi looks like this:

Look familiar? Later, as you might expect, this χ turned into the familiar letter x from the Latin alphabet... and this bastardized abbreviation continues to haunt our textbooks as the most common abbreviation for *stuff*.

The Arabic mathematicians were smart folks, and they chose their abbreviations well. They could do so because they were essentially in the same type of situation we are: in a world without much mathematics, inventing it as they go. Like them, we can always abbreviate things however we want. For example, consider the following two problems. Don't bother doing them. Just stare at them for a few seconds.

1. Here's a description of the f machine:
   $f(x) = x^2 - (5 \cdot x) + 17$
   What does the f machine spit out when we feed it the number 1?

2. Here's a description of the $\circlearrowleft$ machine:
   $\circlearrowleft (\ast) = \ast^2 - (5 \cdot \ast) + 17$
   What does the $\circlearrowleft$ machine spit out when we feed it the number 1?

We don't have to do either of these problems to see that they have the same answer (it's 13, but that's not the point). We're describing the same machine, and we're feeding it the same number in both cases, so we know they have the same answer, even if we didn't bother to figure out what that answer is. Everyone knows that we can abbreviate things however we want. And yet when I'm explaining some piece of mathematics to someone, and I change abbreviations so that we can remember what we're talking about, one of the most common things I hear is "Oh! I didn't know we could do that!" It's important that we practice changing abbreviations, because a lot of ideas in mathematics look scary and complicated when we use one set of abbreviations, but suddenly seem obvious when we use another. We'll see some funny examples of this later.

1.1.5 The Different Faces of Equality

There is another widespread problem with standard mathematical notation that causes tremendous unnecessary confusion to newcomers. That is the need to use different-looking versions of the equals symbol to remind ourselves why things are true.

When we use the normal equals symbol = in this book, we will mean the same thing that all mathematics books mean: $A = B$ means that A and B refer to the same thing, even though they might look different. Therefore, the symbol = just tells you *that* something is true, but it doesn't tell you *why* it's true. We can do better by occasionally using different-looking symbols. In the rest of the book, these three symbols

$$\equiv \qquad \overset{\text{Force}}{=} \qquad \tag{2.17}$$

will all mean the same thing. They all mean "the things on either side of me are the same," but they're different ways of reminding us *why* those two things are the same.

By far the most common alternative version of the equals symbol that I'll use is $\equiv$, and it says that two things are equal because of some abbreviation we're using. A few examples will illustrate what I mean. One of the cases where the symbol $\equiv$ will show up is whenever we're defining something. For example, in the above discussion when we wrote $M(s) = s^2$, we really could have written $M(s) \equiv s^2$. I only used = because we hadn't talked about $\equiv$ yet. The $\equiv$ symbol in the above sentence says "$M(s)$ and s^2 are the same thing, but not because of some mathematics that you missed. We're just using $M(s)$ as an abbreviation for s^2 until we say otherwise."

Now, using the $\equiv$ symbol for definitions isn't unique to this book. Lots of books do that.[6] However, in an attempt to get the most explanatory bang for our notational buck, we'll use this symbol in a slightly more general way. We will use $\equiv$ in any equality that is true simply because of some abbreviation that we're using, and not because of any mathematics that you missed. Just to choose a completely contrived example that refers to absolutely nothing, I might say something like this: Using the fact that $M(s) \equiv s^2$, we can write

$$1 + 5\left(9 - \frac{72}{M(s)}\right)^{1234} \equiv 1 + 5\left(9 - \frac{72}{s^2}\right)^{1234}$$

Just to stress the point, you should be able to understand the above pile of symbols even if you had never heard of addition, multiplication, or numbers! Since it involves $\equiv$, all it is really saying is that the thing on the left and the thing on the right are equal because of some abbreviation we're using, and not because of some mathematics that you missed. As such, whenever you see this kind of equals sign, you're not allowed to be intimidated. There's nothing to be scared about, because equations with $\equiv$ aren't really saying anything. However, we'll see throughout the book how helpful it can be to change back

---

6. Ironically, this seems to be more common in advanced books than it is in introductions, where it's most needed.

and forth between different abbreviations, so it's worth having a special kind of "equals" to remind ourselves when that's all we're doing.

Another way of using equals shows up whenever we're forcing something to be true, and seeing what happens as a result. This is the version of equals that people are using when they say something like "Set yadda yadda equal to zero." This is a strange concept, so it's worth looking at a simple example. When a textbook insists that you "solve $x = x^2$ for x," it's not always clear what that means. The equals sign is clearly being used in an odd way here. First, the sentence $x = x^2$ isn't even true, at least not in general. After all, if the sentence $x = x^2$ were always true, then 2 would equal 4, and 10 would equal 100, and so on. Here's the idea:

> **What they say:** Solve $x = x^2$ for x.
> **What they mean:** Figure out which particular *stuff* makes the sentence $(stuff) = (stuff) \cdot (stuff)$ true. Ignore all the *stuff* that makes it false.

Since this meaning of equals is so different from $\equiv$, we'll write it a different way. How about this:

$$x \stackrel{\text{Force}}{=} x^2$$

To reiterate, all of these different versions of "equals" mean the same thing as the the normal $=$ symbol. The new ones just remind us of why something is true. Even if distinguishing between these different kinds of equals seems unnecessary now, we'll see soon how much easier it makes things.

Attention Reader! This is important! Whatever you do, please don't agonize about learning exactly when you should use which type of equals symbol! And if any teachers are reading this, please, for the love of mathematics, do *not* assign exercises where you make people determine whether $=$ or $\equiv$ or $\stackrel{\text{Force}}{=}$ should be used in some equation or another. This isn't a pedantic distinction we're making because of a compulsive overattention to irrelevant details. It's just a quick and easy way to remind ourselves why something is true. For the same reasons, I'll also occasionally throw a number above an equals sign, like this:

$$Blah \stackrel{(3)}{=} Blee$$

What that means is "*Blah = Blee* because of equation 3." By doing this, each equation can become a way to check and see if you understand an idea, but only if you want to. That is, you can try to figure out why something is true on your own, but whenever you get sick of that, the equals symbol tells you where you can look to find the reason why. I always wished more textbooks would do this. But enough about notation. Time for creation!

1.2 How to Invent a Mathematical Concept

What I cannot create, I do not understand.
—Richard Feynman, from his blackboard at the time of his death

Before we invent calculus, we'll need to know how to invent things in the first place. Specifically, we'll need to know how to invent mathematical concepts. We'll illustrate the creation process with two simple examples: the area of a rectangle and the steepness of a line.[7] It doesn't matter if you already know how to compute both of those things. Everyone has something to gain from a discussion of these issues, whether in their own understanding or in their teaching, because the invention process is so rarely discussed.

When we're inventing mathematics from scratch, we always start with an intuitive, everyday human concept. The process of inventing a mathematical concept consists of attempting to translate that vague qualitative idea into a precise quantitative one. No one can *really* visualize anything in five dimensions, or seventeen dimensions, or infinitely many dimensions, so how do mathematicians define something like "curvature" in a way that allows them to talk about the curvature of higher-dimensional objects? How do human mathematicians arrive at their definitions, when these definitions are often so abstract that it seems as if one would have to be endowed with superhuman capacities of higher-dimensional intuition in order to "see" the truth?

This process is not as mysterious as it seems. Creation is simply translation, from qualitative to quantitative. Hopefully, explicit education in the creation process, at all levels of mathematics, will one day find its rightful place in the curriculum alongside lesser concerns like addition, multiplication, lines, planes, circles, logarithms, Sylow groups, fractals and chaos, the Hahn-Banach theorem, de Rham cohomology, sheaves, schemes, the Atiyah-Singer index theorem, the Yoneda embedding, topos theory, hyper inaccessible cardinals, reverse mathematics, the constructible universe, and everything else we teach students in mathematics from elementary school to the postdoctoral level. It is *so much more important.*

1.2.1 Mining Our Minds: Inventing Area

In this section we'll illustrate the essence of how to invent a mathematical concept by examining the concept of area, in the simplest possible case: the area of a rectangle. The fact that a rectangle with length ℓ and width w has area ℓw is a simple idea, and you almost certainly know it already. But try to forget it. Let's imagine that we have no idea that the area of a rectangle is its length times its width.

---

7. We'll later find that these two concepts form the backbone of all of calculus. The latter is the basis of the "derivative," and the former is the basis of the "integral." These concepts are opposites, and the precise sense in which they're opposites is described by the so-called "fundamental theorem of calculus."

Let's assume that we know roughly what we mean by "area" in a non-mathematical sense. That is, we know it's a word that describes how big a two-dimensional thing is, but we don't know how to tie that concept to anything mathematical. At this point, we can use the abbreviation A to stand for area, and write pointless things like $A = ?$, but that's all we can do. However, based on our everyday non-mathematical concept, we know this for certain:

First thing our everyday concept tells us:
Whatever we mean by the "area" of a rectangle, it somehow
depends on the rectangle's length and width. If someone else has
a definition of "area" that has absolutely nothing to do with
length or width, then that's fine, but it's not what we mean by
"area."

Let's make an abbreviation for all those words. We can express the above sentence in a highly abbreviated form by writing

$$A(\ell, w) = ?$$

instead of simply writing $A = ?$ like we did before. The new stuff inside the parentheses just says, "This *somehow* depends on the length and width, and I'll abbreviate those as ℓ and w. I don't know anything else."

Notice that this abbreviation is similar to our abbreviation for machines from earlier. Either we could say "I'm not talking about machines, I'm just abbreviating," or we could take the analogy with our machine abbreviations seriously and say, "Once we've really said what we mean by area in a precise way, it should be possible to build a machine that spits out the area of a rectangle when I feed it the length and the width. It's this machine that I'm calling A." Either of these interpretations will get us to the same place, so pick your favorite and let's keep going.

Since we're building the precise mathematical concept of area by starting with our intuitive everyday concept, we don't have any numbers to start with. If we don't have anything quantitative to start with, then we have to start with something qualitative. While there are no laws telling us what to do, we want to make sure our precise concept acts like our everyday concept. Toward that end, how's this for another thing our everyday concept tells us:

Second thing our everyday concept tells us:
Whatever we mean by the "area" of a rectangle, if we double the
width without changing the length, then we have two copies of
the original rectangle, so the area should double. If someone else
has a different definition of "area" that doesn't behave like this,
then that's fine, but it's not what we mean by "area."

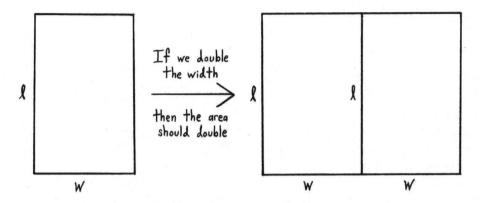

Figure 1.3: Whatever we mean by "area," if we double the width of a rectangle without changing the length, then the area should double.

In case that didn't make sense, look at Figure 1.3. Our vague, intuitive, non-mathematical concept of area isn't enough to tell us that the area of a rectangle is length times width, but it *is* enough to tell us that if we double the width, then the area should double (as long as we keep the length the same). We can abbreviate this idea by writing:

$$A(\ell, 2w) = 2A(\ell, w)$$

For the same reason, if we double the length without changing the width, the area should double too. We can abbreviate this as

$$A(2\ell, w) = 2A(\ell, w)$$

Even more, there's nothing special about the word "double" in this sentence. If we triple the width, then we have three copies of the original thing, so the area should triple. Same for length, and same for quadrupling, or for multiplying by any other whole number. What about numbers other than whole numbers? Well, if we change the length from ℓ to "one and a half ℓ" (without changing the width) then we have one and a half copies of the original thing, so the area should be one and a half of the original. Clearly, whatever we mean by area, sentences like this capture our intuitive concept, no matter what the amount of magnification is. We can abbreviate this infinite bag of sentences all at once by writing

$$A(\#\ell, w) = \#A(\ell, w) \tag{1.1}$$

and

$$A(\ell, \#w) = \#A(\ell, w) \tag{1.2}$$

no matter what the number # is. But if that's true, then we can *trick the mathematics* into telling us the area of a rectangle, by thinking of ℓ as $\ell \cdot 1$, and thinking of—

(*A faint rumbling noise is heard in the distance.*)

Aaah! What was that?!...Was that you?
Reader: Uhh...I don't think so. I think that was on your end.
Author: You sure?
Reader: Yeah, pretty sure.
Author: Hmm...Okay, where were we? Right, equations 1.1 and 1.2 tell us that we can pull numbers outside the Area machine, no matter what those numbers are. But if that's true, then there's nothing preventing us from being tricky and pulling the lengths and widths themselves outside! They're just numbers, after all. Since ℓ is the same as $\ell \cdot 1$ and w is the same as $w \cdot 1$, we can sneakily use the two facts provided by equations 1.1 and 1.2 on the numbers ℓ and w themselves, like this:

$$A(\ell, w) \overset{(1.1)}{=} \ell A(1, w) \overset{(1.2)}{=} \ell w A(1, 1) \tag{1.3}$$

Which says that the area of a rectangle is length times width...times some extra thing? What on earth is that $A(1,1)$ doing there?!

It turns out that equation 1.3 is trying to tell us about the concept of units. It's telling us that we can figure out the area of *any* rectangle, but only once we've decided on the area of a *single* rectangle, the rectangle with a length of 1 and a width of 1 (or any other rectangle). If we're measuring lengths in light-years, then we want $A(1,1)$ to be the area of one square light-year. If we're measuring lengths in nanometers, then we want $A(1,1)$ to be the area of one square nanometer.

We usually get around this by forcing $A(1,1)$ to be 1, but that's just for convenience. We *could* choose to represent $A(1,1)$ by the number 27 if we wanted, and then we'd have formula $A(\ell, w) = 27\ell w$. That may look odd, but it wouldn't be wrong in the least. Instead of forcing $A(1,1)$ to be 1 or some other number, we can interpret equation 1.3 in a different way, by rewriting it like this:

$$\frac{A(\ell, w)}{A(1, 1)} = \ell w$$

What this says is that we don't *have* to talk about units (that is, we don't have to decide what we want $A(1,1)$ to be), but we can no longer talk about the areas themselves. This interpretation tells us that *something* is equal to length times width, but it's not an area. It's a "ratio" of areas, or however many $A(1,1)$'s you can fit into $A(\ell, w)$.

After all that inventing, we see that the mathematics was really smarter than we were — not only did it try to tell us about the concept of units, but it automatically tells us how to convert areas from any system of units to any other (say, from nanometers to light-years). This is one of the many cases we'll see throughout mathematics where inventing a concept for ourselves, even a simple one that we're thoroughly familiar with already, can give us much more insight into the concept itself.

Even better, it's not hard to convince ourselves that this same argument should work in any number of dimensions. Let's imagine that we have a three-dimensional box-type thing, and let's abbreviate its length, width, and height as ℓ, w, and h. For the same reasons as in the area example, if we double the height (say) without changing the length or width, then we've got two of the original box, so the volume should double. Just like before, there's nothing special about the word "double," and the same idea makes sense for any amount of magnification. Same for length and width. So in three dimensions, these three things should be true for any number #, not necessarily the same number in all three sentences:

$$V(\#\ell, w, h) = \#V(\ell, w, h)$$
$$V(\ell, \#w, h) = \#V(\ell, w, h)$$
$$V(\ell, w, \#h) = \#V(\ell, w, h)$$

Just like before, we can use these three ideas on the numbers ℓ, w, and h themselves, and write

$$V(\ell, w, h) = \ell w h \cdot V(1, 1, 1)$$

Now we can do something much more strange and interesting: we can start to say things about higher-dimensional spaces. If n is some large number, then we can't picture anything in n-dimensional space. No one *really* can. At this point, we're not even sure exactly what we *mean* by the phrase "n-dimensional space." That's fine! We're completely free to say "Whatever I mean by n-dimensional space, and whatever I mean by the n-dimensional version of a rectangular box-type thing, they had better behave similarly enough to their cousins in two and three dimensions that we can make the same argument we just made. If they don't behave like that, then that's not what I mean by n-dimensional space right now." Having said that, we can confidently write:

$$V(\ell_1, \ell_2, \ldots, \ell_n) = \ell_1 \ell_2 \cdots \ell_n \cdot V(1, 1, \ldots, 1)$$

where V stands for "whatever we want to call volume in an n-dimensional space" and we've decided not to give all the different directions their own quirky names anymore like we do in two and three dimensions. It's easier to just abbreviate them all by ℓ, and then attach a different number to each one so we can tell them apart (hence: $\ell_1, \ell_2, \ldots, \ell_n$).

Even though we can't even begin to picture what we're talking about, we can still use it to infer other things. For example, if all of the sides of this

n-dimensional box-type thing are the same length (let's call it ℓ), then we'd have an n-dimensional cube-type thing. So if we force that ugly $V(1, 1, \ldots, 1)$ piece to be 1 (just for convenience), then we can infer that the "volume" of this higher-dimensional box is $V = \ell^n$. We can confidently say things about its "n-dimensional volume," even though we can't even begin to visualize what we're saying!

So in summary, we saw that by thinking about our everyday concept of area, and abbreviating our thoughts in a way that expressed infinitely many sentences at once, our vague ideas turned out to *force* the area of a rectangle to be $\ell w A(1, 1)$. We thus found not only the familiar "length times width" formula, but another piece we forgot to consider, though the mathematics was nice enough to remind us about it: the concept of units.

Now, we'll use this simple invention to help us understand and visualize some of the so-called "laws of algebra" in a way that ensures that we'll never have to memorize them ever again. Onward!

1.2.2 How to Do Everything Wrong: A Sermon on the Folly of Memorization

> *But it required a few years before I perceived what a science teacher's job really is. The goal should be, not to implant in the student's mind every fact that the teacher knows now; but rather to implant a way of thinking that will enable the student, in the future, to learn in one year what the teacher learned in two years. Only in that way can we continue to advance from one generation to the next. As I came to realize this, my style in teaching changed from giving a smattering of dozens of isolated details, to analyzing only a few problems, but in some real depth.*
> —E. T. Jaynes, *A Backward Look to the Future*

One of the worst things about many early mathematics courses (at least the ones I had to take) is that the teachers seem to have somehow become convinced that the purpose of mathematics courses is to teach you facts about mathematics. I couldn't disagree more. You might wonder what I think mathematics courses should be about, if not mathematics. This is important, so let's say it once and for all, and put it in a box.[8]

---

8. Why write this in a box with such a grandiose name? Good question! Full disclosure: Sometimes when you're writing a book (as I've learned since I started writing this one), it's fun to insert the occasional homage to something you love. This box is an homage to my favorite textbook: E. T. Jaynes's posthumously published magnum opus *Probability Theory: The Logic of Science*. Perhaps because he died before completing the book — but also because Jaynes was a fiery sort of guy — it's full of Jaynes's strangely heartwarming personal quirks, and various other things one rarely finds in textbooks. One such item is a box called "Emancipation Proclamation" in Appendix B. I always loved that section. Now I'm writing a book of my own, so I get to give Jaynes a well-deserved homage.

Declaration of Independence

The purpose of math courses is not to create students who know things about math. The purpose of math courses is to create students who know how to think.

We'd better be careful here, because at first the phrase "learning how to think" might conjure up an image of a beefy Stalin-esque man in a police uniform, holding a whip and shouting, *"Think this way!"* That's not what I mean at all.

Mathematics is an entire world where nothing is accidental, and where the mind can train itself with an intensity and precision unmatched by any other subject. Moreover, in the course of training your mind, you'll accidentally be learning the subject that just happens to describe everything about the world. It's incredibly useful, but its practicality comes as a side effect of training the mind. Speaking of things that are important enough to write in boxes, here's something else they never tell you:

Mathematics is NOT about

Lines, planes, functions, circles, any of the things you learn about in mathematics courses.

Mathematics IS about

Sentences that look like:
"If this is true, then that is true."

Once we recognize this, we can see two things immediately. First, it's obvious why training the mind in this way is useful, no matter what you are doing. Second, it's obvious that mathematics courses are focused on exactly the wrong things.

Let's look at a particular example of doing the wrong thing. In algebra courses, roomfuls of sleepy humans are told about something called the "FOIL" method, which is an abbreviation for "First, Outer, Inner, Last." It's a way to remember sentences like this:

$$(a + b)^2 = a^2 + 2ab + b^2$$

or more generally,

$$(a + b)(c + d) = ac + ad + bc + bd$$

We can see immediately that the name "FOIL" is a way of helping you to *remember a fact* about mathematics, not a way of teaching you how to reinvent that fact for yourself whenever you need to. What on earth is the point of

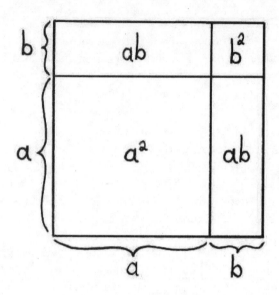

Figure 1.4: This is basically all FOIL is saying.

that? Most students can see what the point of that is better than the teachers can: there isn't one. Let's invent both of these facts in a way that guarantees that we'll never have to remember them again.

If we take a piece of paper and draw a picture on it, drawing the picture doesn't change the area of the piece of paper. This is true whether we draw a house or a dragon or anything else. So suppose we're playing around with the ideas we've invented, and we come across something that looks like $(a + b)^2$. We can think of that as the area of a square. Which square?

Well, if a square has length *blah* on each side, then its area is $(blah)(blah)$, which can be abbreviated as $(blah)^2$. So we can think of $(a+b)^2$ as the area of a square whose sides are all $a + b$ long. Let's draw that square, and then let's draw a picture on it. This is what we're doing in Figure 1.4. The picture we'll draw will look like a weird lopsided + sign, but it's just two straight lines that divide up the sides into a piece that's a long and a piece that's b long. This will give us two ways of talking about the same thing. Since drawing all over the square didn't change its area, we can see that

$$(a + b)^2 = a^2 + 2ab + b^2$$

Now you never have to remember that formula ever again. Let's see if this same type of argument also lets us invent this more complicated sentence:

$$(a + b)(c + d) = ac + ad + bc + bd$$

Any two numbers multiplied together, that is, anything of the general form $(blah) \cdot (blee)$, can be thought of as the area of a rectangle that is $(blah)$ long in

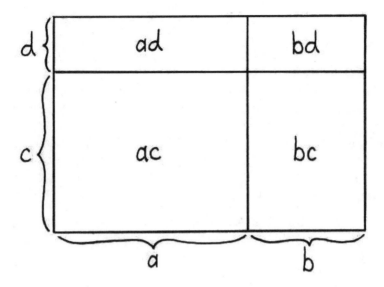

Figure 1.5: This is *really* all FOIL is saying.

one direction and (*blee*) long in the other. Let's draw a picture of this, where (*blah*) is $(a+b)$ and (*blee*) is $(c+d)$. The picture is in Figure 1.5. The picture says that the big rectangle's area is just the area of all the small rectangles added up. So the gist of the picture can be expressed in abbreviated form by the sentence

$$(a + b)(c + d) = ac + ad + bc + bd$$

Now you never have to remember that formula ever again. If you ever forget it, you can invent it for yourself. You shouldn't even *try* to remember either of these formulas. In fact, you should probably try to forget both of them immediately! Every mathematics classroom should have this inscribed over the blackboard:

The First Commandment of Mathematics Education
A mathematics teacher should not
urge students to remember, but to forget.

Since the goal is for you to be able to go through the same process of reasoning yourself, you should not try to memorize the steps in this argument, but rather to understand the argument well enough that if you ever forget either of these formulas (which you *should*), then you can reinvent them for yourself on the spot in a few seconds. When you do this, you'll find that eventually you "memorize" things by accident, just because you understand them so well. The way to check whether you've been successful in this Zen-like process of "learning without remembering" is to see whether you can apply the same process of reasoning in *new* places.

Here's the logic of it: if you can apply the same reasoning in places you've never seen, then it's impossible for you to have just memorized the facts themselves. New contexts act like a sieve that filters out the possibility of memorization. Unfortunately, in an environment that punishes experimentation and failure (e.g., school), trying things out in new contexts becomes a source of anxiety rather than the intellectually gratifying play that it should be. Let's ignore all that and just play.

Inventing Stuff

1. Above, we discussed the silly acronym "FOIL," which stands for "First, Outer, Inner, Last." This method of remembering a fact rather than understanding a process would lead us to come up with infinitely many different "methods," all of which we would have to memorize. For example, it happens to be true that

$$(a + b + c)^2 = a^2 + b^2 + c^2 + 2ab + 2bc + 2ac$$

   That's a really ugly sentence, and no one in their right mind would want to memorize it. If our goal were just to remember it, rather than to learn general strategies of reasoning, then we might use the same approach as the guy who invented the acronym "FOIL" and call this the "LT.MT.RT.15.16.24.26.34.35" method.[9] Let's *not* do that. Instead, use the same strategy we used above (drawing a picture and staring at it) to invent this ugly expression for yourself. Hint: Draw a square and break up each side into three pieces, instead of two like we did.

2. This time let's play in three dimensions, and see if the same type of thinking works there. We really don't want to have to remember an ugly sentence like

$$(a + b)^3 = a^3 + 3a^2b + 3b^2a + b^3$$

   So instead of remembering it, let's invent it in the same way we did above: by drawing a picture and staring at it. Hint: Draw a cube and break up each edge into two pieces. It might help to stare at Figure 1.6. Also, you might want to try the one above this first, because this one can be a bit hard to visualize, so if you get stuck, it's easy to get discouraged and think that you don't understand the idea, even if you really do.

9. This stands for "Left Two, Middle Two, Right Two, First and Fifth, First and Sixth, Second and Fourth, Second and Sixth, Third and Fourth, Third and Fifth" method, and it illustrates the absurdity that the "FOIL" mindset (memorizing facts) leads to if taken just one step further.

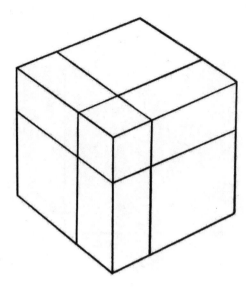

Figure 1.6: This picture might help with invention #2.

Even though this kind of reasoning lets us invent things that other people told us to memorize, there are two things about it that are less than perfect. First, it's not as simple as it could possibly be (I'll say what I mean by this soon). Second, it's basically useless on stuff like $(a + b)^4$ or $(a + b)^{100}$, because the human mind has trouble visualizing things in dimensions higher than three. It turns out that there's a remedy for both of these things.

Instead of using the style of thinking we used above to shatter stuff like $(a+b)^4$ into smaller pieces, let's use something simpler. I know it might seem weird to solve a harder problem by using a simpler method, but this strategy turns out to work all over the place in mathematics. It's really fortunate that things work out that way! Here's the simpler method.

Say we've got a piece of paper, and imagine tearing it into two pieces however you want. Even if we don't know either of the areas numerically, it's clear that the area of the original piece is the area of both torn pieces added together. We can reinvent the "FOIL" method and all of its more complicated friends in any number of dimensions (whether we can picture what's going on or not) just by applying this tearing idea over and over. Let's write down the tearing idea in abbreviated form.

Suppose we're inventing things and we get to a point where we've written down something that looks like $(stuff) \cdot (a + b)$, or maybe $(a + b) \cdot (stuff)$. They're the same thing, so this argument works for either. Just like before, we can picture this as the area of a rectangle with two sides that are $(stuff)$ long and two other sides that are $(a + b)$ long. If we tore this rectangle right along the line in the middle-ish of Figure 1.7, then we'd have a piece with area $a \cdot (stuff)$, and another piece with area $b \cdot (stuff)$. The tearing doesn't change

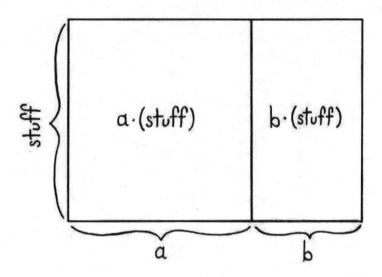

Figure 1.7: The obvious law of tearing things: if you tear something into two pieces, then the area of the original thing is just the areas of the two pieces added together. To say the same thing in abbreviated form: $(a + b) \cdot (stuff) = a \cdot (stuff) + b \cdot (stuff)$. Textbooks usually call this fact the "distributive law."

the total area (because we're not throwing any pieces away), so it has to be true that

$$(a + b) \cdot (stuff) = a \cdot (stuff) + b \cdot (stuff)$$

I'll call this the obvious law of tearing things, but the name isn't important. Call it whatever you want. Textbooks call it the "distributive law," which sounds a bit pretentious, but that name makes sense too. After the next few paragraphs, we usually won't need a name for the idea.

Just like the obvious law, all of the so-called "laws of algebra" can be thought of as abbreviations for simple visual ideas. For example, the fact that multiplication works the same both ways (i.e., $a \cdot b = b \cdot a$) just says that the area of a rectangle doesn't change when we turn in on its side. That's a really simple idea too, so they call it the "commutative law of multiplication" to scare you. But it just means we can switch the order of multiplication anywhere we want. In particular, we can use the obvious law even when $(stuff)$ shows up on the left of $(a + b)$, instead of on the right.

Now, if we wanted to sound like a textbook, we could have written c instead of $(stuff)$ when we wrote down the obvious law. That would be okay too, but I wrote $(stuff)$ to remind ourselves that the law is true no matter what $(stuff)$ looks like. If $(stuff)$ happens to be two things added together (or rather, if we *choose* to think of it that way), then we can replace the $(stuff)$ with something like $(c + d)$, and rewrite the obvious law like this:

$$(a + b) \cdot (c + d) = a \cdot (c + d) + b \cdot (c + d)$$

But then using the obvious law again (on each of the pieces on the right), we get

$$(a + b) \cdot (c + d) = ac + ad + bc + bd$$

which is just the expression that we invented earlier by drawing pictures. The above sentence is also just the "FOIL" method. But since we invented the "FOIL" method using the obvious law, we never have to remember it ever again. Ready, set, forget it forever!

As mundane as the obvious law seems, it turns out to offer us a window into higher dimensions. The visual way of thinking about $(a + b)^3$ required us to picture a three-dimensional object (a cube), and we quickly noticed that this method didn't really help us on $(a+b)^4$ or any larger powers, because we can't visualize four-dimensional objects. However! Even though we're not interested in algebraic tedium like expanding $(a+b)^4$ for its own sake, we might be interested in deeper questions like how to chop up a four-dimensional cube along each of its three-dimensional "surfaces," none of which we can really picture due to the limitations of the human brain. But while we primates run into problems with the visual method, the obvious law has no such limitation. So if we felt like it, we could simply apply the obvious law to something like $(a+b)^4$ repeatedly, and once we had completely unraveled it, the resulting (admittedly long) expression would give us a bit of insight into four-dimensional geometry. For example, the number of pieces in the resulting expression would be the number of different pieces that a four-dimensional cube would be carved into, if we sliced it along each of its three-dimensional faces. I have no idea how to picture what I just said, but it's true! It *has* to be. Simply by using this mundane fact about tearing rectangles in two, we can coax the mathematics into telling us something that reaches far beyond the human brain's powers of visualization.

1.2.3 Unlearning Division / Forgetting Fractions

As we saw above, the small amount of mathematics we've invented so far turns out to be more than enough to reinvent many of the so-called "laws of algebra." In the next few paragraphs, we'll show how two more such laws follow naturally from what we've already done. Understanding where these "laws" come from will allow us to more comfortably forget the things we've been told about the strange nouns known as "fractions," and the correspondingly strange verb "division."

First, you may have been told in the past that we can (warning: jargon ahead) "cancel common factors from the numerator and denominator of a fraction." That is, $\frac{ac}{bc} = \frac{a}{b}$. However, in the universe we've invented thus far, there is no such thing as "division": a symbol like $\frac{5}{9}$ is nothing more than an abbreviation for $(5)(\frac{1}{9})$, which is just multiplication of one number by another. This may seem like cheating, because the symbol $\frac{1}{9}$ certainly looks like it involves division. But division is a concept imported from outside our universe. We're simply using the symbol $\frac{1}{9}$ as an abbreviation for whatever number turns into 1 when we multiply it by 9. To put it another way, we are defining symbols like $\frac{1}{9}$ by their *behavior*, and treating whatever numerical values they might have as a secondary afterthought that we, personally, choose not to focus on. Defining these objects by their behavior also makes it simple to convince ourselves that sentences like $\frac{ac}{bc} = \frac{a}{b}$ are true. Here's how.

We've convinced ourselves that the order of multiplication shouldn't matter, and we also know that $(\#)(\frac{1}{\#}) \equiv 1$, for any number $\#$. The following argument uses only these two ideas to show that the sentence $\frac{ac}{bc} = \frac{a}{b}$ has to be true. Notice that every equals sign below is $\equiv$, except for one. The one that isn't $\equiv$ just involves switching the order of multiplication. Since we can think of that as turning a rectangle on its side (i.e., swapping its length and width without changing its area), I'll write $Turn$ above the equals sign where we use that fact. Here we go:

$$\frac{ac}{bc} \equiv (a)(c)\left(\frac{1}{b}\right)\left(\frac{1}{c}\right) \overset{Turn}{=} (a)(c)\left(\frac{1}{c}\right)\left(\frac{1}{b}\right) \equiv (a)\left(\frac{1}{b}\right) \equiv \frac{a}{b}$$

So the fancy-sounding "law" about "canceling" "common factors" from the "numerator" and "denominator" of a "fraction" really isn't a law at all. Or maybe it is. Or maybe the term "law" isn't really meaningful. Either way, it's really just a consequence of the fact that (i) we decided when we began that the order of multiplication shouldn't matter, and (ii) we're using $\frac{1}{stuff}$ as an abbreviation for whichever number turns into 1 when we multiply it by *stuff*.

Here's another quick one. At some point, we've probably all been told that we can "break fractions apart." That is, someone told us that the sentence $\frac{a+b}{c} = \frac{a}{c} + \frac{b}{c}$ was true, probably without much justification. However, this is just the obvious law of tearing things in disguise. Let's see why. Again, notice that all the equals signs in what follows are $\equiv$, except for one. The one that isn't $\equiv$ is where we use the obvious law, so I'll write $Tear$ above that equals sign. Here we go:

$$\frac{a+b}{c} \equiv (a+b)\left(\frac{1}{c}\right) \overset{Tear}{=} (a)\left(\frac{1}{c}\right) + (b)\left(\frac{1}{c}\right) \equiv \frac{a}{c} + \frac{b}{c}$$

So the ability to break fractions apart isn't some special "law" about fractions. It really has nothing to do with fractions at all. It's just the obvious law of tearing things, written in a slightly unusual way.

The point: Faced with a scary-looking sentence involving a lot of division, we can just rewrite it in the language of multiplication. Surprisingly often, this simple change of abbreviations will make things look a lot simpler, and it also lets us avoid having to memorize all sorts of quirky behaviors about fractions.

Alright! Our inventing muscles still aren't exercised enough yet, so we'll look at one more example of how mathematical concepts are invented, and then we'll end the chapter by summarizing some general principles about this mysterious process of moving from the qualitative to the quantitative.

1.2.4 Of Arbitrariness and Necessity: Inventing Steepness

> *An old definition of the lecture method of classroom instruction: a process by which the contents of the textbook of the instructor are transferred to the notebook of the student without passing through the heads of either party.*
> —Darrell Huff, *How to Lie with Statistics*

When we first hear about the idea of "slope" in mathematics, they usually just tell us that it's "rise over run," briefly say what that means, and then start doing some examples. I never heard anyone explain why it's not "run over rise" or "52 rise over 98 run," and you probably didn't either. Want to know *why* they didn't tell us? Because we *could* define slope to be "run over rise" or "76 rise over 38 run" or any other number of crazy things! It all depends on how much of our vague, everyday concept of "steepness" we want to force our mathematical concept to have, how we choose to do this, and what we think sounds reasonable. What's more, our choice of a formal definition often depends (more than anyone wants to admit) on subjective aesthetic preferences, i.e., what we think is pretty.

The goal of this section is to see why the above paragraph is true by inventing the concept of steepness, or, as they usually call it, "slope." This one is a bit more involved than the invention of area, but don't worry. In both cases, the process of invention follows essentially the same pattern.

We know what the word "steepness" means in an everyday, non-mathematical sense, and we want to use this everyday concept to build a precise mathematical one. We'll focus on straight lines for the moment, just to make things easy on ourselves, and we'll deal with curvy things when we invent calculus in Chapter 2 (essentially just by zooming in until they look straight). So in this section, whenever I talk about "a hill" or "a steep thing," I'm talking about straight lines.

At this point, we can abbreviate steepness by the letter S, but we don't know anything mathematical about it, so we don't know how to write anything other than

$$S = ?$$

But how exactly does our everyday concept work? What properties does it have? What behaviors do we implicitly assign to "steepness" when we reason about the concept in a non-mathematical setting? Before we decide what to do mathematically, we need to explore our everyday concept in some more detail.

Suppose you wake up on another continent. There's no one around. You don't know your latitude, your longitude, or your altitude, even vaguely. You see a hill in the distance, so you decide to walk over to it to see what's on the other side. When you're walking up the hill, you find it to be fairly steep, so you think about turning back and looking for help in the other direction.

The above paragraph reveals something about our everyday concept that we all know intuitively, but which is so obvious that we usually don't bother to mention it, though mentioning such things explicitly will be a huge help in moving from the qualitative to the quantitative. That is, even though you had no idea where you were when you climbed the hill, you still knew that it was steep. A steep thing is equally steep whether we climb it when we're underground or when we're inside of an airplane.

Another way of expressing the same idea is to say that steepness doesn't depend on your vertical or horizontal position *in isolation*. The steepness of a hill isn't an intrinsic property of the horizontal or vertical location where it is located. It is a property of the *change* in vertical location as we walk up the hill. But it's not *just* a property of the change in vertical location. If you walk 10 miles along a not-so-steep sidewalk, you might end up 1000 feet higher in altitude than where you started, but going up 1000 feet would be next to impossible if you only had 10 horizontal feet to do it. So, based on our vague, qualitative, pre-mathematical concept of steepness, we know this:

First thing our everyday concept tells us:
Steepness only depends on *changes* in vertical location and
changes in horizontal location, not on the locations themselves.

Let's come up with some abbreviations for this. We could write the above sentence in abbreviated language by writing

$$S(h, v) = ?$$

The S still stands for "steepness," the new symbols h and v stand for differences in horizontal and vertical location. For example, if you walk 20 feet along the ground and then climb a 10-foot tree, the h between where you started and where you stopped would be 20 feet, and the v would be 10 feet. Note that these abbreviations h and v only make sense once we've decided on *two points*: where to start and where to stop. So which two points were we talking about when we wrote $S(h, v) = ?$ in the sentence above? We aren't saying. We're just playing an abbreviation game at this point. But the sentence $S(h, v) = ?$ stands for the idea that steepness depends only on *changes* in horizontal location (h) and *changes* in vertical location (v), not on the locations themselves.

Now, since h and v are both quantities that compare two points, we need to choose two points on a hill before we can talk about the steepness of it, so (as far as we know at the moment) the steepness of a line might change depending on which two points we choose. But that doesn't seem quite right, because straight lines are straight. At least in an everyday sense of the term, a straight line only has one steepness. It shouldn't depend which points we choose. Let's try to tell the mathematics about this intuition:

Second thing our everyday concept tells us:
Whatever we mean by "steepness," a straight line should have the same steepness everywhere. If someone else has a definition of "steepness" that makes straight lines change steepness in the middle, that's fine, but it's not what we mean by "steepness."

Okay, great! There's something that's *definitely* true about our everyday concept of steepness, and we want to *force* our mathematical concept of steepness to behave like this.

Figure 1.8 lets us visualize this idea. Since steepness is about differences, we need two points to compute it. Imagine that we look at two points on a line that are h apart horizontally and v apart vertically. Looking at two such points essentially leads us to look at the small triangle in the bottom left of Figure 1.8. Now, if we were to look at a *different* pair of points on that same line, then the steepness should be the same. For example, imagine we now look at two points on the same line that are a distance $2h$ apart horizontally (that is, twice the horizontal spacing of the two points we looked at originally). Well, it shouldn't be too hard to see that since we're on a straight line, the vertical distance will double too. That is, it will be $2v$. (Make sure you see why this is true. Figure 1.8 should help.) But because of the "second thing our everyday concept tells us," above, the steepness should be the same in both cases. We can tell the mathematics about this intuition of ours by writing the sentence:

$$S(h, v) = S(2h, 2v)$$

Now, notice that there's nothing special about the number 2 in this argument. If we triple h, then the same reasoning would tell us that v triples, and the steepness stays the same because we're still talking about the same line. We can repeat this argument for any whole number and get $S(h, v) = S(\#h, \#v)$, where $\#$ is any whole number.

What's more, the same idea should work when $\#$ isn't a whole number. For example, if you cut h in half, then v gets cut in half, so $S(h, v) = S(\frac{1}{2}h, \frac{1}{2}v)$. This is an extra fact about our intuitive idea of steepness, and it gets us one step closer to the precise definition we're looking for. Let's write it once and for all:

$$S(h, v) = S(\#h, \#v) \tag{1.4}$$

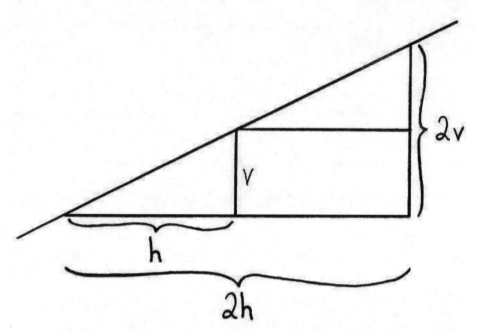

Figure 1.8: An illustration of the second thing our everyday concept tells us about steepness. Whatever we mean by "steepness," a straight line should have the same steepness everywhere. In particular, doubling the horizontal distance between two points also doubles the vertical distance, and we want the steepness to be the same in both cases. Or, in abbreviated form, $S(h, v) = S(2h, 2v)$.

where $\#$ isn't necessarily a whole number. This is really neat, and it tells us a lot about the possible things that we might mean by "steepness." For example, the possible definition $S(h, v) = h$ doesn't work because it's *not* true that $h = \#h$ no matter what numbers $\#$ and h are! For the same reason, the definition of steepness we're seeking can't be $S(h, v) = hv$, or $S(h, v) = h + v$, or $S(h, v) = 33h^{42}v^{99}$, or a lot of other things.

In fact, the longer we stare at equation 1.4, the more it becomes clear how powerful a statement it is. It almost looks like the number $\#$ is "canceling out" of both sides. If we play around for a while, we'll be able to test a bunch of ideas, and make a list of the ones that work (i.e., the ones whose behavior makes equation 1.4 true). Here are a few that work. (Note: In the list below, the symbol $\overset{?}{=}$ just means "these are all definitions we *could* choose, but we haven't chosen any of them yet.")

1. $S(h, v) \overset{?}{=} \frac{v}{h}$ works. This is "rise over run."

2. $S(h, v) \overset{?}{=} \frac{h}{v}$ works too. This is "run over rise."

3. $S(h,v) \overset{?}{=} \left(\frac{v}{h}\right)^2$ works. This is "rise over run" squared.

4. $S(h,v) \overset{?}{=} 3\left(\frac{v}{h}\right) + 14\left(\frac{v}{h}\right)^2 - \left(\frac{h}{v}\right)^{79}$ works. This is crazy.

After playing with this for a while, it becomes clear that any machine that depends only on (h/v) or (v/h) should work.[10] That is, any machine whose description doesn't contain h or v in isolation, but always contains both together in either the form (h/v) or (v/h). Why do we need this? Because it's hard to see how on earth we could get any number to "cancel out," like it has to in equation 1.4 otherwise. There might be some other way to get them to cancel, but we don't care!

1.2.5 Not Your Grandfather's Anarchy

> *Science is an essentially anarchic enterprise: theoretical anarchism is more humanitarian and more likely to encourage progress than its law-and-order alternatives... The only principle that does not inhibit progress is: anything goes.*
> —Paul Feyerabend, *Against Method*

It's worth stepping aside for a moment and reflecting on what exactly we're trying to do. Are we trying to dig deeper and deeper into our intuitive concept, extracting constraints until we can cut down the possibilities to just one? Not necessarily! The choice of "what we're trying to do" is entirely up to us.

By repeatedly translating verbal ideas about our everyday concept of steepness into abbreviated form, we've decided that our mathematical concept of steepness has to (a) depend only on position changes v and h, not the positions themselves, and (b) depend only on $\frac{v}{h}$ and $\frac{h}{v}$. But this still doesn't tell us why the textbooks choose "rise over run" (that is, $\frac{v}{h}$) instead of any multiple of that, like $3\frac{v}{h}$ or $\frac{17}{92}\frac{v}{h}$. Whether and when we should give up and just pick one is a philosophical problem that we have to deal with, and it illustrates a problem that shows up whenever we're inventing a mathematical concept. There are two strategies at this point:

1. **The Soldiering-On Approach.** We could keep trying to prune down the imaginary bag of candidate definitions by (a) thinking of qualitative features of our concept of steepness, (b) abbreviating them, (c) mentally throwing out all the ones that don't work, and (d) repeating this until we get to one and only one possible definition. Of course, once there was only one possibility left, we might not realize it just from looking at the list of requirements we've imposed, so we would have to convince

---

10. Since $h/v = (v/h)^{-1}$, we could have written this sentence by saying "any machine that only depends on the quantity v/h," but we haven't talked about negative powers yet, so we're not acting as if we know they exist. In our universe, they don't yet.

ourselves that there really was only one surviving candidate once we got there. That would be nice, because then we'd know *exactly* where our definition came from, down to every last detail.

2. **The Giving-Up Approach.** Instead, we could decide that we're tired of this process of mining our minds, trying to squeeze the last drop of content out of our intuitions. Maybe our everyday concept isn't specific enough to uniquely determine one and only one definition. So, we could simply give up. "Look at it this way," we could say. "I just wanted *some* definition of steepness that does everything I asked for, and I have several options, so I'm just going to pick one." Who or what gets to determine *which* one we pick? We do, of course. We could just pick the remaining candidate we thought was "prettiest" or most elegant, by any standard. This happens in mathematics more than anyone would like to admit. We're inventing this stuff ourselves. We can do whatever we want. We can conjure up entities ex nihilo and give them life by giving them names. If anarchy exists, this is it!

Okay, wait. That last sentence may have given you the impression that I'm saying there's no such thing as "mathematical truth," because we're making all this stuff up. That's definitely not what I meant to say. Anarchy in the usual sense refers to the absence of human laws, not the absence of physical laws.

In a state of anarchy, there are no "laws," but you still can't fly, because of the "law" of gravity. These are clearly two different concepts. Mathematics outside the confines of a classroom is anarchy in the first sense: we can do anything we want, but we can't make anything be true.

We can choose to define things however we want, and we can choose to play with anything we want, but once we agree on what we're talking about, we find that there is already a preexisting set of truths about our newly invented objects of study, and we have to discover those truths for ourselves.[11]

In summary, at this point we could just decide to give up and choose "rise over run" as the definition we thought was prettiest, and go on to invent calculus. However, it is important to stress that we could *also* give up and choose "run over rise" (the upside-down version) or "42 times (rise over run) cubed" as the definition we thought was prettiest! If we then went on to develop calculus using one of these nonstandard definitions, all of our formulas would *look* slightly different than they do in the standard textbooks, but they would all be saying essentially the same thing as the standard versions.

---

11. **Note to people who know what the terms "Platonist" and "formalist" mean, and who interpreted this section as a defense of either of these views over the other:** It isn't.

1.2.6 Onward! Just for Fun

> *The essence of mathematics lies entirely in its freedom.*
> —Georg Cantor, *Gesammelte Abhandlungen*

Now that we've had that discussion about what is arbitrary and what is necessary in mathematics, let's go on and see what we would have to assume in order to make the standard definition of slope emerge as the one and only possibility left standing.

Let's mine our minds some more, and ask whether our everyday concept of steepness tells us anything else about the properties we want our mathematical concept to have. So far, we have no reason to chose $\frac{v}{h}$ (rise over run) instead of $\frac{h}{v}$ (run over rise). However, while the second of these is a completely fine way to measure steepness, it has one odd property.

The candidate definition "run over rise" says that flat, horizontal things have infinite steepness, and completely vertical cliffs have zero steepness. That is, if the vertical distance between two points is zero (so $v = 0$), then $\frac{h}{v}$ becomes $\frac{h}{0}$, which is infinite (or at least it sort of makes sense to say that it's infinite, because $\frac{1}{tiny} = huge$, and the *huge* piece gets bigger as we make the *tiny* piece smaller). Also, if the horizontal distance between two points is zero (that is, $h = 0$) then "run over rise" is zero. That may not be wrong, but it isn't quite how we usually think. Still, we're the owners of this universe, so we're allowed to impose the intuitive-sounding requirement that flat things have zero steepness. Let's make it official.

Third thing our everyday concept tells us:
Whatever we mean by "steepness," a horizontal line should have
zero steepness.

This rules out a lot more possibilities. For example, it rules out $S(h, v) = (\frac{h}{v})$, it rules out $S(h, v) = 3(\frac{h}{v})^2$, it rules out $S(h, v) = (\frac{h}{v})^{72} - 9(\frac{v}{h})^{12}$, and anything else that isn't zero for horizontal things (i.e., when $v = 0$). This is great! Let's list some possibilities that have survived all of our purges:

1. $S(h, v) \overset{?}{=} \frac{v}{h}$ still works. This is "rise over run."

2. $S(h, v) \overset{?}{=} \left(\frac{v}{h}\right)^2$ still works. This is "rise over run" squared.

3. $S(h, v) \overset{?}{=} 3\left(\frac{v}{h}\right) + 14\left(\frac{v}{h}\right)^2 - \left(\frac{v}{h}\right)^{999}$ still works. This is crazy.

There are still infinitely many candidate definitions, but a lot of them are really weird. We could quit at this point and simply choose our favorite, but let's keep going just to see how much we have to assume in order to arrive at the standard definition.

We still haven't said much about how different hills relate to each other. For example, what does it mean to say that one hill is "twice as steep" as another? We haven't really thought about that yet, so as of right now, there's no correct answer. But we want the concept we invent to make sense *to us*, so let's think about what we want "twice as steep" to mean. Suppose we've got two points, one higher up and to the right of the other, so that the line between them looks like a hill. Then imagine that we grab the higher point and move it up even further, until we've doubled the original vertical distance *without* changing the horizontal distance. That is, imagine we transform one hill into another by doubling the height of the original hill, without changing its horizontal width. Now, it makes a certain amount of sense to say that if two hills are equally wide horizontally but one is twice as tall as the other, then the steepness of the second one should be twice as big. As before, there's no law that forces us to think about things this way, but all the other ways of thinking about it seem even worse: for example, it seems less reasonable to say that doubling the vertical distance should multiply the steepness by 72, because it's not clear why we should prefer this rule over the infinite number of other possibilities. But the idea that doubling height should also double the steepness has a certain simplicity and elegance to it. Let's make it official:

Fourth thing our everyday concept tells us:
Whatever we mean by "steepness," if we double the height of a
hill without making it longer horizontally, then its "steepness"
should double.

How could we abbreviate this idea? Well, we're doubling v without changing h, and we want that to force the steepness to double, so we could abbreviate it like this:

$$S(h, 2v) = 2S(h, v)$$

So, we've managed once again to perform the translation from qualitative to quantitative. Now, as usual, there's nothing special about the 2 in the above argument. The idea we really wanted to convey was more general than that. For example, the same line of reasoning suggests that if you triple the vertical distance without changing the horizontal distance, then the steepness should triple. Let's abbreviate this idea in a way that expresses infinitely many sentences at once, just like we have before:

$$S(h, \#v) = \#S(h, v)$$

Perfect. Now let's look in our imaginary bag of candidate definitions and perform another purge. Which of the remaining possibilities can satisfy this requirement? Let's try some. Well, if we look at the possible definition $S(h, v) \equiv \left(\frac{v}{h}\right)^2$ and we imagine doubling v like we did above, then we get

$$S\left(h, 2v\right) \equiv \left(\frac{2v}{h}\right)^2 = \frac{2 \cdot 2 \cdot v \cdot v}{h \cdot h} = 4\left(\frac{v}{h}\right)^2 \equiv 4S\left(h, v\right)$$

So we doubled the verticalness and the steepness *quadrupled*. That means we can throw this candidate definition away, because it doesn't live up to the fourth thing our everyday concept tells us. Alright, so we just tested $\left(\frac{v}{h}\right)^{\#}$ where $\#$ was 2, but what about when $\#$ is 3 or 5 or 119? Instead of testing each possible power individually (which would take an infinite amount of time), let's test them all at once by remaining agnostic about which power we're testing. By an argument just like the one above:

$$S\left(h, 2v\right) \equiv \left(\frac{2v}{h}\right)^{\#} = 2^{\#}\left(\frac{v}{h}\right)^{\#} \equiv 2^{\#}S\left(h, v\right)$$

But all of this has got to be equal to $2S\left(h, v\right)$, or else it violates the fourth thing our everyday concept tells us. So in order for the steepness to double when we double the height, it has to be the case that $2^{\#}$ is just 2. But that's only true when $\#$ is equal to 1, so this lets us throw out almost all of our remaining possibilities! Let's make a list of some survivors.

1. $S(h, v) \overset{???}{=} \frac{v}{h}$ still works. This is "rise over run."

2. $S(h, v) \overset{???}{=} 3\frac{v}{h}$ still works. This is "3 rise over run."

3. $S(h, v) \overset{???}{=} 974\frac{v}{h}$ still works. This is "974 rise over run."

Basically anything except "some number times rise over run" has been eliminated by one of the requirements we've listed earlier. We're starting to see how much of the reasoning process the textbooks sweep under the rug when they just say "Slope is rise over run." All the remaining definitions we can think of look like $S(h, v) \equiv (number)\left(\frac{v}{h}\right)$, so let's see if our intuitive concept has any of its own opinions about what that $(number)$ should be.

Imagine that gravity changes direction a bit. Then everything that used to be flat would be slightly tilted. If gravity changes its direction by 90 degrees, then stuff that's now horizontal would be vertical, and vice versa. Now, if the direction of "up" changes by 90 degrees, then the steepness of everything is going to change... except one thing: a hill that's halfway between vertical and horizontal. That is, a hill whose horizontal distance is the same as its vertical distance (a hill for which $h = v$) is going to be the *only* thing whose steepness doesn't change when gravity changes like this. Any steepness definition of the form $S(h, v) \equiv (number)\left(\frac{v}{h}\right)$ will assign this special hill a steepness of $(number)$, because the special hill has the property $v = h$. So deciding what we want $(number)$ to be is equivalent to deciding on the steepness of this special hill.

Let's consider some possibilities. Suppose we decided that we want (*number*) to be 5. Then the special hill would have a steepness of 5 both before and after a gravity swap, but other hills would do much weirder things. A hill with $v = 3$ and $h = 1$ would have a steepness of 15 before a gravity swap and a steepness of $\frac{5}{3}$ after. There's nothing wrong with that, but it seems pretty arbitrary, and the steepnesses before and after a gravity swap aren't related to each other in a nice, visually appealing way. However, if for purely aesthetic reasons we choose to assign the special hill a steepness of 1, then *other* hills behave much more nicely. Then, a hill with a steepness of 3 before a gravity swap would have a steepness of $\frac{1}{3}$ after. A hill with steepness $\frac{22}{33}$ beforehand would have a steepness of $\frac{33}{22}$ after. That makes things look a lot simpler. If we make this choice, purely motivated by aesthetics, then we arrive at

$$S(h, v) = \frac{v}{h} = \frac{rise}{run} = \text{The standard definition} \qquad (1.5)$$

as the one and only surviving possibility. Let's write that down:

Fifth thing our everyday concept "tells" us, but not really:
There's only one hill whose steepness is the same before and after a
90-degree gravity swap (i.e., switching v and h). For the sake of elegance and
simplicity, we assign this hill a steepness of 1. This makes all the *other* hills
act nicely under gravity swaps.

Now you know *exactly* how much they weren't telling you in school. As is always the case when inventing a mathematical concept, the definition we finally arrived at was built from a strange blend of translation and aesthetics: some of our definition's behaviors came from our desire to make it behave like our everyday concept, while others came from a desire to make the resulting definition as elegant and simple to deal with as possible, by our own human standards of elegance and simplicity.

1.2.7 Summarizing the Inventing Binge in Words

We've covered the invention process in some detail, because it's important to have at least a few simple examples of the process of inventing a mathematical concept spelled out fully and completely, making clear at every stage what we're making up, what's necessarily true as a consequence of what we've made up, and spelling out the reason behing every step. The invention process is fundamentally important to understand, so let's summarize what we did, first in words, then by listing all the math we invented at once. To save space, we'll abbreviate the phrase "or it's not a good translation" by ONGT. All mathematical concepts are invented like this:

1. You start with an everyday concept that you want to formalize or generalize.[12]

2. You typically have some idea of what you want the concept to do in simple, familiar cases. These simple cases form the basis for your decision about which *behaviors* you want your new concept to have in cases that are less familiar.
   Examples: Whatever "area" means, the area of a rectangle should double if you double its length, ONGT. Whatever "steepness" means, the steepness of a straight line should be the same number everywhere, ONGT. Here's one we didn't do: Whatever "curvature" means, the curvature of a circle or a sphere should be the same number everywhere, and the curvature of straight lines or flat planes should be zero, ONGT.

3. You force your mathematical concept to behave like your intuitive concept in these simple cases, and sometimes in straightforward generalizations from these simple cases.
   Examples: I can't even begin to picture a five-dimensional cube, but its "five-dimensional volume" should be $\ell_1\ell_2\ell_3\ell_4\ell_5$, ONGT. I can't picture a ten-dimensional sphere, but its curvature should be the same number everywhere, ONGT. I can't picture the fifty-two-dimensional version of a "line" or "plane" or whatever, but its curvature should be zero, ONGT.

4. Sometimes you find that all of your vague, qualitative requirements, when written in abbreviated, symbolic language, completely determine a precise mathematical concept.

5. Sometimes, all the intuitive requirements you want to impose may not be enough to single out a single mathematical definition. That's okay! In these cases, mathematicians usually just look into the imaginary bag of candidate definitions that do everything they want, and pick the one they think is prettiest or most elegant. You may be surprised to see these ill-defined aesthetic concepts inserting themselves into mathematics. Don't be.

1.2.8 Summarizing the Inventing Binge in Abbreviations

Finally, let's summarize the inventing binge in symbolic form, to remind us what we did.

---

12. Once we have invented more of mathematics, this set of "everyday concepts" that serves as the raw material for the creation process will come to include simpler *mathematical* concepts that we have invented earlier. This occurs, for example, when we generalize the basic concepts of calculus invented in Chapter 2 to a related set of concepts in a space of infinitely many dimensions, as discussed at the end of the book in Chapter ℵ. The more deeply we explore our mathematical universe, the less clear the distinction between everyday concepts and mathematical concepts becomes.

Inventing Area

Based on our everyday concept of area, we forced the following two properties to be true of the corresponding mathematical concept, in the specific case of a rectangle:

1. $A(\ell, \#w) \overset{\text{Force}}{=} \#A(\ell, w)$ for any number $\#$

2. $A(\#\ell, w) \overset{\text{Force}}{=} \#A(\ell, w)$ for any number $\#$

3. $A(1,1) \overset{\text{Force}}{=} 1$

We then found that this forces the area of a rectangle to be

$$A(\ell, w) = \ell w$$

which is the formula they throw at us in math classes.

Inventing Steepness

Based on our everyday concept of steepness, we forced the following five properties to be true of our mathematical version of the concept, in the specific case of a straight line:

1. Steepness depends only on *changes* in vertical location and *changes* in horizontal location, not on the locations themselves.

2. $S(h, v) \overset{\text{Force}}{=} S(\#h, \#v)$ for any number $\#$.

3. We wanted the steepness of a horizontal line to be zero, so $S(h, 0) \overset{\text{Force}}{=} 0$.

4. If you double the vertical distance of a hill without changing the horizontal distance, this should double the steepness. Also, this property should hold not just for doubling, but for any amount of magnification, so $S(h, \#v) \overset{\text{Force}}{=} \#S(h, v)$ for any number $\#$.

5. When $h = v$, we chose $S(h, v) \overset{\text{Force}}{=} 1$ for purely aesthetic reasons.

We then found that these five requirements together *force* the steepness of a line to be

$$S(h, v) = \frac{v}{h} = \frac{rise}{run}$$

which is the formula they throw at us in math classes.

1.2.9 Using Our Invention as a Springboard

The above discussion might have given you the impression that mathematics is just one big inventing binge and we're not actually discovering anything. In the section "Not Your Grandfather's Anarchy," I tried to explain why this isn't the case, but let's look at a concrete example. We said what we meant by "slope." Now we'll find that by doing so, we have conjured up a world of truth that is independent of us. This world contains truths that we didn't "put in" explicitly, and which may not be obvious to us, but which nevertheless follow from what we've done.

At some point in the past, you may have been told that the "formula" for a line is $f(x) = ax+b$, as if this fact were so simple that it should be self-evident. Notice that we never used this formula in the above discussion, even though we talked about lines the whole time. At least to me, it is neither obvious that machines of the form $f(x) = ax + b$ happen to be lines when we draw them, nor that all lines (except vertical ones) can be represented by machines of that form.

Instead of just accepting the above statement about lines, let's invent it for ourselves. That is, we already invented the concept of steepness, so now let's show that the definition we invented actually *forces* lines to be described by machines that look like $f(x) = ax + b$.

Let's assume that lines can be described by some machine $M(x)$, but we don't want to assume that it looks like $ax + b$, because that's not obvious to us. Instead, we'll just *force* the objects we're describing by the word "line" to have constant steepness. We already made this assumption in the course of inventing steepness earlier, where we called it "the second thing our everyday concept tells us." Let's tell the mathematics that we're assuming this. Suppose x and $\tilde{x}$ are any two numbers. No matter what x and $\tilde{x}$ are, if the machine M describes a line, then we want it to be true that

$$\frac{\text{(One point's vertical position)} - \text{(Other point's vertical position)}}{\text{(One point's horizontal position)} - \text{(Other point's horizontal position)}}$$

$$\equiv \frac{\text{Rise}}{\text{Run}} \equiv \overbrace{\frac{M(x) - M(\tilde{x})}{x - \tilde{x}}}^{\text{Just another abbreviation}} \overset{\text{Force}}{=} \#$$

where the symbol $\#$ stands for "some fixed number that doesn't depend on x or $\tilde{x}$." Okay, that was an awful lot of symbols. We did that so as not to take too many steps at once, but the main message of the stuff above is simply this:

$$\frac{M(x) - M(\tilde{x})}{x - \tilde{x}} \overset{\text{Force}}{=} \# \tag{1.6}$$

We're forcing the steepness to be a fixed number $\#$ everywhere because we want to talk about a line, and equation 1.6 is essentially just our way of "telling the mathematics" that. But now, since we forced equation 1.6 to be true for all x and $\tilde{x}$, it has to be true when $\tilde{x} = 0$. There's nothing special about $\tilde{x} = 0$, and we could have chosen any other number for $\tilde{x}$, or for x. We're just doing this because we're playing around, and equation 1.6 looks a bit simpler when we look at the special case in which $\tilde{x}$ is zero. Alright, so when $\tilde{x}$ happens to be 0, equation 1.6 says:

$$\frac{M(x) - M(0)}{x} = \# \qquad (1.7)$$

no matter what number x is. Since we're remaining agnostic about what x is, that term $M(x)$ in the top left of equation 1.7 is a complete description of our machine! If we could isolate that piece, we'd have succeeded in going from our vague qualitative idea that the steepness of a line should be the same everywhere (part of the definition of steepness we invented earlier) to a precise way of writing down what a line *is* in symbols! Let's try to isolate that description of our machine that's hiding in the top left. Since the two sides of equation 1.7 are equal to each other, they'll still be equal if we multiply them both by the same thing. So if we multiply both sides of equation 1.7 by x, and use the fact that $\frac{x}{x} = 1$, then we see that equation 1.7 is saying the same thing as the sentence $M(x) - M(0) = \#x$. But that's just another way of saying

$$M(x) = \# x + M(0) \qquad (1.8)$$

People who like jargon call the number $M(0)$ the "y-intercept," but we can just think of it as whatever number our machine M spits out when we feed it 0. The symbols $\#$ and $M(0)$ are both just abbreviations for numbers we don't know (or if you prefer, numbers whose specific values we're choosing to remain agnostic about), so we could write the same sentence this way:

$$M(x) = ax + b \qquad (1.9)$$

which is the "textbook" equation for a line. We invented this ourselves, so from now on, we own it.

We did a lot of things in this chapter! Let's remind ourselves what we did. I guess we could just call the next section "Summary." But our own universe deserves a few of its own terms. We're bringing everything we created back together in one place, so how about...

(Author thinks for a moment.)

1.3 Reunion

Okay, in this chapter, we forgot everything we knew about mathematics except addition and multiplication, and started building our own mathematical universe. Occasionally, we left the privacy of that universe to compare the things we'd created to the corresponding concepts in the outside world. We learned:

1. Why textbooks write x as an abbreviation for *stuff* (because you can't say "sh" in Spanish, and because humans are slow to change things that are no longer helpful).

2. Textbooks call our machines "functions." It's not clear why.

3. Textbooks call abbreviations "symbols," and they call sentences "equations" or "formulas." These terms are a bit fancier than the ideas they stand for, but we'll occasionally use them anyway, just to get ourselves used to them.

4. The standard way of describing a machine is to use super-abbreviated sentences that look like $f(x) = 5x + 3$. This sentence contains *three different abbreviations* just on the left side of the equals sign: (i) the name of the machine is f, (ii) the name of the stuff we're feeding it is x, and (iii) the name of the stuff that f spits out when we feed it x is $f(x)$. So the left side is three names rolled into one. The right side of the sentence describes how the machine works.

5. By using different-looking equals symbols like $\equiv$, $\overset{\text{Force}}{=}$, and $\overset{(1.3)}{=}$, we can say that two things are equal, and also remind ourselves why. If this ever confuses you, just pretend I used $=$.

6. The obvious law of tearing things is obvious, and it keeps us from having to remember silly acronyms like "FOIL" or any of its more complicated friends. We can invent them all whenever we need to.

7. Just by using the idea that $\frac{1}{stuff}$ stands for "whatever number turns into 1 when we multiply it by *stuff*," we can invent many of the so-called "laws of algebra" for ourselves, thus forever avoiding the need to memorize them.

8. A mathematical concept is invented like this: We start with an everyday concept that we want to make more precise or more general. There's no single method for doing this. We usually have a few behaviors in mind that we want our mathematical definitions to have, but usually there are lots of candidate definitions to choose from. If our qualitative ideas aren't enough to determine one and only one definition, mathematicians usually just pick the one they think is prettiest or most elegant. Students rarely hear this, so they often end up blaming themselves for not understanding the definitions better.

9. If we *invent* the concept of steepness like we did in Section 1.2, and if we *assume* that an unspecified machine has the same steepness everywhere, then we can *discover* the fact that this machine looks like $M(x) = ax + b$. This non-obvious expression is therefore a straightforward consequence of our intuitive, everyday concepts.

Interlude 1: Dilating Time

[I do not] carry such information in my mind since it is readily available in books... The value of a college education is not the learning of many facts but the training of the mind to think.
—Albert Einstein, on why he did not know the speed of sound, one of the questions included in the "Edison Test," quoted in the *New York Times* on May 18, 1921

There's a lot of things you never see, and you don't know you don't see 'em cuz you don't see 'em. You gotta see something first to know you never saw it, then you see it and say, "Hey, I never saw that." Too late, you just saw it!
—George Carlin, *Doin' It Again*

Things You Never See

There are a lot of things you never see, and this interlude is devoted to some of them. We've all heard of Einstein. Here we'll see how to invent for ourselves one of the many things he's famous for — a mathematical description of how time *actually* works — using mathematics so simple that you'll wonder why they didn't tell you this years ago. In my view, the fact that this short mathematical argument exists but is not presented to every student at some stage of their education is one of the most telling signs that formal education, as it currently exists, has got its priorities all wrong. The simple derivation in the latter part of this interlude beautifully illustrates the strangeness of the universe and the excitement of science, and one can find it floating around mathematics and physics departments, shared between friends like a folk song or an epic poem, but never standardized as part of the body of knowledge that we teach to every member of our society. Yet for decades we've been teaching students everything they need to understand this argument, and then simply not bothering to show it to them. Why? Because special relativity is an "advanced" topic that doesn't quite "belong" in an introductory physics class, nor does it seem to belong in the courses on Euclidean geometry in which students first learn the mathematics needed to understand the argument itself. The argument is homeless. So, lacking any proper home for this beautiful, intuition-shattering argument in the core curriculum, we instead attempt to inspire students about the mysteries of the physical world by focusing on the mathematical description of pendulums, projectiles, and how a ball rolls down a hill. Alright, enough editorializing. Time for some fun!

First, before we see how time really works, we'll see an extremely short argument that makes it obvious why the formula for shortcut distances, usually called the "Pythagorean theorem," is true. Second, we'll see that this is the most complicated mathematical idea you'll need to understand one of the main ideas from Einstein's special theory of relativity: the fact that time slows down when you move. We'll use the phrase "time slows down when you move" as an abbreviated description, but it isn't exactly accurate. More precisely, whenever two objects (people included) aren't moving at the same speed or in the same direction, they begin to observe each other's "time" passing at different rates.[13] As impossible as this idea sounds, it's not just a theory, and it's not just a fact about humans or clocks. It's a fundamental fact about the structure of space and time, and it has received more experimental validation than virtually any other idea in all of science. By the end of this interlude, you'll be able to completely understand the mathematical argument showing that this phenomenon of "time dilation" exists. However, you may come away feeling like you don't understand it, because the conclusion of the argument is *always* surprising, no matter how well you've understood the argument itself. So even though our primate brains have trouble with the concept, the mathematics should make complete sense, and if it doesn't, that's my fault. Ready? Here we go.

Shortcut Distances

Not everything is horizontal or vertical. Things can be tilted in any direction. That's unfortunate, because often the information we're handed comes in the form of facts about two "perpendicular" directions — directions that could (in a vague sense) be thought of as horizontal and vertical. For example, "it's 3 blocks east and 4 blocks north," or "such-and-such is 100 meters tall and 200 meters away." Suppose all we have is information about two such distances: one that we'll call horizontal, and another we'll call vertical. We can discuss this question by drawing a triangle with a horizontal side and a vertical side. To be clear, triangles aren't the point, but this lets us discuss the question abstractly, while remaining agnostic about unimportant details. Let's name the sides of this triangle a, b, and c (see Figure 1.9), and suppose we know the numbers a and b. Using only this information, can we figure out the length of the "shortcut distance," c?

At this point, it's not remotely clear how to figure out the length of c if all we know is a and b. Since we have no idea of how to proceed, our only hope of making progress is to see if we can turn this hard problem into a

---

13. It's difficult to do justice to the idea with a brief verbal description, so if that didn't make sense, don't worry. It'll make more sense soon, after we cover a bit of background.

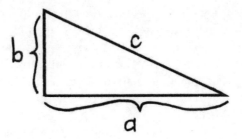

Figure 1.9: This is not a caption.

problem that only involves things we're familiar with. We haven't invented
much mathematics yet, so we're not familiar with many things, but we do
know the area of a rectangle. So a good first shot at the problem might be
to see if we can build a rectangle out of several copies of the triangle above,
and then maybe we'd be able to make a bit of progress (or maybe not, but it's
worth a try). Following this train of thought, the first thing I would personally
think to do is to take two copies of the triangle in the above picture, and stick
them together so we have a rectangle with a width of a and a height of b.
Unfortunately, after a few moments of staring at the resulting picture, we'd
still be confused, since this simplest way of building a rectangle doesn't seem
to tell us much about the shortcut distance. Luckily, the second-simplest way
turns out to be more helpful, as shown in Figure 1.10.

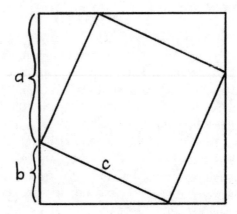

Figure 1.10: Building a square inside a square, using four copies of our triangle and
          some empty space. This lets us talk about something we're *not* familiar
          with (shortcut distances) in terms of something we *are* familiar with
          (the area of a square).

We've built a big square out of four copies of the original triangle, with
the shortcut distances forming a square-shaped region of empty space in the

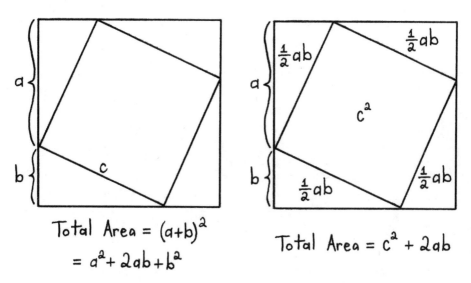

Figure 1.11: By writing the area of the whole thing in two different ways, we can invent a formula for shortcut distances, also called the "Pythagorean theorem" in textbooks.

middle. As in Chapter 1 when we invented the obvious law of tearing things, a surprising amount of knowledge can be squeezed out of the simple fact that drawing a picture on something doesn't change its area. In this case, we've essentially drawn a tilted square inside a larger square. A square's area is one of the few things we know at this point, so this trick lets us form sentences about the shortcut distance using our still very limited vocabulary. We can form such a sentence by talking about the area of the whole thing in two ways. The results are shown in Figure 1.11.

On the one hand, the picture we've drawn is a big square whose length on each side is $a + b$, so its area is $(a + b)^2$. In Chapter 1, we convinced ourselves that $(a+b)^2 = a^2 + 2ab + b^2$ by drawing a picture that made it obvious. That's one way of describing the picture, but we can also describe it in another way. The area of the whole thing is just the area of the empty space (which is c^2) plus the areas of all the triangles. We don't know the area of a triangle, but if we imagine putting any two of the triangles next to each other (like we imagined in our first failed attempt at this problem), we'd have a rectangle with area ab. We have four triangles, so we can build two rectangles out of them. So by carving things up in this way, we can see that the area of the whole picture is also $c^2 + 2ab$. We've described the same thing in two ways, so we can throw an equals sign between the two descriptions, like this: $a^2 + 2ab + b^2 = c^2 + 2ab$.

Now this next part is extremely important, so read carefully. The above mathematical sentence says that one thing is equal to another. If two things really *are* equal (i.e., identical), and we modify both in exactly the same way, then (although the two things will both change individually) they will *still be identical to each other after the modifications*. Two boxes with identical but unknown contents will continue to have identical contents if we perform an identical action to each. This is true no matter what that action is (e.g., "remove all the rocks," or "add seven marbles," or "count up the number of hats in each and double it"), as long as we agree that the two actions are identical. This is why we can now say (in the standard jargon) "subtract the term $2ab$ from both sides of the above equation." Make sure you understand this. This is not a property of mathematics or of equations, and it's not some mysterious "law" of "algebra." It's a simple fact about our everyday concept of two things being identical: identical modifications to identical objects must lead to identical results.[14] If it doesn't, then we cannot have been using the term "identical" very carefully. So, performing this modification, we arrive at this sentence:

$$a^2 + b^2 = c^2$$

This tells us how to talk about shortcut distances in terms of horizontal and vertical bits, so let's call it the "formula for shortcut distances." Textbooks usually call this the "Pythagorean theorem," which sounds like some sort of magical sword, or a disease you catch from drinking unsanitary water.

The Fiction of Absolute Time

> *Events and developments, such as... the Copernican Revolution...*
> *occurred only because some thinkers either decided not to be bound*
> *by certain "obvious" methodological rules, or because they unwit-*
> *tingly broke them.*
> —Paul Feyerabend, *Against Method*

> *A few lines of reasoning can change the way we see the world.*
> —Steven E. Landsburg, *The Armchair Economist*

Get some popcorn, dear reader, and prepare yourself, because you're about to see one of the most beautiful arguments in all of science. The conclusion isn't easy for our primate brains to accept, so you won't be able to understand it viscerally. No one can. But even the simple mathematics we've already invented offers us a way to circumvent and move beyond certain inherent limitations of the primate brain. This section will go a bit more quickly than we have been so far, but don't worry. The derivation below is logically independent from the rest of the book, so even if you don't understand anything in what follows, you won't be behind when we move on to inventing calculus in

---

14. An understanding of this simple fact and its consequences would allow us to skip a large proportion of a typical introductory course in algebra.

Chapter 2. With that in mind, sit back and enjoy yourself. We'll need three things to get where we're going:

1. How far you go = (How fast you're going)·(How long it takes), as long as your speed doesn't change along the way. We all know this intuitively, but we can easily forget that we know it when it's phrased in this abstract form. It's just saying: (a) if you go 30 miles per hour for 3 hours, you will have gone 90 miles, and (b) there's nothing special about the numbers we used in item (a). Let's write this as $d = st$ to stand for "(distance) equals (speed) times (time)."

2. The formula for shortcut distances that we invented above (i.e., the "Pythagorean theorem").

3. A strange fact about light.

The strange fact about light is not a mathematical fact, but a physical fact, and it's so completely ridiculous that you shouldn't be surprised if it doesn't make sense. The full absurdity of this fact about light is best expressed by seeing how it's different from something we all know. So first, the thing we all know: if you throw a tennis ball at 100 miles per hour and then immediately (by some superpowers) run up behind it at 99 miles per hour, then it will look like the tennis ball is moving away from you at 1 mile an hour, at least until it falls to the ground. Not a mystery.

Here's the seemingly impossible fact about light: if you "throw" some light (e.g., if you let a few light particles, or "photons," out of a flashlight while you're standing still) and then you immediately run up behind it at 99% of the speed of light, then it *won't* look like the light is moving away from you at 1% of the speed of light! Rather, it will still look like the light is moving away from you at the full speed of light — the same speed it *would have* looked like it was moving away from you if you hadn't bothered chasing after it.

If that seems impossible, good! That means you're paying attention. Rather than trying to understand this fact by worrying about how it could possibly be true, we'll try to understand it by playing a game similar to the one Einstein played in 1905. We'll say, "Okay, this fact seems impossible, but the evidence suggests that it's true, so how about we just ask ourselves: *if it were true, what else would have to be true?*"

To begin, let's imagine a strange device that I'll call a "light clock." To build a light clock, just imagine holding two mirrors some small distance apart. Since light bounces off mirrors, this imaginary device will keep the light trapped, bouncing back and forth between them. As we all know, we can measure time in seconds, hours, days, or however else we want, so let's choose to define our unit of time to be: however long it takes the light to bounce from one mirror to the other. We could give this amount of time a name, like "schmeconds" or something, but we don't need to.

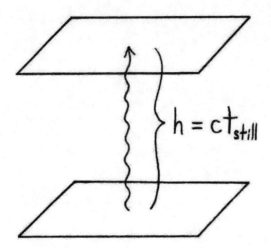

Figure 1.12: Our imaginary light clock consists of two mirrors, one raised a height
        h above the other, with a particle of light bouncing back and forth
        between them.

Now for some abbreviations. For strange historical reasons, people usually use the letter c to stand for the speed of light. Basically, c is the first letter of the Latin word for "swiftness," and the speed of light is quite literally the fastest anything in our universe is able to go, so aside from the Latin, it sort of makes sense.

So, c will stand for the speed of light. Let's use h to stand for the height difference between the two mirrors, and let's use t_{still} to stand for the amount of time it takes for the light to go from one mirror to the other (we'll see in a minute why we're calling this t_{still} instead of just t). I'll draw the light clock in Figure 1.12.

Now, at the beginning of this section, we convinced ourselves that (How far you go) = (How fast you're going)·(How long it takes) as long as your speed isn't changing. So using all the abbreviations we just defined, we can write $h = ct_{still}$, or to say the same thing in a different way:

$$t_{still} = \frac{h}{c} \tag{1.10}$$

Now, let's imagine two people looking at the same light clock. We imagine that one of them is on a rocket moving horizontally, holding the light clock. The other person will be on the ground, watching the rocket and light clock fly by at some speed, which we'll abbreviate as s. This is shown in Figure 1.13.

Okay, so the argument above where we wrote $h = ct_{still}$ should describe what the guy on the rocket sees. You might be confused about why we're using the word *still* to talk about this situation, since after all, the guy on the rocket is

Figure 1.13: Our light clock is on a rocket, moving past an observer on the ground at some speed s.

"moving." We're describing it this way because the guy on the rocket is not moving *relative to the light clock*, since he's holding it, so it's "still" compared to him. As we'll discuss later, "moving" doesn't really mean anything unless you say "moving relative to such and such." Okay, now what will the guy on the ground see? Well, for him, the light trapped in the light clock will still be bouncing up and down, but it's also moving past him horizontally, so the light particle will appear to be bouncing diagonally in a kind of sawtooth pattern, as shown in Figure 1.14.

Recall that in the argument above (where we concluded $h = ct_{still}$) we were thinking about the time it takes for the light to bounce from one mirror to the other. Let's do that again, this time from the point of view of the guy on the ground. We can write t_{mov} as an abbreviation for the amount of time it takes for the light to bounce from one mirror to the other, as seen by the guy on the ground. The subscript *mov* reminds us that this guy sees the light clock moving. Now, you might wonder why we would give two different names to this amount of time. After all, they're clearly the same. But don't be so confident! We already saw that light behaves in a very strange way, and Einstein took seriously the possibility that these times might not be the same. For now, let's give them two different names just in case. If they actually *are* the same, we'll discover that later. If they're not, we'll discover that too.

Now, just focusing on the light's path from Figure 1.14, we can figure out the distance that the light travels in one "clock tick" as seen by the guy on the ground. This is shown in Figure 1.15. The vertical distance between the

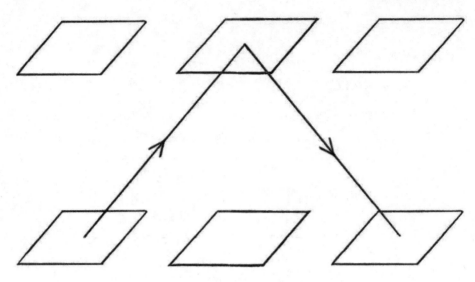

Figure 1.14: Three snapshots of our light clock as the light moves up and down, as seen by the guy on the ground. From his perspective, the light is moving diagonally, since it is bouncing up and down between the mirrors, while the light clock itself is moving past him from left to right. As he watches the rocket go past, the light particles will trace out a kind of sawtooth pattern.

mirrors is still h, and the distance the light travels horizontally is st_{mov}, since the rocket is going at speed s and we're thinking about what happens during a time t_{mov}.

Here's where we use the strange fact about light: that however fast you're traveling, light always *looks* like it's moving the same speed. Because of this, both our characters see the light traveling at the same speed c. But the guy on the ground sees the light traveling diagonally, and the distance that the light travels along the diagonal in a time period t_{mov} is still just "speed times time," or ct_{mov}. This is weird. For example, if the light were any normal bouncy object bouncing back and forth between the mirrors, then the diagonal speed of the ball seen from the ground would be *faster* than the vertical speed of the ball as seen by the guy on the rocket. So we've "told the mathematics" that we're assuming the strange fact about light. Now we can see what else has to be true as a result.

Here's where we use the formula for shortcut distances. Since "horizontal" and "vertical" are perpendicular to each other, the picture in Figure 1.15 tells us:

$$h^2 + (st_{mov})^2 = (ct_{mov})^2$$

We eventually want to compare the times t_{still} and t_{mov}, and we already have an expression for t_{still} from earlier, so let's try to isolate t_{mov} in the above

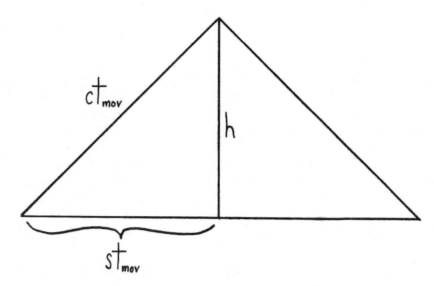

Figure 1.15: Drawing all the distances. Let's think about the time it takes for the light to go from the bottom to the top, from the point of view of the guy on the ground. The vertical distance between the mirrors is still h. The horizontal distance traveled is st_{mov}, and the diagonal distance traveled is ct_{mov}, because of the strange fact about light from earlier.

equation, and then maybe we can see if these two times are the same. If we want to isolate t_{mov}, it might help to throw everything involving t_{mov} onto one side of the equation, like this:

$$h^2 = (ct_{mov})^2 - (st_{mov})^2$$

Now, since the order of multiplication doesn't matter, it has to be the case that $(ab)^2 \equiv abab = aabb \equiv a^2b^2$, no matter what a and b are. Toward the goal of isolating t_{mov}, let's try to rewrite the above equation like this:

$$h^2 = c^2 t_{mov}^2 - s^2 t_{mov}^2$$

But then each piece on the right has a t_{mov} attached, so we can turn this into:

$$h^2 = \left(c^2 - s^2\right)(t_{mov})^2$$

Or to say the same thing another way:

$$\frac{h^2}{(c^2 - s^2)} = (t_{mov})^2 \tag{1.11}$$

Now remember earlier we found that $t_{still} = \frac{h}{c}$, and the left side of the equation above *almost* has a piece in it that looks like $\frac{h}{c}$. The trouble is that obnoxious

$-s^2$. If that piece didn't exist, then we'd have $\frac{h^2}{c^2}$ on the left side, which is just t_{still}^2, so the two times would be equal. However, that s^2 is getting in our way. So let's perform a sneaky mathematical trick: lying, and then correcting for the lie. Here's the idea. We want to compare the times t_{still} and t_{mov}, because we feel so strongly that they should be the same. If they're not, then that means "time" in the everyday sense doesn't exist — an unsettling thought! We could compare these two times if only that $-s^2$ weren't there. We can't just get rid of the $-s^2$, because that would be lying, and then our conclusions wouldn't be right. However, if we lie and then correct for the lie, then we'll have a correct answer, so let's do that. We want to rewrite equation 1.11 so that it looks like this:

$$\frac{h^2}{(c^2 - s^2)} = \frac{h^2}{c^2 (\clubsuit - \spadesuit)} = (t_{mov})^2$$

Now, we have absolutely no idea what the symbols $\clubsuit$ and $\spadesuit$ are! Our job is to figure out what they have to be, in order to make that sentence true. Why would we do this? Well, if we can dream up some values for $\clubsuit$ and $\spadesuit$ that would make the sentence true, then we could turn the h^2/c^2 bit in the above equation into a t_{still}^2 by using equation 1.10, which would let us compare the times, so then we could see how time really works. Our goal is make this sentence true:

$$c^2 (\clubsuit - \spadesuit) = (c^2 - s^2)$$

but when we frame the problem like that, it's not too hard. We want the symbol $\clubsuit$ to turn into c^2 when we multiply it by c^2, so we can just choose $\clubsuit$ to be 1. We want $\spadesuit$ to turn into s^2 when we multiply it by c^2, so we can just choose $\spadesuit$ to be s^2/c^2, so that the c^2 on the bottom kills the c^2 on the top.

Most math books would avoid all this stuff about $\clubsuit$ and $\spadesuit$, and instead just say "factor out c^2." We'll do that too, once we're comfortable with the idea. However, saying that at this stage might make it sound like "factoring" is something that we have to spend time learning. We don't. While the end result of this whole process could be described as "factoring," that isn't a good description of the thought process we went through. What really happened is that we *wanted* something to be true (namely, we wanted there to be a c^2 on the bottom), so we *lied* in order to make it true (i.e., we just wrote a c^2 where we wanted it to be), and then we *corrected for the lie* so that we'd still get the right answer.

Even more importantly, the phrase "factor out a c^2" makes it sound like there already has to be a c^2 inside. There doesn't! If we ignore the concept of "factoring" and instead think about lying and correcting, then it's perfectly clear that we can pull anything out of anything; we can pull a c out of $(a + b)$, a term that doesn't even have a c inside. How? Same logic as with $\clubsuit$ and $\spadesuit$ above. If we happen to want a c outside of $(a + b)$, just follow that same logic, and you'll end up rewriting it as $c \cdot (\frac{a}{c} + \frac{b}{c})$. Okay, sorry for the sermon, but it wasn't just a random change of topic. This stuff is fundamentally important, and I can think of no better time to mention it. Anyways, we've now figured out that

$$\frac{h^2}{c^2 \left(1 - \frac{s^2}{c^2}\right)} = \left(t_{mov}\right)^2$$

Taking the square root[15] of both sides, and using the fact that $t_{still} = h/c$, we get

$$t_{mov} = \frac{t_{still}}{\sqrt{1 - \frac{s^2}{c^2}}} \tag{1.12}$$

Equation 1.12 may look complicated, but let's first ignore most of the complexity and just mention the most important part of it: the times t_{mov} and t_{still} are not the same unless s is zero! This tells us that whenever two objects are moving at different speeds, then their light clocks desynchronize, and start "ticking" at different rates. We can rephrase equation 1.12 by throwing all the time-related stuff over to one side (i.e., dividing both sides by t_{still}). The only reason we might feel like doing this is because then the right side would only depend on the speed s. Of course it also depends on the speed of light c, but that's just a number that never changes (this is that strange fact about light from earlier). However, the speed s is something that we can change. This lets us visualize this strange time-slowing phenomenon a bit better, which is what we're doing in Figure 1.16. The figure tells us how the quantity t_{mov}/t_{still} changes as we change s, the speed of the rocket. We can think of that quantity as telling us how many times bigger t_{mov} is than t_{still}. The larger this quantity becomes, the more our everyday concept of time breaks down.

Now, it turns out that equation 1.12 isn't just a fact about light clocks, or even about clocks in general. It's a fact about the fundamental structure of space and time, and it has been experimentally tested too many times to count since Einstein discovered it in 1905. Why don't we notice this effect in our everyday lives? That is, if you and I are hanging out, and then I drive to the store and back, we don't tend to think that we've lived through two genuinely different amounts of time. However, as a glance at Figure 1.16 shows, the times we experience are equal when we're moving at the same speed with respect to

---

15. We haven't talked in depth about square roots yet, although later we'll see that they arise as part of a strange process whereby a formerly content-free abbreviation gains new life and becomes a genuine idea. If you didn't understand the step where we took the square roots of both sides, don't worry. We'll talk about them soon. For now, we're just using the symbol $\sqrt{stuff}$ to stand for whatever (positive) number turns into *stuff* when you multiply it by itself. That is, $\sqrt{stuff}$ stands for whichever number (?) makes the sentence $(?)^2 = stuff$ true. You're definitely not expected to know how to compute the square roots of any particular numbers. As long as you get the general idea, that's more than enough for now.

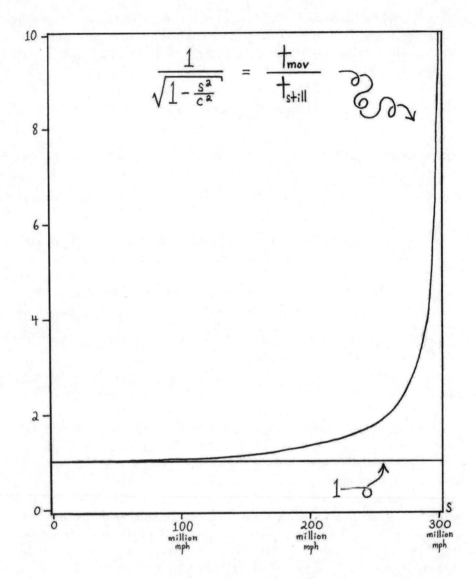

Figure 1.16: Visualizing time dilation. The horizontal axis is speed, and the vertical axis is the quantity t_{mov}/t_{still}, which tells us how much bigger t_{mov} is than t_{still} (i.e., how much our everyday concept of time has broken down). In our everyday life, we feel like time is a universal thing, which is to say we think $t_{mov} = t_{still}$, or equivalently $t_{mov}/t_{still} = 1$. This is the horizontal line in the figure. The curvy line is reality: when you are moving relative to someone, their time appears to be moving more slowly. For speeds that are small compared to the speed of light, our everyday concept of time is very close to being correct, but it increasingly breaks down as this relative speed more closely approaches the speed of light (roughly 300 million meters per second).

each other, and they're extremely close to being equal when we're moving at speeds that are small compared to the speed of light.

But even this tiny quantitative difference, as insignificant as it may be in our daily lives, requires a large qualitative change in how we think about the universe. The world we're used to, in which there is a single absolute notion of time, is simply a helpful approximation: a lie that happens to be useful, as long as we're not moving too fast relative to the objects around us. But as useful as our everyday concept of time may be, it is a startlingly poor description of the fundamental nature of reality.

Even worse, it turns out that we're not completely justified in using the subscripts $_{mov}$ and $_{still}$ when we wrote t_{mov} and t_{still}. More careful consideration of this issue shows that as long as both guys are moving with some fixed speed and direction (i.e., neither one of them is speeding up, slowing down, or changing direction), then we're not justified in saying that either of them is "still." We're used to using words like "moving" and "still" because we live on a gigantic rock covered in air, and whenever we're on or near the surface of the Earth (i.e., pretty much always), there's a special frame of reference that *appears* to be "not moving" — namely, standing still with respect to the Earth. However, this frame of reference isn't really "still" in any universal sense, and if we imagine two people floating past each other in outer space, the issue becomes more clear. Each person might think it's the other one who is really moving, and he who is standing still. Or he might think that he's still and the other guy is moving. Or he might think that they're both moving. All of these ways of thinking would be equally right and equally wrong.

The more we consider arguments like the one we just made, the more we see that it makes no sense to say that my real speed is such-and-such. It only makes sense to say that my speed is such-and-such compared to some arbitrary other thing that I'm defining to be "still." Because of this, the conclusions of the above argument are much weirder than they might have seemed at first. In our light clock example, it's not as if person A observes person B's time to be slowed down, and person B observes person A's time to be sped up, so that everyone can agree. The reality is far stranger than that. They would *each* see the other person's time slowed down, and — as long as neither person changed speed or direction — neither of them would be wrong! Are you wondering who would be older if one of two twins left Earth on a near-light-speed rocket ship while the other stayed home, and then they eventually met back up and compared watches? Good! Look up the twin paradox. The universe is crazy. Let's learn some more.

2 The Infinite Power of the Infinite Magnifying Glass

Calculus is the most powerful weapon of thought yet devised by the wit of man.
—W. B. Smith, *Infinitesimal Analysis*

2.1 Turning Hard Stuff into Easy Stuff

2.1.1 Oh, There I Am!

First, a joke about mathematicians. I didn't come up with this joke, but I don't know who did. Ready, set, joke:

> As a psychology experiment, a mathematician is placed in a room with a sink, a cooking pot, and a stove. He is asked to boil a pot of water. He takes the empty pot, fills it with water from the sink, sets it on the stove, and turns it on. Next he is led into a room with a sink, a pot *full* of water, and a stove. Again he is asked to boil a pot of water. He takes the pot and dumps the water out in the sink. He then announces, "I have reduced the problem to the previously solved problem."

As goofy as the mathematician's behavior is, the joke makes an important point. It shows us that there are two ways to solve any problem — not just mathematical problems, but problems in anything we're trying to do. Here are the two ways to solve a problem:

1. Solve the problem from scratch.

2. Solve a small part of the problem, and then notice that the rest of the problem is like something you already know how to do. Then do that.

To put it another way, problems are hard only when we don't know what to do. Once we know what to do, we can put ourselves on autopilot and just relax until we're done. For example, at some point each of us has been lost in some place we're not familiar with and trying to get home. How did you get home? Usually you don't just suddenly stumble into your own backyard and say, "Oh, I'm here." That is, you don't solve the problem of being lost all at once. What usually happens is that you stumble upon some *other* place that you're familiar with. You say, "Oh! There's that methadone clinic with the stained-glass windows! I know how to get to Grandma's house from here." By stumbling upon a familiar location, you've managed to reduce the problem to one you've already solved in the past, and the rest is easy. That's all of mathematics! One of the best ways to gain a visceral understanding of this is to invent calculus. Let's do that now.

2.2 Inventing Calculus

2.2.1 The Problem: Curvy Stuff Is Baffling

In Chapter 1, we exercised our inventing muscles by inventing the concepts of area (for rectangles) and steepness (for straight lines). Obviously, straight lines are straight, and rectangles are built from straight lines. Neither of these things is really "curvy." However, as long as there was no curviness involved, we still found that we could talk about some pretty deep and interesting things. For example, when we invented the concept of area, we saw that it's not hard to confidently talk about n-dimensional objects, or to convince ourselves that it makes sense to define the "n-volume" of an "n-dimensional cube" to be ℓ^n, where ℓ is the length of one side (even though we can't visualize anything in more than three dimensions).

What about curvy stuff, like circles? A circle is a *lot* easier to picture than an n-dimensional cube! And yet who among us can look at a circle and just intuitively "see" what a reasonable formula for its area is? *None of us.* You've probably been *told* what the area of a circle is. At some point in your life, someone told you that the area of a circle with radius r is πr^2, where π is some bizarre number that's slightly bigger than 3. But forget that. We haven't invented that fact, and it's not intuitively obvious to anyone. In fact, a fairly popular book called the Bible says that π is equal to 3, so it would appear that even deities have problems with curvy things.[1] That's why we're approaching the subject in this backward way, inventing calculus before its prerequisites: because most of those prerequisites somehow involve curvy stuff, and curvy stuff is nearly impossible to deal with before we've invented calculus (and especially before we've learned how to invent things). So despite all we've done, we still haven't got the faintest clue how to deal with curvy stuff on its own terms. Let's put this to rest once and for all.

2.2.2 The Embarrassing Truth

Here's the central insight behind all of calculus. It's kind of embarrassing how simple it is.

---

1. The line is 1 Kings, chapter 7, verse 23: "And he made a molten sea, ten cubits from the one brim to the other: it was round all about, and his height was five cubits: and a line of thirty cubits did compass it round about."

All of Calculus

If we zoom in on curvy stuff, it starts to look more and more straight.

What's more, if we were to zoom in "infinitely far" (whatever that means), then any curvy thing would look *exactly* straight. But we know how to deal with straight stuff! At least a little bit. If we can learn to make sense of this idea of zooming in infinitely far — if we can invent an "infinite magnifying glass" — then we'll be able to turn any problem involving curvy stuff (a hard problem) into a problem involving straight stuff (an easy problem). If we could do that, then maybe we'd be able to go back and reinvent for ourselves all of those unexplained facts they taught us in high school. Then we could forget them forever, and just reinvent them if we ever needed to.

Dear Reader,

Stop and take a breath.

This is where math starts to get interesting.

2.2.3 The Infinite Magnifying Glass

The essence of mathematics is not to make simple things compli-
cated, but to make complicated things simple.
—Stan Gudder

When we invented the concept of steepness for lines, we needed to pick two points, so that we could compare their horizontal and vertical positions. It didn't matter *which* two points we picked, but we had to pick two. But for curvy things, picking two random points doesn't seem to make sense, because if the steepness is constantly changing (as is typical for curvy things), then it seems like we'd get different answers depending on which two points we picked. That would make for a really ugly definition. What's more, our brains seem to somehow know what steepness means at a single point. If we forget about mathematics and just stare at a curvy thing (say, this squiggle ∿∿∿), it's very clear that some places are more steep than others, even though we don't know how to use numbers to say how steep the different spots are. Is there any way to make sense of the idea of steepness at a *single point* of some general curvy thing? Well, if we had an infinite magnifying glass at our disposal, we could reduce this hard problem to an easy problem by zooming in on curvy things until they look straight.

Our Problem: If we have a curvy thing (for example, the graph of some machine M that isn't just a line), is there any way of saying what we mean by the "steepness" of this curvy thing at a single point x?

So someone hands us a machine M and a number x, and we need to make sense of the concept of "steepness" at the point they handed us. Well, here's one idea. Let's look at the graph of M near x. That is, if we visualize x as some number on the horizontal axis, and if we visualize $M(x)$ as some number on the vertical axis, then the point with horizontal coordinate x and vertical coordinate $M(x)$ will live on the graph of the machine M. We can write this point as something like $(x, M(x))$, or however else we want. Now let's look really hard at that point. If we had an infinite magnifying glass, then we could center it over this part of M's graph and zoom in infinitely far. Then we'd see a straight line. Since we already invented the concept of steepness for lines, we could just apply that old concept to two points that are infinitely close to each other. What does it mean for two points to be infinitely close to each other? I don't know! Let's decide.

Let's write *tiny* to stand for a number that's infinitely small. It's not zero, but it's also smaller than any positive number. If you're worried about that idea, let's talk about it in this footnote.[2] Okay, time for some abbreviations. Let's say the point where we zoomed in has horizontal coordinate x and vertical coordinate $M(x)$, while the point infinitely close to it has horizontal coordinate $x + tiny$ and vertical coordinate $M(x + tiny)$. To say it another way:

$$\text{Steepness of } M \text{ at } x$$

$$\equiv \frac{\text{Tiny Rise}}{\text{Tiny Run}} \equiv \frac{\text{Vertical Distance}}{\text{Horizontal Distance}} \equiv \frac{M(x + tiny) - M(x)}{(x + tiny) - x}$$

All of these are just different abbreviations for the same idea, but the far right is the most important. Notice that the bottom right side of this long string of equations is $(x + tiny) - x$. The two x's cancel each other, so we can rewrite this as:

$$\text{Steepness of } M \text{ at } x \equiv \frac{M(x + tiny) - M(x)}{tiny}$$

There's a picture of this idea in Figure 2.1.

---

2. You're right to be worried! It's not clear that this idea of infinitely small numbers makes any sense, but if we're worried about it, we can just imagine that *tiny* is 0.00(*etc*)001, where there are maybe 100 or 1000 or 10,000 zeros between the decimal point and the 1. Then instead of using an infinite magnifying glass, we'd just be using an extremely powerful magnifying glass. After zooming in like this, curvy things won't be exactly straight, but they'll be so close to straight that we could act as if they were straight and our answers would all be so accurate that we'd never notice the difference. Indeed, all of calculus could be done this way, so — resting assured that we always have this safer but less elegant method to fall back on if our "infinite magnifying glass" approach runs into problems — we can simply forge ahead with our more risky way of thinking, always knowing that we have a safety net.

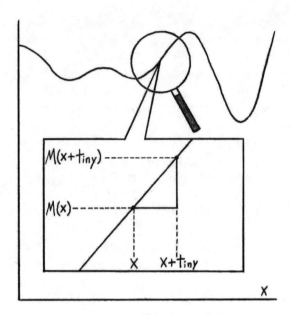

Figure 2.1: Pick any point on a curvy thing and zoom in infinitely far. Once we're
zoomed in, we can just treat it like a straight line. For example, we can
define the steepness (at the point we zoomed in) just by looking at the
"rise over run" of two points that are infinitely close to each other.

2.2.4 Does Our Idea Make Sense? Testing It on Some Simple Examples

All of this is getting a bit abstract, and we've just been making this stuff up
as we go, so let's stop for a reality check to make sure we haven't gone off the
rails. Whenever we invent a new concept, it's always a good idea to test it on
some simple example where we know what we expect.

The idea of infinitely small numbers lets us talk about the steepness of
curvy things, but we're not really sure it makes sense. However, it had better
reproduce what we already know about straight things. If it doesn't, then
either our new concept is broken or else we didn't invent what we meant to.
Let's see if it gives us what we expect.

Trying It Out on the Simplest Machines

First let's test the idea on a really simple machine: $M(x) \equiv 7$. This is a
machine that spits out 7 no matter what we feed it. If we "graph" this machine,
it's just a horizontal line, so its steepness is zero. Since we know what to
expect, let's compute its steepness using infinitely small numbers and see if
we get zero. As before, we'll use the abbreviation *tiny* to stand for a really
tiny number, either infinitely small or just "as small as we want it to be,"

depending on our philosophical preferences. Since M always spits out 7, we have:

$$\text{Steepness of } M \text{ at } x \equiv \frac{M(x + tiny) - M(x)}{tiny} = \frac{7 - 7}{tiny} = \frac{0}{tiny} \equiv 0 \left(\frac{1}{tiny} \right) = 0$$

Notice that we didn't use any special facts about the number 7 in the above argument, so the same argument should work for any machine that always spits out the same number no matter what we feed it. Alright, so for all the machines that look like $M(x) \equiv \#$, the idea of infinitely small numbers gave us what we expected. Onward!

Trying It Out on Lines

Let's test our idea on another simple kind of machine: straight lines. In Chapter 1, we found that straight lines can be described by machines that look like $M(x) \equiv ax + b$. Let's see if computing their steepness using infinitely small numbers gives us the answer we expect (namely, a).

$$\text{Steepness of } M \text{ at } x \equiv \frac{M(x + tiny) - M(x)}{tiny}$$

$$\equiv \frac{[a \cdot (x + tiny) + b] - [ax + b]}{tiny} = \frac{a \cdot (tiny)}{tiny} = a$$

Perfect! Our strange idea hasn't let us down yet. Let's see how it works in a less familiar situation.

Trying It Out on a Genuinely Curvy Thing

Okay, now let's try this zooming in idea on something where it actually *matters* whether we zoom in or not: the Times Self Machine. This was the machine $M(x) = x^2$ that we talked about in Chapter 1. Whatever number we feed it, it multiplies that number by itself and hands back the result. Let's see what these infinitely small numbers give us when we try to compute the steepness at some point x, whose particular value we'll choose to remain agnostic about.

$$\text{Steepness of } M \text{ at } x \equiv \frac{M(x + tiny) - M(x)}{tiny} = \frac{(x + tiny)^2 - x^2}{tiny}$$

We can use the obvious law from Chapter 1 to unwrap $(x + tiny)^2$ into the form $x^2 + 2x(tiny) + (tiny)^2$. Then the above sentence becomes

$$\text{Steepness of } M \text{ at } x = \frac{\overbrace{x^2 + 2x(tiny) + (tiny)^2 - x^2}^{\text{The two } x^2 \text{ pieces kill each other}}}{tiny}$$

$$\overbrace{}^{\text{Cancel }(tiny)\text{ from the top \& bottom}}$$
$$= \frac{2x(tiny) + (tiny)^2}{tiny}$$

$$= 2x + tiny$$

Since *tiny* stands for a number that's infinitely small, we'll never be able to tell the difference between the answer we got, which is $2x + tiny$, and an answer that's infinitely close to it, namely, $2x$. So if we accept this odd way of reasoning, we can write:

The Result of Zooming In Infinitely Far:

If M is the Times Self Machine, $M(x) = x^2$,
then the steepness of M at x is $2x$.

2.2.5 What Just Happened? Infinitesimals vs. Limits

To this day calculus all over the world is being taught as a study of limit processes instead of what it really is: infinitesimal analysis. As someone who has spent a good portion of his adult life teaching calculus courses for a living, I can tell you how weary one gets of trying to explain the complex and fiddling theory of limits.
—Rudy Rucker, *Infinity and the Mind*

Sometimes it is useful to know how large your zero is.
—Author unknown

If this idea of infinitely small numbers scares you a little, you're not alone! After Isaac Newton invented calculus, people racked their brains for more than a century trying to figure out how arguments like this kept giving the right answers. After all, they seemed so obviously ridiculous. Either the number *tiny* is zero or it isn't! How can we act like it's not zero, and then a few lines later act like it is zero?

Over the years, people have invented all sorts of mathematical Rube Goldberg contraptions — ways of "formalizing" calculus — in order to help them make sense of exactly what is going on here. That's good! It's always helpful to make sure the crazy ideas we're inventing make sense, and we should be happy that people have done it. But the idea of infinitely small numbers is so beautiful that it's a shame to hide them under all sorts of contraptions, especially since they actually *work*! In fact, physicists tend to be less scared of using the concept of infinitely small numbers directly. They do their calculations just like we did, and they get the same answers that mathematicians do, but often with much less work.[3] Now, some of these contraptions take the idea of infinitely small numbers seriously, while others try to avoid the idea altogether. The second type is *much* more common, so I'll mention the first type first, just for the sake of heresy.

One of the contraptions that can be used to formalize the idea of infinitely small numbers is called an "ultrafilter." Ultrafilters are pretty complicated, and never mentioned in the introductory textbooks. Even though we won't talk about them after this, it's good to know that they exist, because it means that there's at least one precise way of making sense of calculus that takes the idea of infinitely small numbers seriously.

However, the contraption you'll see in all the standard intro textbooks is called a "limit," and it's much simpler, but it serves the same purpose: it lets mathematicians get all the *benefits* of using infinitely small numbers, without giving the idea of infinitely small numbers any of the credit.

Here's the basic idea of the "limit" contraption. Instead of thinking of the number *tiny* as a number that's infinitely small, let's just think of it as a number that is as small as we want it to be. That is, by not writing down a particular number like *tiny* = 0.00001, we can instead choose to remain agnostic about its value and just run through the same calculation. To put it another way, we can think of the number *tiny* as having a knob attached to it that we can turn at will, choosing to make it as small as we want, as long as it's not *exactly* zero. But notice that this entire discussion assumed that the steepness of M at x could be figured out by pretending it was a perfectly straight line. Now, that assumption is going to be true only when we've turned the "*tiny*" knob all the way down to zero (i.e., once we've zoomed in infinitely far). So instead of what we wrote, you would see something like this in a standard textbook. Stare at all of these strange new symbols for a moment, and then I'll try to explain how it's essentially the same as what we did.

$$M'(x) \equiv \lim_{h \to 0} \left[\frac{M(x+h) - M(x)}{h} \right] = \lim_{h \to 0} \left[\frac{(x+h)^2 - x^2}{h} \right] = \lim_{h \to 0} \left[\frac{2xh + h^2}{h} \right]$$

$$= \lim_{h \to 0} \left[2x + h \right] = 2x$$

What's going on here? Let's translate it.

First, they're using the letter h instead of the word *tiny*. I'm not sure why they use h, but I assume it stands for "horizontal," since a tiny change in x will be a tiny change on the horizontal axis. Second, instead of writing the phrase "Steepness of M at x" like we did, the textbooks abbreviate this as $M'(x)$. That's fine, and it's a lot quicker to write.[4] Third, there's the weird thing on

---

3. This difference gets even bigger as we move to more advanced contexts, as we'll see throughout the book.

4. Basically, the apostrophe is just an abbreviation that means "Zoom in on $M(x)$ and find the slope as if it were a line." They pronounce $M'(x)$ by saying "M prime of x."

the left of each piece that looks like this:

$$\lim_{h \to 0}$$

This is the contraption that allows us to avoid thinking about infinitely small numbers if we want to. The symbol above is pronounced "the limit as h goes to 0," and it means something like this:

> **What the abbreviation $\lim_{h \to 0}[stuff]$ means:**
> Calculate everything inside me (the *stuff*) as if h were a regular everyday number, not an infinitely small number. Then, when we've gotten rid of all the h's on the bottom of *stuff* (so that we don't have to worry about what it means to divide by zero), we imagine turning the h knob down so that h gets smaller and smaller and smaller. For example, something like $3 + h$ will get closer and closer to 3 as you turn the h knob closer and closer to zero. A big complicated thing like $79x^{999} + 200x^2 h + h^5$ will get closer and closer to $79x^{999}$ as you turn the h knob down to zero.

This is a perfectly fine way to do the same thing we were doing when we imagined zooming in infinitely far, but it can be pretty confusing if they don't explain why they're doing it. In one sense it's making things simpler because we don't have to worry about the meaning of an infinitely small number. But in another sense it makes things harder because it's not always obvious to students why they have to learn all about these odd things called "limits," especially since we're usually taught about limits *before* we hear about ideas like reducing curvy problems to straight problems, and defining the slope of curvy things by zooming in. That is, they teach us the behavior of these limit things before we have any reason to care, and before they tell us the reason that limits were invented in the first place. So it's not surprising that people think calculus is confusing.

It's less confusing if we realize that limits are just one of several (optional!) contraptions that let us avoid worrying about the meaning of infinitely small numbers *if we want to*. Throughout the book, we'll occasionally use limits, and we'll occasionally use infinitely small numbers, just so you get used to both. Fortunately, we'll always get the same answers using either method, so you're free to pick whichever one you prefer.

2.2.6 A Laundry List of Abbreviations

In the previous section, we used the phrase "Steepness of M at x" as an abbreviation for the process of zooming in on a curvy thing and computing its steepness as if it were a line. That's a lot of words. Let's look at some common ways of abbreviating the idea. All of these mean the same thing:

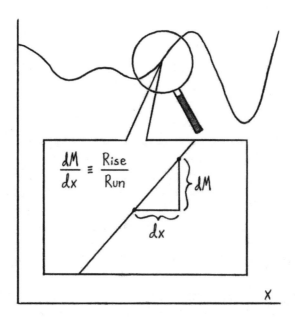

Figure 2.2: The steepness (or "derivative") of a machine M is sometimes abbreviated by writing $\frac{dM}{dx}$. This is why.

1. The Steepness of M at x

2. The Derivative of M at x
   This is definitely the most common name for the concept. It's a noun, and the corresponding verb is "differentiate," which means "figure out the derivative."

3. $M'(x)$
   This abbreviation emphasizes the fact that we can think of the steepness as a machine in its own right. $M'(x)$ stands for a machine that works like this: when we hand it some number x, the machine $M'(x)$ spits out the *slope* of the original machine M at the same point x.

4. $\frac{dM}{dx}$
   As much as I complain about the abbreviations that textbooks usually use, this one is pretty good once you get used to another abbreviation that's not as good. Just for a moment, let's rename our machine V to stand for "vertical," and let's use H instead of x to talk about the stuff it eats. So we'll write $V(H)$ instead of $M(x)$, but only for the next few paragraphs. We're doing this because we're thinking of drawing the graph of the machine, so that its output is drawn in the vertical direction and its input is drawn in the horizontal direction. In the first chapter, we wrote h and v to stand for the differences in horizontal and vertical

location between two points. Our h and v would normally be called something like ΔH and ΔV in the standard textbooks, where $\Delta stuff$ stands for "a difference in *stuff* between one place and another." This is why (when the machine V happens to be a line) you'll sometimes see its slope written like this:

$$\frac{\Delta V}{\Delta H}$$

This is just "rise over run," or what we called $\frac{v}{h}$ in the first chapter. Now, the symbol Δ is the Greek letter d (it's really more like D, but stay with me), so it sort of makes sense to use Δ as an abbreviation for a "difference" between two things, or a "distance" between two points, which is what the textbooks do: ΔV stands for the vertical distance between two places, and ΔH stands for the horizontal distance between them. So all of the following are different ways of abbreviating the steepness of the machine V, as long as it's just a straight line:

$$\text{Steepness of V} \equiv \frac{\text{Rise}}{\text{Run}} \equiv \frac{\text{Change in } V}{\text{Change in } H} \equiv \frac{\Delta V}{\Delta H}$$

Now, while $\Delta stuff$ stands for a change in *stuff* between two points, it basically always stands for a regular change involving regular numbers, not an infinitely small change involving infinitely small numbers. But now that we've started to invent calculus, we suddenly find that we'd like to distinguish between normal changes (when we're *not* zoomed in) and infinitely small changes (when we *are* zoomed in). Here's an occasion where the standard notation does something really nice: to change an expression involving regular numbers to an expression involving infinitely small ones, just change the Greek alphabet to the Latin alphabet (i.e., change Δ to d). So if we use the abbreviation $d(stuff)$ to stand for an infinitely small change in *stuff* between two infinitely close locations, then we can write a string of equations similar to the ones above, but this time they'll also apply to curvy things like $V(H) = H^2$, instead of only applying to straight lines:

$$\text{Steepness of } V \equiv \frac{\text{Tiny Rise}}{\text{Tiny Run}} \equiv \frac{\text{Infinitely small change in } V}{\text{Infinitely small change in } H} \equiv \frac{dV}{dH}$$

This is why you'll often see the derivative of a machine M written as $\frac{dM}{dx}$. See Figure 2.2 for a picture of this idea. Similarly, we'll see later that when the textbooks pass from the letter Σ (the Greek S, which stands for the word "sum") to its corresponding Latin letter S (they actually write $\int$, which kind of looks like an S), they're doing a similar trick. In both cases, the passage from Greek letters to their Latin equivalents signifies the passage from a mundane expression involving regular numbers, to an expression involving numbers that are infinitely small. To be clear, Greek letters in mathematics *certainly* don't always have this nice interpretation. They're used for all sorts of different things. But at least in the above two cases (unlike others we'll see later) the standard notation was designed extremely well.

In summary, all of the above abbreviations are referring to exactly the same idea, so it might seem like we have more than we need. But we'll soon see how switching back and forth between different abbreviations has a surprising ability to make complicated things seem simple, and vice versa.

2.3 Understanding the Magnifying Glass

In this section we'll try to get better at using our infinite magnifying glass. Even though we invented it ourselves, it's not entirely clear what we invented or how it behaves. That is, even though we invented the *idea*, we haven't had a lot of practice actually *using* it on specific machines. In this section we'll get a lot of practice by playing with a bunch of different examples. The only machines we know about are ones that can be completely described in terms of addition and multiplication, so that's the only kind of machine we'll play with at this point.

2.3.1 Back to the Times Self Machine

Okay, so we've tested our infinite magnifying glass idea on machines that look like $M(x) \equiv \#$, and we got $M'(x) = 0$. We tested it on different machines that look like $M(x) \equiv ax + b$, and we got $M'(x) = a$. Up to that point, we had only rediscovered things that we *could have* discovered with the normal slope formula and no infinite zooming in.

Then we tested our infinite magnifying glass on our first curvy thing: the machine $M(x) \equiv x^2$, and we got $M'(x) = 2x$. Before we move on to any different examples, let's make sure we understand what this is saying by looking at it in two different ways.

2.3.2 The Usual Interpretation: The Machine's Graph Is Curvy

The first way of looking at this is the usual way: "graphing" $M(x) \equiv x^2$ and seeing that the graph is curvy. This is what we're doing in Figure 2.3.

We imagine laying a bunch of Times Self Machines next to each other along the horizontal axis. We feed them the numbers on the horizontal axis, and what they spit out we draw on the vertical axis. The machine next to the number 3 is fed $x = 3$, and it spits out $M(x) = 9$, which is why we drew the height of the curvy thing at $x = 3$ to be 9. All the other places on the graph say essentially the same thing.

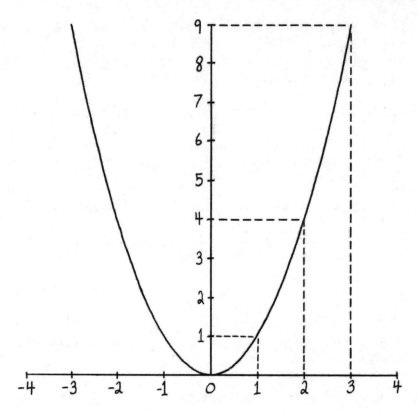

Figure 2.3: Visualizing the Times Self Machine $M(x) \equiv x^2$. The numbers in the horizontal direction are different things we can feed it, and the numbers in the vertical direction (the height) tell us what it spits out. Since the graph is curvy, it has different amounts of steepness at different places, so while the sentence $M(x) \equiv x^2$ tells you the height at x, the sentence $M'(x) = 2x$ tells us the steepness at x. The graph is flat in the middle (at $x = 0$), so the steepness should be zero there. Fortunately, $M'(x)$ tells us this too, because $M'(0) = 2 \cdot 0 = 0$.

So we chose a random point with horizontal position x and vertical position $M(x)$. Then we zoomed in infinitely far on the curve at this point and found its steepness there using the simple old "rise over run" business that we invented in Chapter 1. When all the dust settled, we found that the steepness there was $M'(x) = 2x$. Since we chose to be agnostic about which particular number x was, we really did infinitely many calculations at once. So the sentence $M'(x) = 2x$, in just a few symbols, manages to express an infinite number of sentences. Let's see what some of them say.

One of these sentences says $M'(0) = 2 \cdot 0 = 0$. This tells us that the steepness of the curve is zero when $x = 0$. If we look at Figure 2.3, this makes more sense. The graph is flat and horizontal there, so the steepness is zero. What about some of the other sentences hiding inside the infinite sentence

$M'(x) = 2x$? Well, some other ones are:

1. $M'(1) = 2 \cdot 1 = 2$, so at the place where $x = 1$, the steepness of the curve is 2.

2. $M'(\frac{1}{2}) = 2 \cdot \frac{1}{2} = 1$, so at the place where $x = \frac{1}{2}$, the steepness of the curve is 1.

3. $M'(10) = 2 \cdot 10 = 20$, so at the place where $x = 10$, the steepness of the curve is 20.

We could keep going, but all the sentences are saying basically the same thing. At a location with horizontal position h, the steepness of this curve is exactly $2h$. So the steepness is always twice the horizontal distance away from 0. Just from this last sentence, we can see why the graph of M has to keep going up faster and faster. As the horizontal distance from zero gets bigger, the steepness is steadily increasing at every step.

2.3.3 The Reinterpretation Dance: The Machine Has Nothing to Do with Curviness

Okay, so in the previous section we talked about the usual interpretation of the sentence "The machine $M(x) \equiv x^2$ has derivative $M'(x) = 2x$." This interpretation involved graphing the machine M, noticing that it had different steepnesses in different places, and interpreting the derivative as telling us what the steepness was at different points.

But I also promised that we'd be able to look at it a second way, so let's do that now. We'll arrive at all the same conclusions by thinking about this machine differently. To start, notice that we don't have to visualize the Times Self Machine by graphing it. We can also visualize it by thinking of $M(x) \equiv x^2$ as the area of a square whose length on each side is x. Since we're thinking about it differently now, let's use the abbreviation A instead of M, and ℓ instead of x. Then we can write $A(\ell) \equiv \ell^2$, and we'd still be talking about the exact same machine, but we're not thinking of it as talking about anything curvy now.

As always in calculus, we've got a machine, and we're asking a question like, "If I change the stuff I'm feeding it by a tiny amount, how does the machine's response change?" Since d is the first letter of "difference" and ℓ is the first letter of "length," let's use the abbreviation $d\ell$ to stand for some tiny difference in length. We're thinking of $d\ell$ as a change we're making to the side length of a square, so $\ell_{after} \equiv \ell_{before} + d\ell$. When we change the length a little bit, we can ask how the area changed. The area before the change is ℓ^2, and the area after the change is $(\ell + d\ell)^2$. Let's write all this in a box.

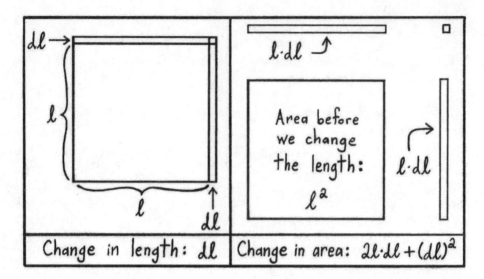

Figure 2.4: Another way of understanding the derivative of the Times Self Machine
$A(\ell) \equiv \ell^2$. Originally, we have a square that is ℓ long on all sides. Then
we change ℓ a tiny bit to make it $\ell + d\ell$. Then we see how the area
changed. From the picture, the change in area will be $dA \equiv A_{after} -
A_{before} = 2(\ell \cdot d\ell) + (d\ell)^2$. So before shrinking $d\ell$ down to zero, we have
$\frac{dA}{d\ell} = 2\ell + d\ell$. Then shrinking $d\ell$ down to zero (or if you prefer, thinking
of it as "infinitely small" from the beginning), we get that the derivative
is $\frac{dA}{d\ell} = 2\ell$. We can think of the sentence $\frac{dA}{d\ell} = 2\ell + d\ell$ as saying that
the two long thin rectangles shrink to two lines (hence the 2ℓ), while the
tiny square shrinks down to a point (hence the $d\ell$), which only adds an
infinitely small length to either of those lines, so we can ignore it and
just write $\frac{dA}{d\ell} = 2\ell$.

Making Tiny Changes to a Square:

Length before we change it: ℓ
Length after we change it: $\ell + d\ell$
Change in length: $d\ell = \ell_{after} - \ell_{before}$

Area before we change it: ℓ^2
Area after we change it: $(\ell + d\ell)^2$
Change in area: $dA = A_{after} - A_{before}$

So we changed what we're feeding the machine by a tiny amount. How does
the stuff it spits out change? Let's draw a picture. Our picture is in Figure
2.4, and it shows how the area changes when we change the side lengths a little
bit. Here are different ways we could abbreviate the resulting change in area:

$$dA \equiv A_{after} - A_{before} \equiv A(\ell + d\ell) - A(\ell)$$

The picture makes it clear that this change in area will be

$$dA = 2(\ell \cdot d\ell) + (d\ell)^2$$

This is just another way of saying

$$\frac{dA}{d\ell} = 2\ell + d\ell$$

Then, shrinking $d\ell$ down to zero (or if you prefer, thinking of it as infinitely small from the beginning, so that $2\ell + d\ell$ will be indistinguishable from 2ℓ), we find that the "derivative" of the Times Self Machine $A(\ell) \equiv \ell^2$ is

$$\frac{dA}{d\ell} = 2\ell$$

This is the same sentence we found earlier, when it was written as $M'(x) = 2x$. In both cases, it was saying that the "derivative" or "steepness" or "rate of change" of the Times Self Machine at any number we feed it is two times as big as the number itself.

2.3.4 What We've Done So Far

We're still only thinking about machines that can be completely described using addition and multiplication. So we still don't know anything about any of those bizarre machines you might have heard of, like $\sin(x)$, $\ln(x)$, $\cos(x)$, or e^x. We still have no idea what the area of a circle is, we don't know what π means, and so on. We basically don't know much except addition, multiplication, the idea of a machine, how to invent a mathematical concept, and a bit of calculus. The first mathematical concepts we invented were "area," in the easy case of rectangles, and "steepness," in the easy case of straight lines. Then we noticed that curvy stuff turns into straight stuff if you zoom in infinitely far, so we invented the idea of an infinite magnifying glass. This lets us talk about the steepness of any machine M whose graph is curvy, just by finding the "rise over run" of two points that are infinitely close to each other, like this:

$$\frac{dM}{dx} \equiv \frac{\text{Tiny Rise}}{\text{Tiny Run}} \equiv \frac{M(x + dx) - M(x)}{dx}$$

Even though we feel like this idea should work for any machine we can describe at this point, we still haven't played around with our infinite magnifying glass much yet. So far, we've only used it on constant machines, lines, and the Times Self Machine. Of these, only the last one was curvy, so we've really only played with the full power of our infinite magnifying glass in one example. Let's play some more to try to get used to it.

2.3.5 On to Crazier Machines

The Machine $M(x) \equiv x^3$

If we want to play around with our infinite magnifying glass some more, we've got to think of some machines to use it on. Let's try $M(x) \equiv x^3$. So far, we've used the two abbreviations *tiny* and dx to stand for a tiny (possibly infinitely small) number. We also used $d\ell$, but that's the same type of abbreviation as dx. We could use either of those, but both of them are fairly clunky. Let's use the one-letter abbreviation t to stand for *tiny*. Just like before, t is a tiny (possibly infinitely small) number. So we want to feed the machine x, and then feed it $x + t$, and see how its response changes. Let's use dM as an abbreviation for the change in its response, or $dM \equiv M(x+t) - M(x)$. Then

$$dM \equiv M(x + t) - M(x) = (x + t)^3 - x^3 \tag{2.1}$$

The $(x + t)^3$ piece will have a x^3 hiding inside it, so that's going to kill the negative x^3 on the right side of dM, but we can't see what the leftovers will look like unless we do the tedious job of breaking $(x + t)^3$ apart. (Soon we'll discover a way to avoid this.)

$$(x + t)^3 \equiv (x + t)(x + t)(x + t) = x^3 + 3x^2t + 3t^2x + t^3 \tag{2.2}$$

This is a pretty ugly sentence, and no one would want to memorize it. Fortunately, we saw how to invent the above sentence in Chapter 1, either by drawing a picture and staring at it, or by using the obvious law of tearing things a few times. As ugly as it is, equation 2.2 gives us another way of writing equation 2.1. That is:

$$dM = M(x + t) - M(x) \overset{(2.1)}{=} (x + t)^3 - x^3 \overset{(2.2)}{=} 3x^2t + 3t^2x + t^3$$

where the numbers above the equals signs tell you which equation to look back to if you aren't sure what we did. Everything has at least one t attached to it, so if we divide by that, we get

$$\frac{M(x + t) - M(x)}{t} = 3x^2 + 3xt + t^2$$

Notice that $3x^2$ is the only piece that doesn't have any t's still attached. But remember that t was our abbreviation for an infinitely tiny number, or if you prefer, a number that we can imagine making smaller and smaller until we don't notice it anymore. So we can rewrite this as

$$\frac{M(x + t) - M(x)}{t} = 3x^2 + GonnaDie(t)$$

The term $GonnaDie(t)$ is an abbreviation for all the stuff that's gonna die when we turn the tiny number t all the way down to zero, and the piece on the far left is just the "rise over run" between two extremely close points, so it will turn into the derivative of M when we turn t down to zero. Let's do that. Shrinking t down to zero gives

$$M'(x) = 3x^2$$

Did it work? Well, we don't know yet. We're on our own here. Let's keep moving, and maybe we'll eventually figure out whether what we just did makes sense. (Don't worry. It does.)

The Machines $M(x) \equiv x^n$

The hardest part of the example above wasn't the calculus — that just involved throwing away all the pieces with a t attached. The hardest part was the tedious business of expanding $(x + t)^3$. Let's see if we can find the derivative of (i.e., use our infinite magnifying glass on) the machine $M(x) \equiv x^4$ *without* having to expand everything. Just like before, we want to calculate this:

$$\frac{M(x + t) - M(x)}{t} \equiv \frac{(x + t)^4 - x^4}{t} = ?$$

where t is some tiny number. We need to figure out a way to get rid of the t on the bottom[5] so that we can just go ahead and turn t down to zero.

How can we avoid having to expand $(x + t)^4$? Why avoid it at all? Well, expanding it all out wouldn't be a very intelligent process. We'd waste a lot of time, and more importantly, we wouldn't *learn* anything that would actually help us if we ever had to deal with $(x + t)^{999}$, or $(x + t)^n$. So let's avoid expanding it, but still try to learn, in a vague sense, what the expanded answer looks like. The term we want to avoid expanding is $(x + t)^4$, which is an abbreviation for

$$(x + t)(x + t)(x + t)(x + t)$$

This can be thought of as four bags, each of which has two things inside it: an x and a t. If we took the time to expand all this nonsense, we'd get a result that was a bunch of pieces added together. We don't care about the full expanded result. We just want to get a feel for what the individual pieces of the result would look like, so we can get a vague idea of how the final result might look without having to actually expand it all out. Suppose we use an abbreviation like this:

---

5. Why do we need to figure out a way to get rid of the t on the bottom? Because $(tiny/tiny)$ isn't really a tiny number! Similarly, if $(stuff)$ is some normal number, not assumed to be infinitely small, then $(stuff)(tiny)/(tiny)$ isn't infinitely small either. It's just equal to $(stuff)$. So the reason we need to get rid of the t on the bottom is because it prevents us from seeing which pieces are really infinitely tiny, and which are normal numbers like 2 or 78.

$$(x+t)(x+t)(x+t)(x+t) \equiv (4 \text{ bags}) \equiv (x+t)(3 \text{ bags})$$

Imagine applying the obvious law of tearing things to the far right side, and then focusing in on one of the two pieces. Like this:

$$
\begin{aligned}
(4 \text{ bags}) &\equiv (x+t)(3 \text{ bags}) \\
&= x(3 \text{ bags}) + t(3 \text{ bags}) \\
&= x(3 \text{ bags}) + \ldots
\end{aligned}
\tag{2.3}
$$

In the three lines above, we essentially tore open one bag, got two pieces out, and then picked one to focus on. Remember, we don't care about every single term. We just want to see what an arbitrary piece in the end result looks like, to get an intuitive feel for how the end result is built. So since our goal is pretty modest, we can just keep ignoring all but one piece, like we did above, and keep on tearing the bags apart. Which side should we choose to focus on each time we tear? It doesn't matter. Let's just choose randomly. We just picked the x from the first bag, so let's pick the t from the second, the t from the third, and the x from the fourth: x, t, t, x. Each time we unwrap, tear, and pick one term, I'll write which one we picked above the equals sign. Each line below will be an abbreviation for three lines of reasoning just like those in equation 2.3: unwrap, tear, and focus. If that seems complicated, here's all I mean:

$$
\begin{aligned}
(4 \text{ bags}) &\stackrel{x}{=} x(3 \text{ bags}) + \ldots \\
&\stackrel{t}{=} xt(2 \text{ bags}) + \ldots \\
&\stackrel{t}{=} xtt(1 \text{ bag}) + \ldots \\
&\stackrel{x}{=} xttx + \ldots
\end{aligned}
\tag{2.4}
$$

That wasn't much work, but it turns out to give us exactly the information we wanted. We still don't know what the full "expanded" result is, but we can now see that each term in the fully expanded version will be built from a choice. Or rather, four choices: a choice of one item from each bag. We just made our choice of $xttx$ at random, so each other choice we *could have* made has to be in the final result as well. So without even expanding $(x+t)^4$, we know that the fully expanded version will have an $xttx$, but there will also be an $xxxx$, and a $tttt$, and a $xttt$, and so on.

That's really all we need to know. This simple knowledge makes it incredibly easier to compute the derivative of x^4, or even x^n for that matter. Let's return to the problem with this new bit of knowledge. We want to compute this:

$$\frac{M(x+t) - M(x)}{t} = \frac{(x+t)^4 - x^4}{t} = ? \tag{2.5}$$

Here's the thought process:

1. One of the pieces in $(x+t)^4$ will be $xxxx$, which is x^4, but that gets killed by the negative x^4 in the equation above (equation 2.5).

2. Some of the pieces in $(x+t)^4$ will have only one t attached. That single t has to come from one of the four bags, so there should be four pieces with only a single t. That is, $txxx$, $xtxx$, $xxtx$, and $xxxt$. The single t in each will get killed by the t on the bottom of equation 2.5, turning each of those four terms into xxx, which is x^3. That's four copies of x^3, or $4x^3$. Now, since these pieces won't have any more t's attached once we've canceled the t on the bottom, they won't die when we turn all the t's down to zero. Let's write all of these pieces as $t \cdot Survivors$ to emphasize that they all only have one t hanging on to them, so when we divide them by t, they turn into $Survivors$, where $Survivors$ is a bunch of stuff that doesn't have any t attached. In this case, $Survivors$ is just $4x^3$, but writing it in this more general way will help us make the argument for powers other than 4.

3. The rest of the terms will have more than one t attached, like $ttxx$ or $txxt$. The t on the bottom of equation 2.5 will kill one of these, but they'll all have at least one t still attached in the end, so these pieces will all die when we turn the t knob down to zero. Let's call all these pieces $t \cdot GonnaDie(t)$ to emphasize that they still have t's hanging on even after we divide them by t, thus turning them into $GonnaDie(t)$.

So, thinking about the problem this way, we can rewrite equation 2.5 using all the abbreviations we just decided on. Then we get:

$$\frac{M(x+t) - M(x)}{t} = \frac{x^4 + [t \cdot Survivors] + [t \cdot GonnaDie(t)] - x^4}{t} \qquad (2.6)$$

The x^4 pieces kill each other, and we can cancel the t's to get

$$\frac{M(x+t) - M(x)}{t} = Survivors + GonnaDie(t) \qquad (2.7)$$

Now let's imagine turning t down closer and closer to zero. This does nothing to the $Survivors$ piece, but it kills $GonnaDie(t)$. The left side turns into the derivative of M, which we'll write as $M'(x)$. Summarizing all that, we've got

$$M'(x) = Survivors \qquad (2.8)$$

Hey! So the derivative of M is just that term we called $Survivors$, which in this case was $4x^3$. But in that whole argument, starting at equation 2.5, we didn't really use the fact that the power was 4. At least not in any important sense. So the same type of argument should work for the more general version $(x+t)^n$, no matter what number n is.

The benefit of making the strange argument we just made — looking for general patterns, since we were too lazy to expand $(x+t)^4$ in the usual tedious way — is that it instantly lets us figure out the derivative of x^n, no matter what number n is! Here's how. To figure out the derivative of x^n while remaining agnostic about n, we need to compute:

$$\frac{(x+t)^n - x^n}{t} = ?$$

(2.9)

Now, just like before, let's think in terms of bags. This time we have n of them:

$$(x+t)^n = \underbrace{(x+t)(x+t)\cdots(x+t)}_{n \text{ times}}$$

(2.10)

Each piece of $(x+t)^n$ will be n things multiplied together, with various numbers of x's and t's. For example, one piece will be the guy with only x's, or

$$\underbrace{xxxx\cdots xxxx}_{n \text{ times}}$$

That's x^n, and it gets canceled by the negative x^n in equation 2.9. Another one of the pieces will be the guy with a t in his second slot, a t in his final slot, and all the rest x's, like this:

$$\underbrace{xtxx\cdots xxxt}_{n \text{ things. Only 2 } t\text{'s}}$$

But this term has two t's, so even after we cancel one of them against the t on the bottom of equation 2.9, there's still at least one t left, so this term will die when we turn t down to zero, as will all the others with two or more t's.

So we don't have to expand $(x+t)^n$ using some complicated formula.[6] As before, the derivative will only be affected by all the pieces with a single t, since those are the ones that will survive when their single t gets canceled by the t on the bottom of equation 2.9. One such piece will be $txxx\cdots xxxx$, another one will be $xtxx\cdots xxxx$, and so on. But each of these is just tx^{n-1}. There are n bags that a single t could come from, so the same term is just showing up n times.

$$t\cdot Survivors = \underbrace{tx^{n-1} + tx^{n-1} + \cdots + tx^{n-1}}_{n \text{ times}}$$

(2.11)

We can rewrite the same thing this way:

$$t\cdot Survivors = ntx^{n-1}$$

(2.12)

---

6. The complicated formula textbooks use is called the "binomial theorem." It's basically a complicated way of talking about bags, like we did, but without saying that that's what we're doing, all while using a strange notation that looks almost like a fraction, but not quite. We can definitely do without it.

Since there's one t on both sides, we can kill it off to get

$$Survivors = nx^{n-1} \tag{2.13}$$

But for the same reason as before, *Survivors* turns out to be the derivative of the machine we started with, which was $M(x) \equiv x^n$, so we can summarize this entire section by saying

---

What We Just Invented

If $M(x) \equiv x^n$,

then $M'(x) = nx^{n-1}$

Now, you might trust the $n = 2$ calculation and the $n = 3$ calculation more than that crazy business about *Survivors* that we just did in this section, but notice that we can use our old results to check our new ones! If we were right that the derivative of machines like $M(x) \equiv x^n$ is really $M'(x) = nx^{n-1}$ no matter what number n is, then this formula has to reproduce our old results, or else it's wrong. That is, our new formula has to predict[7] that the derivative of x^2 is $2x$, and that the derivative of x^3 is $3x^2$. In fact, it does! When $n = 2$, the expression nx^{n-1} turns into $2x$, and when $n = 3$, it turns into $3x^2$. Now, you could still argue, "We shouldn't feel completely certain that the above argument worked just because it gave us the right answers in the two cases where we knew what to expect!" You'd be right, but it does (and should) give us more confidence that we're on the right track, and that we didn't make any slip-ups in reasoning. In this process of inventing mathematics, it's always up to us to decide when we're convinced, since our mathematical universe doesn't come pre-filled with books where we can simply look up the answer.

2.3.6 Describing All of Our Machines at Once: Ultra-agnostic Abbreviations

Having played with our infinite magnifying glass by testing it on a bunch of specific machines, where do we go next? At the moment, the only machines we know about are ones that we can completely describe in terms of addition and multiplication. We've said the previous sentence so many times that it's worth asking ourselves exactly what we mean. What kinds of machines *can* be "completely described" just "in terms of" addition and multiplication, that is, in terms of what we know? Well, of course it depends what we mean by that. For example, do we allow division in the description? It might seem like the answer should be no, because division doesn't really exist in our universe. But it sort of does. The number $\frac{1}{s}$ in our universe is just an abbreviation for

---

7. Or rather, "postdict"... Or maybe "retrodict"... Or something like that.

whichever number turns into 1 when we multiply it by s. But then what does it mean to describe something only in terms of addition and multiplication?

This is really just an issue of how we use words, so let's not worry too much about it for now. We won't allow "division" in the description, at least not yet, which just means we're excluding machines like $m(s) \equiv \frac{1}{s}$ for the moment. What sorts of machines can we describe? Here are some:

1. $m(s) \equiv s^3$

2. $r(q) \equiv q^2 - 53q + 9$

3. $f(u) \equiv 5u^2 + 7u^3 - 92u^{79}$

We don't really have any reason to be interested in any of these machines, but we like our infinite magnifying glass, and we'd like to get better at using it. As we list more and more machines that can be described in terms of addition and multiplication, it starts to seem like there's a pattern. All the machines we can describe at this point have a common structure. They're all made out of pieces that look like this:

$$(Number) \cdot (food)^{number}$$

where "*food*" is whatever abbreviation we're using for what the machine eats (what textbooks call a "variable") and where *number* isn't necessarily the same as *Number*. All of the machines we listed above, and essentially any machine that we can completely describe in terms of what we know, is going to be a bunch of things that look like this, all added together. Why not "added and multiplied" together? Good question! That's because two things that look like $(Number) \cdot (food)^{number}$, when multiplied together, will be of the same form. That is, multiplying together something like ax^n and bx^m gives $(ab)x^{n+m}$, which is just $(Number) \cdot (food)^{number}$ again. Okay, so any machine that we can completely describe in terms of addition and multiplication will be built from adding up pieces that look like this. Let's come up with some abbreviations so that we can talk about any and all of our machines at once.

We need different abbreviations for a *lot* of different numbers. We only know what $(stuff)^{\#}$ means when $\#$ is a positive whole number,[8] so in our universe, there's no such thing as a negative or fractional power (yet!). So, since we're only thinking about whole number powers at this point, we can write all of our machines like this:

$$M(x) \equiv \#_0 + \#_1 x^1 + \#_2 x^2 + \cdots + \#_n x^n \tag{2.14}$$

where n is the biggest power that shows up in the description of that particular machine, and the symbols $\#_0$, $\#_1$, $\#_2$, $\ldots$, $\#_n$ are just abbreviations for

---

8. Remember $(stuff)^{\#}$ is just an abbreviation for $(stuff)(stuff)\cdots(stuff)$, where the $\#$ is the number of times *stuff* shows up. If we're being consistent about this, then $(stuff)^1$ should just be another way of writing $(stuff)$.

whatever numbers show up out front. We put subscripts on them so that we can tell them apart, and so we don't have to use a different letter for each one. Also, we started the subscripts at 0 instead of 1 to make the rest of the subscripts match up with the powers, so that each piece (except the first one) looks like $\#_k x^k$ instead of $\#_k x^{k-1}$. The only reason we did this was because it's slightly prettier. If we just use x^0 as an abbreviation for the number 1, then every single piece, including the first one, will look like $\#_k x^k$. (There's actually a better reason than this for choosing $x^0 = 1$. We'll see what that is in the next interlude.) At present, our choice to write x^0 as an abbreviation for the number 1 is entirely a matter of aesthetics, and we'll only use it so that every piece in equation 2.14 can be written to look like $\#_k x^k$. For the moment, we have no reason to suspect that zero powers really *are* equal to 1 in any principled sense. We're just abbreviating.

Okay, so equation 2.14 is an abbreviation that describes *all* of the machines that we can talk about at this point. Now, it's fairly tedious to keep writing all the dots (these things $\longrightarrow \cdots$) in our description of these machines. So let's invent a shorter way of saying the same thing. As an abbreviation for the right side of equation 2.14, we could write something like

$$Add\left(\#_k x^k\right) \qquad \text{where } k \text{ starts at 0 and goes to } n$$

but this is a bit clunky too. We don't need all those words on the right. We just need to remind ourselves where k starts and where it ends. So we could abbreviate the same thing this way:

$$Add\left(\#_k x^k\right)_{k=0}^{k=n}$$

The way this is usually written in textbooks is

$$\sum_{k=0}^{n} \#_k x^k$$

Actually, they usually use a letter like c (c stands for "constant") instead of our number symbol $\#$. They use a Σ (the Greek letter S) because S is the first letter of "sum," and "sum" means "add." This way of writing things can look scary before you're used to it, but it's just an abbreviation for the right side of equation 2.14, where we express the same thing using the "$\cdots$" notation. If you don't like sentences with Σ, you can always just rewrite them by using the dots.

So we've built an abbreviation with enough agnosticism in it to let us talk about all of our machines at once. Since we've come up with a way of abbreviating them all, let's come up with a name for them all too, so we don't have to keep saying "machines that can be completely described using only addition and multiplication." Let's call any machine of the above form a "plus-times machine." Based on everything we've talked about so far, you might have guessed that textbooks have some unnecessarily fancy name for this simple type of machine. In fact, they do! They call them "polynomials," which isn't the best term.[9]

Anyways, if we could figure out how to use our infinite magnifying glass on any plus-times machine, then we would have truly progressed to a new level of skill in using it. Let's try that, but first let's summarize the above discussion, and make it official by writing it in its own box.

An Ultra-Agnostic Abbreviation

The strange symbols on the left are just an abbreviation for the stuff on the right:

$$\sum_{k=0}^{n} \#_k x^k \equiv \#_0 x^0 + \#_1 x^1 + \#_2 x^2 + \cdots + \#_n x^n$$

Why: We're using this abbreviation because we want to be able to talk about *any* machine that can be completely described using what we know: addition and multiplication.

P.S. The number symbols (i.e., $\#_0, \#_1, \#_2, \ldots, \#_n$) are just normal numbers like 7 or 52 or 3/2. Writing symbols like $\#$ (instead of specific numbers like 7) lets us remain agnostic about which specific numbers they are. This way, we can secretly talk about infinitely many machines at once.

Name: We'll call machines that can be written like this "plus-times machines." Textbooks outside our universe usually call them "polynomials."

2.3.7 Breaking Up the Hard Problem into Easy Pieces

At this point, we only have two ways of figuring out the derivative of a machine. The first way is just to use the definition of the derivative — that is, to apply our definition of slope from Chapter 1 to two points that are infinitely close to each other. So if someone hands us a machine M, we can try to find its derivative by using this:

$$M'(x) \equiv \frac{M(x + tiny) - M(x)}{tiny}$$

---

9. Granted, the term "plus-times machine" isn't the best either. It's a bit awkward and clunky. But at least it reminds us what we're talking about.

where *tiny* stands for some infinitely small number (or if you prefer, some tiny number, not necessarily infinitely small, which we imagine turning all the way down to zero after we've gotten rid of the *tiny* on the bottom).

The only other way we have of figuring out the derivative of a machine is to ask ourselves, "Have we figured out its derivative yet?" For example, we've figured out that the derivative of any machine that looks like $M(x) \equiv x^n$ is just $M'(x) = nx^{n-1}$, where n is any whole number. We could call this the "power rule," like textbooks do, but of course that's just a name for something we figured out ourselves, using the definition of the derivative. So I guess we really only have one way of figuring out the derivatives of things: the definition. As in all areas of mathematics, these "rules" (like the "power rule" textbooks talk about) aren't really *rules* after all. They're just names for things we figured out earlier... using the definition... The definition of a concept we created ourselves... starting from a vague, qualitative, everyday concept and falling back on aesthetics and anarchic whim whenever we didn't quite know what to do. Huh... mathematics is weird...

Anyways, so we've written down an abbreviation that is large enough and agnostic enough to describe any machine in our universe. As a test of how far our skills have progressed in mastering the art of the infinite magnifying glass, let's see if we can use it on *any* plus-times machine. Unfortunately, even though we managed to capture all our machines in a single abbreviation, it's a pretty hairy one. If we just plugged the expression

$$M(x) \equiv \sum_{k=0}^{n} \#_k x^k \qquad (2.15)$$

into the definition of the derivative, we'd get a big ugly mess, and we probably wouldn't know what to do. So let's try to break apart this hard question into several easy questions that we can answer more easily.

Well, we've already made a helpful observation: any plus-times machine is just a bunch of simpler things added together. Our abbreviation in equation 2.15 says that these simpler things each look like $\#_k x^k$, where k is some whole number, and $\#_k$ is any number, not necessarily a whole number. Maybe if we could figure out how to differentiate any machine that looks like $\#x^n$, and if we could think of a way to talk about the derivative of a bunch of things added together in terms of the derivatives of the individual things, then we would have discovered a way to use our infinite magnifying glass on *any* plus-times machine. At that point, there would be no corner of our universe that we hadn't conquered.

The Pieces: $\#x^n$

We already figured out how to use our infinite magnifying glass on x^n. Its derivative is just nx^{n-1}. So our job at this point is to ask what we should do with that number $\#$ when we differentiate the pieces $\#x^n$. Let's define

$m(x) \equiv \#x^n$ and try to figure out its derivative. Using the familiar definition of slope for two points that are extremely close to each other:

$$\frac{m(x+t) - m(x)}{t} \equiv \frac{\#(x+t)^n - \#x^n}{t} = \#\left(\frac{(x+t)^n - x^n}{t}\right)$$

Recall that t is the tiny number that we're thinking of as having a "knob" attached to it. When we turn the tiny number t all the way down to zero, the far left side of this becomes the definition of the derivative, $m'(x)$. What about the far right side? Well, $\#$ is just a number, so it stays the same when we shrink t, and the piece to the right of $\#$ is just the derivative of x^n, which we figured out already. So as we might have hoped, the number $\#$ simply hangs along for the ride, and the derivative of $m(x) \equiv \#x^n$ will just be

$$m'(x) = \#nx^{n-1}$$

But wait... except for that last step, we didn't use any special properties of the machine x^n. Could we make this same argument in a more general way? Let's see if we can make a similar argument when $m(x)$ looks like (some number) times (some other machine), or to say the same thing in abbreviated form, when $m(x) \equiv \#f(x)$. That is, is there a relationship between the derivatives of two machines that are almost the same, except one is multiplied by some number? Let's make exactly the same argument we did above.

$$\frac{m(x+t) - m(x)}{t} \equiv \frac{\#f(x+t) - \#f(x)}{t} = \#\left(\frac{f(x+t) - f(x)}{t}\right)$$

Now, exactly like we did before, let's turn the tiny number t all the way down to zero. The far left side of the above equation turns into the definition of $m'(x)$. What does the right side turn into? Well, just like before, $\#$ is just a number, and it doesn't depend on t, so it doesn't change when we shrink t down to zero. The piece to the right of $\#$, however, will turn into the definition of $f'(x)$. So we just discovered a new fact about our infinite magnifying glass, and how it interacts with multiplication! We're really proud of what we just invented, so let's give it its own box, and write the same idea in several different ways.

What We Just Invented

Is there a relationship between the derivatives of any two machines that are almost the same, except one is multiplied by a number? Yes!

$$\text{If } m(x) \equiv \#f(x), \text{ then } m'(x) = \#f'(x).$$

Let's say the same thing in a different way:

$$[\#f(x)]' = \#f'(x)$$

Let's say the same thing in a different way... again!

$$\frac{d}{dx}[\#f(x)] = \#\frac{d}{dx}f(x)$$

This is getting crazy, but let's do it again!

$$(\#f)' = \#(f')$$

You still there? Okay, one more time!

$$\frac{d(\#f)}{dx} = \#\frac{df}{dx}$$

Combining the Pieces

In our attempt to figure out how to use our infinite magnifying glass on any plus-times machine[10] we realized that we might be able to break this hard question into two easy questions. The first was how to find the derivative of $\#x^n$, where n is a whole number. We just played around with that question, answered it, and then along the way we figured out that the same type of argument could answer an even bigger question. This led us to the discovery that we can pull numbers outside of derivatives, or $(\#f)' = \#(f')$.

Now let's try to answer the second of our two "easy" questions. If we have a machine that's made out of a bunch of smaller machines added together, can we talk about the derivative of the whole machine if we only know the derivatives of the individual pieces?

Let's imagine that we have a machine that is secretly two smaller machines sitting right next to each other, but someone put a metal case over both of them, so it looks like one big machine. Imagine we feed the big machine some number x. If the first spits out 7, and the second spits out 4, all we see is a large box spitting out 11. If you think about any complicated machine (a modern computer, for example), it makes sense to think of it in both ways at once: as one big machine, and also as lots of smaller machines bundled.

Why is this important? It's important because, in mathematics, as in our daily lives, the question "*Exactly* how many small machines is a big machine made of?" is kind of a meaningless question. If it helps us to think of a big machine as two simpler machines stuck together, then we're free to think of it that way. Let's abbreviate this idea by writing $M(x) \equiv f(x) + g(x)$. Since we haven't specified what any of these three machines are, we're not really saying anything specific, but sentences like this let us express the idea that we *can* think of a machine as being made from two "simpler" machines, if we want

---

10. In textbook language: "find the derivative of any polynomial."

to. Following this train of thought, let's see if we can say anything about the derivative of the "big" machine M just by talking about the derivatives of its parts. Given that we don't know anything about the smaller machines f and g, we had better go back to the drawing board, that is, to the definition of the derivative. This gives:

$$\frac{M(x+t) - M(x)}{t} \equiv \frac{[f(x+t) + g(x+t)] - [f(x) + g(x)]}{t}$$

$$= \frac{f(x+t) - f(x) + g(x+t) - g(x)}{t}$$

$$= \frac{f(x+t) - f(x)}{t} + \frac{g(x+t) - g(x)}{t}$$

In the first equality we used the definition of M. In the second and third equalities, we were just shuffling things around, trying to segregate the f stuff from the g stuff. We did that because the whole point of all this is to break a hard problem into a bunch of easy problems, so it would be helpful to be able to talk about the derivative of a big machine M just by talking about the derivatives of its parts f and g. It turns out that we've managed to do exactly that. If we now imagine shrinking t down closer and closer to zero, the top left of the above equation will turn into $M'(x)$. Similarly, the bottom is made up of two pieces, and when we shrink t down to zero, one of them will morph into $f'(x)$, and the other will morph into $g'(x)$.

So! We just discovered a new fact about our infinite magnifying glass, one that will be helpful in breaking large problems into simpler parts. Let's give this its own box like we did before, and write the same idea in several different ways.

What We Just Invented

Is there a relationship between the derivatives of a big machine $M(x) \equiv f(x) + g(x)$ and the derivatives of the parts $f(x)$ and $g(x)$ out of which it is built? Yes!

If $M(x) = f(x) + g(x)$, then $M'(x) = f'(x) + g'(x)$

Let's say the same thing in a different way:

$$[f(x) + g(x)]' = f'(x) + g'(x)$$

Let's say the same thing in a different way... again!

$$\frac{d}{dx}\Big(f(x) + g(x)\Big) = \Big(\frac{d}{dx}f(x)\Big) + \Big(\frac{d}{dx}g(x)\Big)$$

This is getting crazy, but let's do it again!

$$(f + g)' = f' + g'$$

You still there? Okay, one more time!

$$\frac{d(f + g)}{dx} = \frac{df}{dx} + \frac{dg}{dx}$$

This might work for any number of machines, but we don't know yet:

$$(m_1 + m_2 + \cdots + m_n)' \overset{???}{=} m_1' + m_2' + \cdots + m_n'$$

So we discovered how to deal with two machines added together. But what if we have 3 or 100 or n machines added together? Do we need to figure out a new law for each possible number from 2 up? If we think about it for a moment, we can see that this problem goes away as soon as we recognize the power of an idea we mentioned earlier. That idea was this:

> The question "How many machines are there *really* in a big machine?" is a meaningless question. If it helps us to think of a big machine as two smaller machines taped together, then we're free to think of it that way.

The same philosophy clearly applies to any number of pieces, not necessarily two. If we acknowledge that a bunch of machines taped together also counts as a single "big machine," then we can pull an interesting trick. Imagine that we have a big machine that we're thinking of as three pieces added together. What's the derivative of the big machine? We only know that "the derivative of a sum is the sum of the derivatives" for the simple case of two parts, that is to say:

$$(f + g)' = f' + g'$$

But if we think of the sum of a bunch of machines as a machine in its own right, then we can just apply the simple two-parts version twice, like this:

$$[f + g + h]' = [f + (g + h)]' = f' + (g + h)' = f' + g' + h'$$

The first equality just says "we're thinking of $(f + g)$ as a single machine, just for a moment." In the second and third equalities, we used the simpler version that we know already: you can distribute the apostrophes when there are only two parts. First we used it on f and $(g + h)$, thinking of $(g + h)$ as a single machine. Then we used it on g and h, no longer thinking of them as a single machine. So the way we were thinking about $(g + h)$ completely changed in the middle of the equations above. It's not hard to convince ourselves that this same kind of reasoning can take us as high as we want. Try to convince yourself that if we have n machines, then the following is true:

$$(m_1 + m_2 + \cdots + m_n)' = m_1' + m_2' + \cdots + m_n'$$

and that this comes from exactly the same type of reasoning we used to get from two machines to three. We just have to imagine applying the same argument over and over. The consequences of this are fairly amazing, because just by figuring out the "two-parts" version $(f + g)' = f' + g'$, we automatically got a more general version. By using the "two-parts" version and then repeatedly changing our minds about whether we think of something as two machines or as one big machine with two parts, we were able to *trick the mathematics* into thinking that we had figured out the bigger "n-parts" version!

(*A faint rumbling noise is heard in the distance.*)

Aaah! It happened again!
Reader: Why do you keep doing that?
Author: It's not me!
Reader: (*Skeptically*) ... Are you *sure*?
Author: Yes! I don't know what's going on any more than you do... Where were we? Oh, I think we're done with this section! Onward, dear Reader!

2.3.8 The Last Problem in Our Universe... for Now

In the last section we invented the following facts:

$$(\#M)' = \#(M') \tag{2.16}$$

and

$$(f + g)' = f' + g' \tag{2.17}$$

where $\#$ is any number, and M, f, and g are *any* machines, not necessarily plus-times machines (though we know of no other kind at this point). Also, we found earlier that the derivative of x^n is

$$(x^n)' = nx^{n-1} \tag{2.18}$$

and we saw that the "two-parts" version $(f + g)' = f' + g'$ was just as powerful as the "n-parts" version:

$$(m_1 + m_2 + \cdots + m_n)' = m_1' + m_2' + \cdots + m_n' \tag{2.19}$$

We've got all the ingredients in place to figure out how to use our infinite magnifying glass on any plus-times machine (or if you prefer, on all possible plus-times machines at once). Let's do that! We convinced ourselves earlier that the abbreviation

$$M(x) \equiv \sum_{k=0}^{n} \#_k x^k \tag{2.20}$$

was large enough and agnostic enough to let us talk about any plus-times machine. Now we know how to talk about the derivative of the whole thing in terms of the derivatives of the pieces, and we know how to figure out the derivatives of each of the pieces. So all we have to do is combine these two discoveries, and we will have mastered our invention as much as it can be mastered... at least at this point. That is, at the beginning of the chapter, we invented the infinite magnifying glass, and now if we can manage to solve this last problem of ours, we will have figured out how to apply that invention to all the machines that currently exist in our universe.

In the following argument, we'll see how helpful it can be to have different kinds of equals signs in a long mathematical sentence. As usual, I'll use $\equiv$ whenever something is true just because we're reabbreviating, so you don't need to worry about why those parts are true. The argument below may look fairly complicated, but a surprising number of steps are just reabbreviation. Three steps won't be, though. I'll write 2.19 above the equals sign where we're using equation 2.19, and I'll do the same thing for equations 2.16 and 2.18. These are the only things we'll need. Take a deep breath, and try to follow along. Here we go...

$$[M(x)]' \equiv \left[\sum_{k=0}^{n} \#_k x^k \right]' \equiv \left[\#_0 x^0 + \#_1 x^1 + \#_2 x^2 + \cdots + \#_n x^n \right]'$$

$$\overset{(2.19)}{=} \left[\#_0 x^0 \right]' + \left[\#_1 x^1 \right]' + \left[\#_2 x^2 \right]' + \cdots + \left[\#_n x^n \right]'$$

$$\overset{(2.16)}{=} \#_0 \left[x^0 \right]' + \#_1 \left[x^1 \right]' + \#_2 \left[x^2 \right]' + \cdots + \#_n \left[x^n \right]'$$

$$\equiv \sum_{k=0}^{n} \#_k \left[x^k \right]'$$

$$\overset{(2.18)}{=} \sum_{k=0}^{n} \#_k k x^{k-1}$$

Done! In retrospect, most of those steps weren't entirely necessary, but I wanted to go as slowly as possible to avoid getting lost in all the symbols. Now that we get the idea, though, we could have made the exact same argument without so many reabbreviations. Here's what that argument would look like if we made it more casually. We want to differentiate a machine that looks like this:

$$M(x) \equiv \sum_{k=0}^{n} \#_k x^k \qquad (2.21)$$

This is just a sum of a bunch of things. We know from equation 2.19 that the derivative of a sum is just the sum of the individual derivatives, and because of equations 2.16 and 2.18 we know that the derivative of $\#_k x^k$ is just $\#_k k x^{k-1}$. So, knowing both of those things, we can make the above argument all in one step, and conclude that the derivative of M is

$$M'(x) = \sum_{k=0}^{n} \#_k k x^{k-1}$$

And we're done again. We're essentially done with the chapter as well. At least the meat of it. So, since the chapter began, we've come up with the idea of an infinite magnifying glass, used it to define a new concept (the derivative), and figured out how to apply that concept to every machine that currently exists in our universe.

Before we officially end the chapter, let's spend a bit of time in semi-relaxation mode. First, we'll spend a few pages talking about a new ability we now have as a side effect of our magnifying glass expertise. Then, we'll briefly discuss the role of "rigor" and "certainty" in mathematics.

2.4 Hunting Extremes in the Dark

Extremes are interesting. Watching an Olympic gold medalist sprint or swim or throw a javelin is generally more captivating than watching a randomly chosen person from your neighborhood doing the same thing. We enjoy watching the superlative performances of people who are the best at some particular activity. On the other extreme, the worst performances in any given category have a similar power to draw our attention. The principle that extremes are interesting also tends to be true in the world of mathematics. It may come in handy to be able to locate those extremes — places where a given quantity is largest or smallest, highest or lowest, best or worst — and to do so simply by manipulating symbols, since we won't always be able to visualize the objects we're studying.

Without even realizing it, we gained a surprisingly powerful ability when we invented the concept of the derivative: the ability to hunt down the locations where a machine achieves its extremes, even if we can't even begin to picture what the machine looks like! Here's the idea.

Since a machine's derivative tells us the slope of the machine at that point, we can make use of the following convenient fact: all flat points of a machine are places where that machine's derivative is zero. As such, we can find the flat points of a machine m by forcing its derivative to be zero

$$m'(x) \stackrel{\text{Force}}{=} 0$$

and then trying to figure out which numbers x make that sentence true. If we can figure that out, then we will have found the flat points of the machine. Then we just have to check a small number of cases to see where the extremes are. Importantly, we can do this even if we can't visualize what the machine looks like.

Let's look at a few simple examples of this. Back in Figure 2.3, we drew a picture of the Times Self Machine $m(x) \equiv x^2$. It's apparent from the picture that this machine has no maximum (it keeps getting bigger as x gets further from zero, in either direction), but it clearly has a minimum at $x = 0$. Now, even if we couldn't picture this machine, we could still coax the mathematics into telling us where the minimum is. Since $m(x) \equiv x^2$, we already know that $m'(x) = 2x$. Now we can just write down the sentence "this machine's derivative at x is zero" in symbolic form, like this:

$$m'(x) = 2x \stackrel{\text{Force}}{=} 0$$

I used $\stackrel{\text{Force}}{=}$ because the sentence $2x = 0$ isn't always true, so the $\stackrel{\text{Force}}{=}$ helps us remember that it's something we're forcing to be true (i.e., something we're insisting on), in order to see which particular x's actually make the sentence true. Which x's are those? Fortunately, it's not too hard to see that $2x = 0$ only when $x = 0$, which tells us that the machine $m(x) \equiv x^2$ has one and only one flat point, and it lives at $x = 0$. We knew this in advance (from the pictures we drew earlier), but it's always nice to test our new ideas on familiar cases to make sure they give what we expect. That's how we can check what we've done, in our own private universe, without requiring the help of a textbook or an authority figure.

Similarly, what if we looked at the machine $f(x) \equiv (x-3)^2$? In a sense, this is just like the previous example. Anything that looks like $(stuff)^2$ is going to be positive unless $stuff = 0$, so we might expect that this machine would have one and only one flat point, at $x = 3$, and that this point would be a minimum, just like last time. Now, that's all correct, but suppose we were handed this same machine in a form that made its similarity to $(stuff)^2$ less clear, namely:

$$f(x) \equiv x^2 - 6x + 9$$

This is exactly the same machine as $f(x) \equiv (x-3)^2$, but it's less clear from looking at the above equation that it will have one and only one flat point at $x = 3$. Nevertheless, we can get the mathematics to tell us this fact using the same idea we used above — compute the derivative, and force it to be equal to zero:

$$f'(x) = 2x - 6 \stackrel{\text{Force}}{=} 0$$

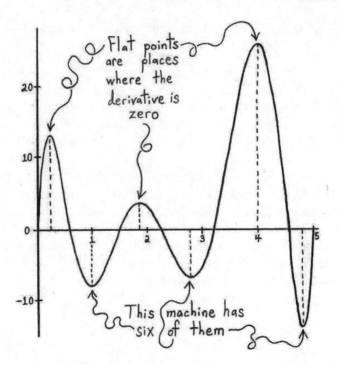

Figure 2.5: This machine has six flat points. If we describe them by their horizontal
coordinates, then the flat points occur when x is 0.25-ish, 1, 1.8-ish, 2.7-
ish, 4, and 4.8-ish. Not all of the flat points are extremes, but both of
the extremes are flat points. That is, the maximum of this machine in
the region pictured is at the flat point $x = 4$, while the minimum of this
machine in the region pictured is at the flat point $x \approx 4.8$.

Now, the sentence $2x - 6 = 0$ is saying the same thing as the sentence $2x = 6$,
which is saying the same thing as the sentence $x = 3$. So just as we expected,
we were able to figure out that this machine has one and only one flat point,
and its location is at $x = 3$. And the fact that we got the same result in two
different ways provides further evidence that our ideas make sense.

It is important to notice that this idea will not always spit out the extremes
(i.e., the maximum spots and minimum spots) of a given machine. However,
this is not because of a failure of the mathematics, but rather because of a
few obvious facts we overlooked. To get the idea, let's look at a few examples.
Suppose we wanted to find the maximum of the machine $g(x) \equiv 2x$, and we
proceed using the ideas above. We find its derivative and force it to be equal
to zero, which gives

$$g'(x) = 2 \stackrel{\text{Force}}{=} 0$$

The "force the derivative to be zero" method has just spat out the ridiculous
sentence $2 = 0$. Does this mean that 2 is really equal to 0? Hopefully not! Does
it mean that the "force the derivative to be zero" method of finding extreme

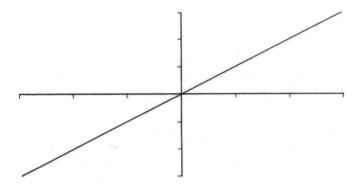

Figure 2.6: This machine has no flat (i.e., horizontal) points. When we try to force the mathematics to tell us where they are, it tells us that $2 = 0$. Don't worry, it's just messing with us. This is how the mathematics typically lets us know that one of our assumptions was wrong.

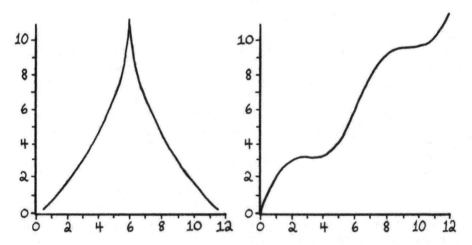

Figure 2.7: Above, we discussed the strategy of finding the highest and lowest parts of a machine by figuring out where its derivative is zero. In a perfect world, that strategy would always work, but there are some obnoxious situations where it doesn't. I should at least mention them briefly, though we won't run into them after this. The machine on the left has a maximum, but it occurs at an infinitely pointy place, where the derivative won't spit out zero, so the "force the derivative to be zero" method will miss it. Fortunately, these infinitely pointy machines don't show up unless we ask them to, so we won't have to deal with finding the extremes of any of them in this book. The machine on the right has two flat points in the region pictured, but neither of them is an extreme point (i.e., neither is a maximum or a minimum). As such, the "force the derivative to be zero" method would spit out the locations of these more boring flat points, which are located somewhere near $x \approx 3$ and at $x \approx 9$ (the squiggly equals sign means "roughly").

points has broken down? Not really. The graph of the machine $g(x) \equiv 2x$ is a tilted line, and straight lines have no flat points, unless the whole line is horizontal (I drew a picture of this in Figure 2.6, but it's hardly necessary). This failure is not a flaw of the "force the derivative to be zero" method. Rather, by spitting out something impossible like $2 = 0$, the mathematics is simply telling us that we assumed something impossible. Not a mystery.

Although unfortunate examples like those in Figure 2.7 won't be showing up much in the book, a brief mention of them is important if you want to understand the odd way that mathematicians write. Mathematicians tend to be somewhat obsessed with "counterexamples" — rare, bizarre exceptions to simple rules — and this obsession makes their theorems much harder to read. For example, in describing the "derivative equals zero" method, a mathematician might say: "Let $f : (a, b) \to \mathbb{R}$ be a function and suppose that $x_0 \in (a, b)$ is a local extremum of f, where f is differentiable at x_0. Then $f'(x_0) = 0$." That probably sounds like gibberish, but what they're really trying to say is pretty simple. The translation would be something like: "Picture the highest (or lowest) point on the graph of some machine. The machine has to be flat there. Oops, I mean, unless it's infinitely pointy, like on the left side of Figure 2.7. But that doesn't happen very often." The exception on the right side of Figure 2.7 shows up in the above theorem too, but it's more hidden. It's the reason why the gibberish is saying:

"*If* we're at a max or min, *then* the derivative is zero"

rather than

"*If* the derivative is zero, *then* we're at a max or min"

The second sentence would be true if it weren't for flat points like the ones on the right side of Figure 2.7 (where the derivative is zero, but we're *not* at a max or min). If such unfortunate examples didn't exist, then they could say the second sentence instead of the first, and that would be much more convenient, since the point of all this "derivative equals zero" business is usually to figure out where those max and min points are.

Okay! So, even though there are exceptions, we will continue to run into this idea throughout the book: the extremes of a given machine can usually be found by figuring out where the derivative of that machine is zero. As we generalize the notion of the derivative to more exotic types of machines, the way we will have to state this idea will change slightly, but the central idea will remain the same. For machines that eat one number and spit out one number, for machines that eat two numbers and spit out one number, for machines that eat N numbers and spit out one number, and finally for machines that eat infinitely many numbers and spit out one number, the basic principle that "extremes are usually places where the derivative equals zero" will continue to apply no matter how far we go, and no matter how strange our mathematical universe becomes.

2.5 A Brief Sermon on Rigor

> *Give me the fruitful error any time, full of seeds, bursting with its*
> *own corrections. You can keep your sterile truth for yourself.*
> —Vilfredo Pareto

In this chapter, we discussed (and made use of) the idea of infinitely small
numbers, an idea that some consider to be mathematically taboo. Before the
end of the chapter, I want to make a general point about the place of rigor
and certainty in mathematics. Sometimes in mathematics, you find yourself
making an odd argument like the ones we made above, and you're not sure
if it's "right," or if it might lead to contradictions somewhere down the road.
That's okay! It is not a sin to simply forge ahead, and attempt to make
sense of what you've done later. Much of the mathematics that is written
in modern textbooks was discovered in just this way, and "cleaned up" later,
often long after the original pioneers were dead. If someone else prefers the
job of formalizing ideas and cleaning things up, that's perfectly fine too! As
they say: To each ~~his//hshi///shers~~ own *god I hate English pronouns.*

If you ever run into a macho mathematician who thinks that this approach
renders our discussion undeserving of the name "real mathematics," kindly
ask them to remember the name Leonhard Euler, or for that matter, virtually
any other mathematician before the days of Bourbakism. Leonhard Euler was
one of the greatest mathematicians of all time, and one day he wrote down
this bizarre expression:

$$\sum_{n=0}^{\infty} 2^n = -1$$

The thing on the left is a sum of infinitely many positive terms. It's essentially

$$1 + 2 + 4 + 8 + 16 + \cdots (forever)$$

How could one of the best mathematicians of all time have thought this was
equal to -1?! Well, it turns out he was following a surprisingly reasonable
chain of argument. He wasn't crazy. When you see the argument for the first
time, it's easy to become convinced that he was right! The point of this is to
say that the ultra-cautious "how do I know if I'm right?" feeling we all develop
in mathematics classes is a feeling that we should, at least partially, learn to
abandon. When we're inventing mathematics (or anything) for ourselves, we
don't know if we're right. No one ever does. We might be uncertain of a given
argument, and only later think of a different argument that arrives at the same
results. This later argument might appear more convincing to us, and to the
extent that we can find independent ways of getting to the same conclusion,
we can become more convinced that the conclusion makes sense. But we can

never be unquestionably sure that what we're doing makes sense.[11]

I understand the desire for rigor, I really do. For several years I planned to go into mathematical logic, focusing especially on the foundations of mathematics. There are astonishingly few modern mathematicians who focus primarily on foundations. As the logician Stephen Simpson has said in his phenomenal textbook *Subsystems of Second Order Arithmetic*, "Regrettably, foundations of mathematics is now out of fashion." In spite of its unpopularity, however, I was always drawn to the field. During that time, my obsession with rigor was stronger and more totalizing than that of most. Only later did I realize how much this mindset was killing my mathematical creativity. When I began to read some of the most famous papers by the field's early pioneers — Kurt Gödel, Alonzo Church, Alan Turing, Stephen Kleene, and modern-day giants like Harvey Friedman and Stephen Simpson — I found that the field's greatest minds reasoned and spoke in a surprisingly informal way about the formal languages and formal theories they studied. Their proofs were entirely rigorous by the standards of mathematics, but none of them appeared to confine themselves to that level of rigor when thinking about their field. This is not a shortcoming, but a virtue. The old saying among physicists appears to be true: too much rigor can, and does, lead to rigor mortis.

2.6 Reunion

Let's remind ourselves what we did in this chapter.

1. We noticed that curvy stuff is generally harder to deal with than straight stuff. However, we realized that if you zoom in on curvy stuff, it starts to look more and more straight. We noticed that if we could zoom in "infinitely far" (whatever that means), then any curvy thing would be exactly straight. That is, if only we had an infinite magnifying glass, we could potentially reduce curvy problems to straight problems. Rather than being sad that we didn't have an infinite magnifying glass, we just pretended that we had one.

2. We used this idea of an infinite magnifying glass to define the steepness of curvy things. We did this by computing the steepness using two points that were "infinitely close to each other" (though we weren't entirely sure what that meant).

3. We briefly discussed a few contraptions that people have invented over the years to allow them to avoid the idea of infinitely small numbers. We will occasionally use the contraption called a "limit," but we'll usually

---

11. I'll resist the temptation to mention Gödel's second incompleteness theorem here, but I won't resist the temptation to mention apophasis.[ω]

ω. I'll also resist the temptation to explain the joke(s?) in the above footnote, but I won't resist the temptation to mention that the recursive footnote is a much less popular literary device than it deserves to be.

just use the idea of infinitely small numbers directly. We'll get the same answers with both methods, so it's okay to switch back and forth.

4. We discussed lots of different names and abbreviations that textbooks use to talk about these ideas. Textbooks usually call the steepness of M at x "the derivative of M at x." Some common abbreviations for this idea are (i) the notation $M'(x)$, which emphasizes that the derivative of M can be thought of as a machine in its own right, and (ii) the notation $\frac{dM}{dx}$, which emphasizes that the derivative of M can be thought of as the "rise over run" between two points that are infinitely close to each other.

5. We proceeded to test our new idea on two non-curvy examples: constant machines and straight lines. We did this to make sure that our new idea gave sensible answers in simple cases where we already knew what to expect.

6. Then we tested our idea on some curvy machines, and eventually figured out how to use it on any machine that currently exists in our universe: the plus-times machines, or "polynomials."

7. We discussed how the derivative often allows us to find the extremes of machines, namely, the places where they achieve their highest and lowest values. We explained why such extremes can usually be found by figuring out which points x make the derivative $m'(x)$ equal to zero, and when this idea breaks down.

Interlude 2: How to Get Something from Nothing

Ex Nihilo Redux: From an Abbreviation to an Idea?

Soon after we began our adventure, we introduced the concept of powers. Using the term "concept" is a bit of a stretch, since powers weren't really an idea. They were just a meaningless abbreviation that we invented for the sake of convenience. Currently, the symbol $(stuff)^n$ is nothing more than a abbreviation for

$$\underbrace{(stuff)\cdot(stuff)\cdots(stuff)}_{n \text{ times}}$$

which is just repeated multiplication. There's nothing to know about powers that isn't already true of multiplication, so powers don't really have a life of their own. However, if you leave our private mathematical universe and spend some time stumbling around in the world, you'll occasionally hear people talking about strange things like "negative powers" or "fractional powers" or "zeroth powers," along with a variety of mysterious sentences like $(stuff)^0 = 1$. Now, given how *we* defined powers, it's not at all clear that $(stuff)^0$ should be 1, because if we translate that abbreviation back into English, it says that $(stuff)^0$ is $(stuff)\cdot(stuff)\cdots(stuff)$, where there are 0 copies of $(stuff)$ in a row. At first glance, it seems like this should be 0, not 1, right? We used $x^0 \equiv 1$ as an abbreviation earlier, but its only motivation was aesthetics. It allowed us to write the expression for an arbitrary plus-times machine in a simpler way, but we had no reason to assume that the sentence $x^0 \equiv 1$ made any sense beyond that.

Although we have no idea what a power like 0 or -1 or $\frac{1}{2}$ could possibly mean, we do have a bit of experience inventing things. So instead of simply trusting other people's assertions about how x^0 is 1 and so on, let's see if we can *invent* a way to extend the idea of powers that makes sense to us. That is, let's try to generalize our own definition of $(stuff)^n$ to situations where n could be *any* number, not just a whole number.

Whenever we try to extend a familiar definition to a new context, we immediately run into the fact that there are an infinite number of ways we could do this. However, even though there are lots of ways we *could* generalize the concept of powers, the vast majority of such generalizations would be sterile, boring, and useless. For example, we could extend our definition of $(stuff)^n$ by saying that $(stuff)^\#$ means the same thing it always meant when $\#$ is a positive whole number, but whenever $\#$ isn't a positive whole number, then we define $(stuff)^\#$ to be 57. That's not illegal, but it seems pretty boring. However, it's consistent with our old definition, so its merit (or lack thereof) is to be judged only by the fact that it doesn't help us at all, and we personally don't think it's interesting.

Faced with this infinite abundance of choices, how should we choose to generalize the concept of powers? Obviously, any generalization is useless unless it is useful. Let's use this obvious fact to move beyond the familiar. We'll define our generalization not directly but indirectly, as "whatever keeps the definition useful," and as always, it's up to *us* to say what we mean by "useful." That is, we need to find some property or behavior of the abbreviation $(stuff)^n$ that we think is useful, or which makes it easier to deal with things that look like $(stuff)^n$.

So let's find a way of saying something that is *useful* about the familiar definition of $(stuff)^n$. You might think, "Well, everything we would ever want to know about $(stuff)^n$ is contained in the definition itself, so if we're looking to say something that's useful about $(stuff)^n$, why not just give its definition: $(stuff)^n \equiv (stuff) \cdot (stuff) \cdots (stuff)$?" That's all perfectly correct, but there's one problem. The definition

$$(stuff)^n \equiv \underbrace{(stuff) \cdot (stuff) \cdots (stuff)}_{n \text{ times}}$$

doesn't give us any hint about how we can define something like $(stuff)^{-1}$ or $(stuff)^{1/2}$, because this old definition relies on the idea of *how many times* $(stuff)$ shows up. What could it possibly mean for $(stuff)$ to show up half a time, or a negative number of times? That's clearly not the way we want to think about things if we want to make sense of negative or fractional powers. Our goal is to choose some behavior of the familiar definition that makes *just as much sense* for powers that aren't whole numbers. Here's something more useful.

If n is some whole number (say, 5), then we can always write sentences like $(stuff)^5 = (stuff)^2(stuff)^3$, because if we translate all the abbreviations on both sides back into English, then the left is saying "5 copies of *stuff* in a row" and the right is saying "2 copies of *stuff* in a row, then 3 more copies of *stuff*." It's clear that these are the same thing. For the same reason, whenever n and m are positive whole numbers, the sentence $(stuff)^{n+m} = (stuff)^n(stuff)^m$ is true, because both sides are just different ways of saying "$n + m$ copies of *stuff* in a row." That's certainly useful, and it expresses the same idea that the original definition did, but it doesn't necessarily require n and m to be whole numbers! So maybe we can use this as the raw material to build a more general concept of powers. Now comes the important part.

At this point, we just choose to say

I have no idea what $(stuff)^{\#}$ means when $\#$ isn't a whole number...

But I really want to hold on to the sentence

$$(stuff)^{n+m} = (stuff)^{n}(stuff)^{m}$$

So I'll force $(stuff)^{\#}$ to mean:

whatever it has to mean in order to make that sentence keep being true.

Take a moment to think about what we're saying in that box. It's not a complicated idea, but it is wildly different from the way we're normally taught to think about mathematics in school. Yet as unfamiliar as this style of thinking might be, it is an infinitely more honest representation of real mathematical reasoning than any number-juggling calculation could ever be. This style of thought lies at the core of mathematical invention, and a surprising number of mathematical concepts are invented in exactly this way. Generalizing ideas like this is really nice for two reasons. First of all, it lets us import things we're already familiar with into unfamiliar territory. That is, rather than simply surrendering to the sad fact that new things are unfamiliar, we can instead use a clever conceptual hack, and choose to define new things in a way that guarantees that we will already be familiar with them. That hack is simply to define things *indirectly*, not by what they are, but by how they behave. Second, instead of having to remember the meaning of strange things like $(stuff)^{0}$ or $(stuff)^{-1}$ or $(stuff)^{1/2}$ we can figure out what they mean for ourselves! Let's do that.

Why This Forces Zero Powers to Be 1

We don't know what $(stuff)^{\#}$ means, but we're forcing it to mean *whatever it has to mean* in order to let us write $(stuff)^{a+b} = (stuff)^{a}(stuff)^{b}$, where a and b are any numbers, not necessarily whole or positive. Let's use this unusual way of thinking to see what $(stuff)^{0}$ has to mean. Using the idea in the box above, we can write:

$$(stuff)^{\#} = (stuff)^{\#+0} = (stuff)^{\#}(stuff)^{0}$$

This is telling us that $(stuff)^{0}$ has to be whatever number doesn't change things when you multiply by it. So I guess $(stuff)^{0}$ has to be 1. Hey, that finally makes sense! Let's write it down.

> **Our indirect definition forces it to be true that:**
>
> $$(stuff)^0 = 1$$

Why This Forces Negative Powers to Be Handstands

We don't know what $(stuff)^{-\#}$ means, but let's try the same strategy as before. Notice that the sentence we used as the basis for our generalization only involves the *addition* of powers: $(stuff)^{a+b} = (stuff)^a(stuff)^b$. What about subtraction of powers, like $(stuff)^{a-b}$? Well, by writing $a - b$ in the odd form $(a) + (-b)$, we can trick the original sentence into letting us talk about subtraction in the language of addition. Using this idea, together with our recently acquired knowledge that $(stuff)^0 = 1$, we can do this:

$$1 = (stuff)^0 = (stuff)^{\#-\#} = (stuff)^{\#+(-\#)} = (stuff)^{\#}(stuff)^{-\#}$$

Now, if we divide both sides by $(stuff)^{\#}$, we can build a sentence that tells us how to translate negative powers into the language of positive powers. Let's write it down:

> **Our indirect definition forces it to be true that:**
>
> $$(stuff)^{-\#} = \frac{1}{(stuff)^{\#}}$$

In textbooks, a term that looks like $\frac{1}{x}$ would usually be called the "reciprocal" of x. We won't need a name for this concept very often, so it doesn't really matter what we call it. Still, the term "reciprocal" is a bit unclear, so let's use the term "handstand," because $\frac{1}{x}$ is just x upside down.

Why This Forces Fractional Powers to Be n-Cube Side Lengths

Alright, we've unraveled zero powers and negative powers. What about powers that aren't whole numbers, like $(stuff)^{\frac{1}{n}}$? First let's assume that n itself is a whole number, and see what we get.

$$(stuff) = (stuff)^1 = (stuff)^{\frac{n}{n}} = (stuff)^{\frac{1}{n}+\frac{1}{n}+\cdots+\frac{1}{n}}$$

$$= \underbrace{(stuff)^{\frac{1}{n}} \cdot (stuff)^{\frac{1}{n}} \cdots (stuff)^{\frac{1}{n}}}_{n \text{ times}}$$

This is strange. We have an unfamiliar thing (namely, the $\frac{1}{n}$ power of *stuff*) and an unfamiliar thing (namely, *stuff*). Usually when we use the word "explanation," we're describing an unfamiliar thing in terms of one or more things that are more familiar. This does the opposite! It describes *stuff* in terms of a bunch of unfamiliar things: n copies of its $1/n$ power. Let's say this more simply.

Our indirect definition forces it to be true that:

$(stuff)^{\frac{1}{n}}$ is any number that can say the following sentence without lying:
"Multiply me by myself n times and you get $(stuff)$."

In a sense, this is the opposite process of finding the "volume" of an n-dimensional cube. Earlier, when we invented the concept of area, we convinced ourselves that the n-dimensional volume of an n-dimensional cube should be ℓ^n, where ℓ is the length of each edge. Now, the definition of $(stuff)^{\frac{1}{n}}$ we just obtained sort of seems to be talking about the volumes of n-dimensional cubes... but in reverse. It is not saying the usual thing:

From Lengths to n-Volumes: n^{th} powers
If you hand me the length ℓ of an n-dimensional cube, then the volume is ℓ^n

Rather, it's saying the opposite. Something like:

From n-Volumes to Lengths: $(1/n)^{th}$ powers
If you hand me the volume V, then the length of each side is $V^{\frac{1}{n}}$, whatever that is.

I assume this is where the funny terms "square root" and "cube root" came from. If all we knew was the area of a square (which we can abbreviate as A), then how might we figure out the length of its sides? Well, we may not know how to figure out an exact *number* for the length if A is something crazy like 9235, but that's fine. Numbers aren't the point. Ideas are. Here's the idea: We know that the "side length" is whatever number turns into A when we multiply it by itself. That is, the side length is any number (?) that makes the sentence $(?)^2 = A$ true. Which number is that? I don't know, but that number is called $A^{\frac{1}{2}}$. So even though we have no idea how to compute specific numbers for $7^{\frac{1}{2}}$ or $59^{\frac{1}{2}}$, we know how these numbers behave and what they mean: if a square's area is A, then its side lengths are $A^{\frac{1}{2}}$. The same type of story works for cubes in the normal three-dimensional sense of the word, or even for those weird n-dimensional cubes we've talked about before (we can't picture these, but again, that doesn't matter). This is why $(stuff)^{\frac{1}{n}}$ is often called the "n^{th} root of $(stuff)$."

The word "root" isn't really necessary: an nth root of something is just a $(1/n)^{th}$ power. Giving "roots" their own name (not to mention their own $\sqrt{strange}$ notation) often seems to give people the impression that the term "root" refers to a different (and more mysterious) concept than powers. But there's nothing mysterious about them, and understanding the concept *certainly* doesn't require you to know how to compute the arbitrary root of an arbitrary number! We'll need to invent some more calculus before we figure out how to do that. But again, computing specific numbers isn't the point. The point is the ideas and how they're created. And in the case of powers, as is always the case in mathematics, the fundamental ideas are extremely simple.

3 As If Summoned from the Void

> And every science, when we understand it not as an instrument of power and domination but as an adventure in knowledge pursued by our species across the ages, is nothing but this harmony, more or less vast, more or less rich from one epoch to another, which unfurls over the course of generations and centuries, by the delicate counterpoint of all the themes appearing in turn, as if summoned from the void.
> —Alexander Grothendieck, *Récoltes et Semailles*

3.1 Who Ordered That?

3.1.1 From an Abbreviation to an Idea... By Accident

In Interlude 2, we discovered something quite strange. Originally we introduced $(stuff)^n$ as an abbreviation. That is, we simply defined

$$(stuff)^n \equiv \underbrace{(stuff) \cdot (stuff) \cdots (stuff)}_{n \text{ times}}$$

As long as this was all we meant by raising something to a power, we could be confident that there was nothing about the concept of powers that we didn't understand. After all, powers were not really a *concept* to begin with. They were a completely vacuous and content-free *abbreviation*. There was simply nothing to know about them. However, in extending this (non-)concept to powers other than positive whole numbers, we performed a strange feat of mathematical creation. We said:

I have no idea what $(stuff)^\#$ means when $\#$ isn't a whole number...

But I really want to hold on to the sentence

$$(stuff)^{n+m} = (stuff)^n (stuff)^m$$

So I'll force $(stuff)^\#$ to mean:

whatever it has to mean in order to make that sentence keep being true.

By attempting to extend a vacuous non-concept to a non-vacuous concept, we found that we had unwittingly summoned from the void an idea about which there *was* something to know, and with which we were *not* familiar. Having performed what appeared to be nothing but a harmless act of generalization and abstraction, we found that we had created a new and unexplored part of our universe. After a brief exploration of it, we proceeded to discover three simple facts about our accidental conceptual progeny.

We discovered that we had forced the following to be true:

$$(stuff)^0 = 1$$

$$(stuff)^{-\#} = \frac{1}{(stuff)^{\#}}$$

$$\underbrace{(stuff)^{1/n} \cdot (stuff)^{1/n} \cdots (stuff)^{1/n}}_{n \text{ times}} = (stuff)$$

We have no reason to expect that the above constitutes a complete map of the world we accidentally invented. For example, by creating negative and fractional powers, we also accidentally created a unexplored swarm of new machines.

New machines that must now exist,
as a consequence of our invention:

$$M(x) \equiv x^{-1}$$

$$M(x) \equiv x^{1/2}$$

$$M(x) \equiv x^{-1/7} + 92x^{21/5} - \left(x^{-3/2} + x^{3/2}\right)^{333/222}$$

The third example above makes it clear just how large this new and unexplored corner of our universe really is. And just when we were starting to feel accomplished! Recall that at the end of Chapter 2, we not only thought of a way to abbreviate all of the machines that our universe contained at that point — the plus-times machines — but we also managed to figure out how to differentiate any (and thus all) of them, by pulling the clever trick of differentiating our abbreviation that stood for an arbitrary one of them (thus differentiating all of them simultaneously).

To stress just how much larger our world has suddenly become, notice that presumably all of these machines have derivatives. At this point we know none of them, even though it was we who conjured up this world. Mathematics is strange... Well, let's take a nap to try to wrap our minds around the immensity of what we've done, and when we wake up, maybe we'll have the energy to play with a few of these machines for a while and see if we want to acknowledge them, or if we would prefer to simply pretend that they don't exist.

3.1.2 Visualizing Some of These Beasts

In keeping with the spirit of creating mathematics from the ground up, we're making direct use only of facts we have discovered for ourselves (though mentioning others along the way), deciding on our own abbreviations and terminology (occasionally allowing some orthodox terminology to tag along), and drawing all of our pictures by hand. But against that background, how can we possibly hope to picture these new beasts that we've unwittingly conjured up? We have no idea at this point how to calculate a specific number for quantities like $\frac{1}{53}$ or $7^{\frac{1}{2}}$, so how on earth are we supposed to visualize something like the graph of $\frac{1}{x}$ or $x^{\frac{1}{2}}$? Though our powers of visualization are limited in this new part of our universe, this lack of visual acuity need not stand in the way of our getting a visceral feel for the behavior of these new machines. Let's spend a moment using what we know in order to get an idea of what some of these machines look like. Rather than trying to picture arbitrarily complicated machines like

$$M(x) \equiv x^{-1/7} + 92x^{21/5} - \left(x^{-3/2} + x^{3/2}\right)^{333/222}$$

let's instead focus on the simplest representatives of the new machines we've created. In the previous interlude, we invented three new types of powers: zero, negative, and fractional. Since zero powers just turned out to be 1, we really only have two genuinely new phenomena to deal with: negative powers and fractional powers. The simplest representative of negative powers seems to be

$$M(x) \equiv \frac{1}{x} \equiv x^{-1}$$

while the simplest representative for fractional powers seems to be

$$M(x) \equiv \sqrt{x} \equiv x^{\frac{1}{2}}$$

Let's start with the first one. It doesn't really matter that we don't have the patience or even the knowledge to calculate a specific number for 7^{-1} or $(9.87654321)^{-1}$ or anything else. All we want to know is the general, big-picture behavior of the machine $M(x) \equiv x^{-1}$, so that we can gain an intuitive feel for what it looks like and how it behaves. Basically everything we know about this machine can be summarized by the following four mathematical sentences, which are all really just one sentence expressed in different ways.[1]

---

1. Note: In the past, we've used *tiny* to stand for an infinitely small number, but here it just stands for a regular old number like 0.000(bunch of zeros)0001. Similarly, *huge* is a just a regular number too, not an infinitely big one.

$$\frac{1}{tiny} = huge$$

$$\frac{1}{huge} = tiny$$

$$\frac{1}{-tiny} = -huge$$

$$\frac{1}{-huge} = -tiny$$

For example, $\frac{1}{100,000,000}$ is extremely tiny, while $\frac{1}{0.00000001}$ is huge, and by making x smaller and smaller, we can make $\frac{1}{x}$ as huge as we want. We don't need to do any tedious arithmetic to see this. It just follows from our view that there's no such thing as division. Since the beginning, we've been saying that $\frac{a}{b}$ is just an abbreviation for $(a)(\frac{1}{b})$, and the funny symbol $\frac{1}{b}$ is just an abbreviation for whichever number turns into 1 when we multiply it by b. Having defined this symbol not by *what it is* but by *how it behaves*, it follows naturally that $\frac{1}{huge}$ should be a tiny number: $\frac{1}{huge}$ stands for whichever number turns into 1 when we multiply it by *huge*, but if something only manages to get magnified to a size of 1 after we multiply it by a huge number, then the original number must have been fairly tiny. No arithmetic, just reasoning. As simple as this idea is, it's the only idea we need in order to draw Figure 3.1.

Okay, so we still can't compute a specific number for $\frac{1}{7}$ or $\frac{1}{59}$, but getting a general understanding of $\frac{1}{x}$ for *all* values of x somehow wasn't as hard as we expected. Mathematics can be strangely backwards like that. What about the other new type of machine? Can we figure out a way to visualize $m(x) \equiv x^{\frac{1}{2}}$, also known as $\sqrt{x}$? Well, just as before, we have no idea how to compute a specific number for $\sqrt{7}$ or $\sqrt{729.23521}$ or infinitely many other things. However! We *do* know how to do basic multiplication, so we know how to square whole numbers. For example,

$$1^2 = 1 \qquad 2^2 = 4 \qquad 3^2 = 9 \qquad 4^2 = 16$$

and so on. Now, using the definition of fractional powers that we invented in the previous interlude, we can express all of the above sentences in a slightly different language:

$$1 = 1^{\frac{1}{2}} \qquad 2 = 4^{\frac{1}{2}} \qquad 3 = 9^{\frac{1}{2}} \qquad 4 = 16^{\frac{1}{2}}$$

What about numbers less than 1? Well, as a simple example, we know that $\frac{1}{4} = \frac{1}{2 \cdot 2} = \frac{1}{2}\frac{1}{2} = \left(\frac{1}{2}\right)^2$, which is just another way of saying:

$$\left(\frac{1}{4}\right)^{\frac{1}{2}} = \frac{1}{2}$$

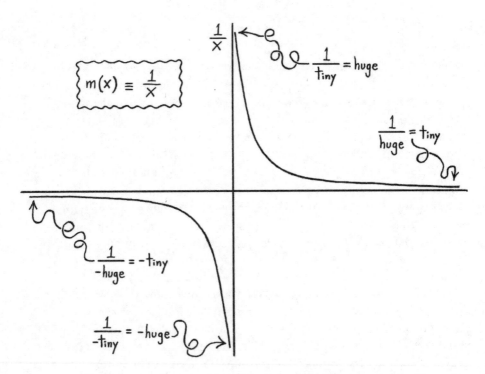

Figure 3.1: We know that (i) 1 over a tiny number is a huge number, (ii) 1 over a
huge number is a tiny number, and (iii) in the two previous phrases, the
meaning of "tiny" can be made as tiny as we want, for huge enough values
of "huge," and vice versa. This lets us build a general understanding
of what the machine $m(x) \equiv \frac{1}{x} \equiv x^{-1}$ looks like, even though we are
completely incapable of computing a specific number for quantities like
$\frac{1}{7}$, as well as largely uninterested in doing so.

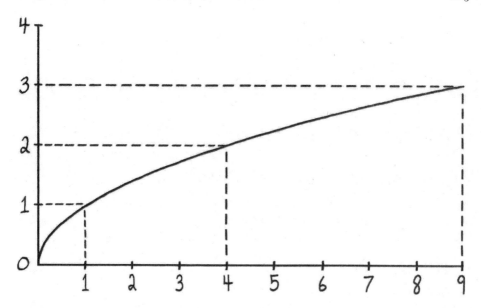

Figure 3.2: Using what little we know to picture the one-half-power machine $m(x) \equiv \sqrt{x} \equiv x^{\frac{1}{2}}$.

So whereas taking the $\frac{1}{2}$ power of a number bigger than 1 makes it smaller, taking the $\frac{1}{2}$ power of a number *smaller* than 1 seems to make it *bigger*. Using everything we just wrote down, we can build a reasonably detailed picture of what the machine $m(x) \equiv x^{\frac{1}{2}}$ looks like. Our attempt is shown in Figure 3.2. Just as before, we managed to develop a general feel for the machine's behavior for *all* values of x, even though we still have no idea how to compute a specific number for arbitrary quantities like $7^{\frac{1}{2}}$. In all of mathematics, contrary to what we are led to believe, this is the norm.

3.1.3 Playing with Our Accidents and Getting Hopelessly Stuck

> *We have a habit in writing articles published in scientific journals to make the work as finished as possible, to cover all the tracks, to not worry about the blind alleys or to describe how you had the wrong idea first, and so on. So there isn't any place to publish, in a dignified manner, what you actually did in order to get to do the work, although, there has been in these days, some interest in this kind of thing.*
> —Richard Feynman, *Nobel Prize Lecture, December 11, 1965*

Okay, let's see if our magnifying glass idea continues to make sense when we use it on these new things. Let's try it on the simplest of our new fractional power machines: $m(x) \equiv x^{\frac{1}{2}}$. Here we go:

$$\frac{M(x+t) - M(x)}{t} \equiv \frac{(x+t)^{1/2} - x^{1/2}}{t} = Uh\ldots$$

We have absolutely no idea what to do with this. Let's give up on this one for a minute and see if our magnifying glass is any easier to use on one of those negative-power machines. Again, let's try the simplest example: $m(x) \equiv \frac{1}{x}$. Ready, set, go:

$$\frac{M(x+t) - M(x)}{t} \equiv \frac{\frac{1}{(x+t)} - \frac{1}{x}}{t} = \ldots ?$$

Stuck again. We have no clue what to do with any of these ugly new beasts. But despite our almost total lack of familiarity with them, we do know *something* about them. We invented them, after all! We generalized powers by saying "they're anything that behaves this way," where "behaves this way" meant "obeys the break-apart-the-powers formula." Because of this, one of the fundamental properties of these unfamiliar new machines is that we can turn them into something more familiar by banging them into other machines with multiplication. For example, no matter what n is, we invented the fact that $(x^n)(x^{-n}) = x^{n-n} = x^0 = 1$, or more briefly:

$$(x^n)(x^{-n}) = 1 \tag{3.1}$$

Remember in the last chapter, we figured out that for *any* machines f and g (not necessarily plus-times machines, but possibly something much more exotic), it had to be the case that $(f + g)' = f' + g'$. That is, we found a way to talk about the derivative of two things added together in terms of the derivatives of the individual pieces (or to say the same thing differently, the derivative of a sum is the sum of the derivatives). We were surprised to find how easy it was to convince ourselves of this using only our definition of the derivative, even if we didn't know anything at all about the particular individual personalities of f and g. We have no idea what the derivative of x^{-n} is, but the above idea together with equation 3.1 suggests a promising approach.

If we could figure out how to talk about the derivative of two things multiplied together in terms of the derivatives of the individual pieces, then we might be able to ambush these new untamed machines by using equation 3.1 in two different ways. Let's say we write down an abbreviation for what we can call a "trick machine." It could be something like $T(x) \equiv (x^n)(x^{-n})$. We're using the letter T because we're attempting to *T*rick the mathematics into telling us something. This machine T is js u t...

(The familiar rumbling noise is heard again.
The intensity of the rumbling increases for several seconds
and then immediately stops without warning.)

Ugh, not again! Stupid noise made me spell "just" wrong. Sorry, Reader, that must be one of those famous California earthquakes. Just ignore them if you can. I doubt they'll turn out to be important. Where were we? Right! The trick machine $T(x) \equiv (x^n)(x^{-n})$ is just a fancy way of writing the number 1, so we know that the derivative of T is zero.

However, if we had in our possession a way to talk about the derivative of a "product" (two things multiplied together) in terms of the derivatives of its pieces, then we could switch hats, think of the machine T as a product of the two machines x^n and x^{-n}, and use this as-yet-undiscovered method of talking about the derivative of the whole thing in terms of its pieces. We would then have done the same thing in two ways. On the one hand, we know the derivative of T is zero, since the machine T is just the constant machine 1 in disguise. On the other hand, we'd have another way of expressing T's derivative in terms of a thing we know (the derivative of x^n) and a thing we want to know (the derivative of x^{-n}). Then, if we happened to get extremely lucky, we might be able to manipulate this complicated expression, and somehow isolate the piece that we wanted (namely, the derivative of x^{-n}). It's a total shot in the dark, and it probably won't work, but at very least it's a creative idea.

And hey! If we could find a general formula for the derivative of two things multiplied together, then we might be able to tackle the fractional power machines too! We completely failed on $M(x) \equiv x^{1/2}$ earlier, but because of the way we invented the idea of powers, we know that

$$\left(x^{1/2}\right)\left(x^{1/2}\right) = x \tag{3.2}$$

So just like before, if we could somehow talk about the derivative of a product in terms of the individual derivatives, then we might be able to use the above expression to *trick the mathematics* into telling us what the derivative of $x^{1/2}$ is!

> *(As the section comes to an end,*
> *no more rumbling is heard.*
> *An unsettling silence*
> *has fallen over*
> *the book*
> *...)*

3.1.4 A Shot in the Dark

This idea seems a bit far-fetched, and we're not sure if it'll work, but we may as well try. After all, there's not much else to do, and there's no one around to punish us if we fail. Let's play around a bit and see if we can make any progress on this hopeful delusion of a problem.

Richard Feynman famously said that the first step in discovering a new physical law is to guess it. It was a joke, but he wasn't really joking. Discovering new things is an inherently anarchic process. Reality doesn't care how we stumble upon its secrets, and guessing is as good a method as any. Let's take that approach. We already know that $(f + g)' = f' + g'$. That is, we know that the derivative of a sum is the sum of the derivatives. So, the first natural thing to guess is that maybe the derivative of a product is the product of the derivatives. Sounds reasonable, right? Let's write it down, making sure to remind ourselves it's just a guess:

$$(fg)' \stackrel{Guess}{=} f'g' \tag{3.3}$$

Alright, we've got a guess. The above guess was inspired by another fact we *know* is true, namely, $(f + g)' = f' + g'$, so we're not just guessing randomly. If our guess is really correct, though, it has to be consistent with what we already know, so let's see if it reproduces things we discovered earlier. We know that the derivative of x^2 is $2x$, and if the above guess is true, then we could also write the derivative of x^2 like this:

$$\left(x^2\right)' \equiv (x \cdot x)' \stackrel{???}{=} (x)'(x)' = 1 \cdot 1 = 1$$

Well that didn't work. It's definitely *not* true that $2x$ is always equal to 1 no matter what x is. Oh well. Our guess was wrong, but I guess we learned something. And hey! In retrospect, we should have been able to see that our guess was wrong to begin with. After all, if it were true that $(fg)' = f'g'$, then the derivative of *everything* would be zero. Why? Well, any machine f is the same as 1 times itself, so if our guess had been true, then we could always just write this:

$$(f)' = (1 \cdot f)' \stackrel{Oops!}{=} 1' \cdot f' = 0 \cdot f' = 0$$

Now that our first guess failed, it's not really clear what to do. I guess there's not much we can do except go back to the drawing board. What was the drawing board? Well, I guess the definition of the derivative is as far back as we can go. Let's just imagine that f and g are any machines, not necessarily plus-times machines, and then let's define $M(x) \equiv f(x)g(x)$. So M spits out the product of whatever f and g spit out. Then using the definition of the derivative, and using the abbreviation t to stand for a tiny number, we have

$$\frac{M(x+t) - M(x)}{t} \equiv \frac{f(x+t)g(x+t) - f(x)g(x)}{t} = \ldots ? \tag{3.4}$$

And we're stuck again. That didn't take long to unravel! What now?

Well, it would be nice if we could just lie and change the problem, because it *almost* looks like a problem we could deal with. I mean, if only that $f(x+t)$ piece were really $f(x)$ instead, then we might be able to make some progress. Let's give up on the real problem for a moment and examine the version we'd get if we were allowed to lie. I'll write $\stackrel{Lie!}{=}$ at the spot where we change the problem. The $\overset{Lie!}{=}$ symbol is where we're lying in order to make the problem easier. That's fine, as long as we remember that we're not actually solving the original. This may turn out to be pointless, but who knows? Lying in order to make progress might give us some ideas of what to do on the problem that defeated us. Here we go:

$$\frac{f(x+t)g(x+t) - f(x)g(x)}{t} \overset{Lie!}{=} \frac{f(x)g(x+t) - f(x)g(x)}{t}$$

$$= f(x)\left(\frac{g(x+t) - g(x)}{t}\right)$$

Nice! We already made it past where we got stuck before. Now turning t down to zero as usual makes the far right turn into $f(x)g'(x)$, and the far left is just the derivative of $M(x) \equiv f(x)g(x)$, so we get

$$[f(x)g(x)]' \overset{Lie!}{=} f(x)g'(x)$$

Does this help? Well, we got *an* answer, but it's not the real answer, because we lied. But maybe if we go back to the original problem and perform the same lie, but then *correct for the lie*, then we'll be able to make a similar kind of progress *without* changing the original problem. How do we lie and correct? Well, let's add zero... but not just any zero. Instead of actually modifying the problem like we did above, let's just add the piece we *wish* were there, and then subtract the exact same piece so that we don't change the problem.

So sort of like before, the lie is to add $f(x)g(x+t)$ to the top of the original problem, but now we'll also have to subtract $f(x)g(x+t)$ back off again, to avoid changing anything. The end result is that we're doing nothing. We're just adding zero. But saying "now add zero," in the middle of a long string of equations doesn't really express what we're thinking, and it certainly doesn't do justice to how strange and clever this idea is. Really, we're just anarchically doing *whatever we want*, in order to make progress, and then apologizing for our recklessness by adding an equal and opposite antidote to undo whatever we did. This mixture of poison and antidote doesn't change the problem, but the problem grows a kind of scar of the form "$(stuff) - (stuff)$" as a result of what we did, and if we constructed our lie carefully, this gives us something to grab on to in order to pull ourselves forward. Let's see if this idea actually works. Ready? Here we go:

$$\frac{f(x+t)g(x+t) - f(x)g(x)}{t}$$

$$= \frac{f(x+t)g(x+t) - f(x)g(x) + \overbrace{[f(x)g(x+t) - f(x)g(x+t)]}^{\text{Lie \& correct}}}{t}$$

$$= \frac{\overbrace{[f(x)g(x+t) - f(x)g(x)]}^{\text{Piece we wanted}} + \overbrace{[f(x+t)g(x+t) - f(x)g(x+t)]}^{\text{Leftovers}}}{t}$$

$$= f(x)\left(\frac{g(x+t) - g(x)}{t}\right) + \left(\frac{f(x+t) - f(x)}{t}\right)g(x+t)$$

That worked out much better than we expected. In correcting for the lie, we ended up with some extra leftovers that we didn't necessarily want. However, in a surprising stroke of luck, the same nice thing ended up happening with the leftovers as with the piece we wanted, because we found that we could peel off the $g(x+t)$ piece and take it outside.

We've got four pieces in the last line of the equations above. Now, once we turn t down to zero, every one of these pieces will turn into either one of the two machines or one of their derivatives. The first piece will stay $f(x)$, the second will morph into $g'(x)$, the third will morph into $f'(x)$, and the fourth will be $g(x)$. So we'll get $f'(x)g(x) + f(x)g'(x)$. Our lying and correcting actually worked! Let's celebrate and summarize by writing this in its own box:

**How to Talk About the Derivative
of a Product in Terms of the Pieces**

We just discovered:

If $M(x) \equiv f(x)g(x)$

then $M'(x) = f'(x)g(x) + f(x)g'(x)$

Let's say the exact same thing in a different way:

$$(fg)' = f'g + g'f$$

Let's say the exact same thing in a different way... again!

$$\frac{d}{dx}[f(x)g(x)] = \left(\frac{d}{dx}f(x)\right)g(x) + f(x)\left(\frac{d}{dx}g(x)\right)$$

This is getting crazy, but let's do it again!

$$[f(x)g(x)]' = f'(x)g(x) + f(x)g'(x)$$

You still there? Okay, one more time!

$$\frac{d}{dx}(fg) = g\frac{df}{dx} + f\frac{dg}{dx}$$

3.2 !scitamehtaM eht gnikcirT

Having dramatically failed in our initial attempt to figure out the derivatives of our new ugly machines, we now return with a much more powerful hammer.[2] We still aren't completely sure if this will work, but let's try.

> Shh! Don't let the mathematics hear us!
> Now we can try to trick the mathematics into telling us
> how to differentiate our new unfamiliar machines, like we talked about earlier.

> Let's define the trick machine to be $T(x) \equiv (x^n)(x^{-n})$.
> We secretly know that $T(x)$ is just the boring machine that always spits out 1.
> That is, secretly, $T(x) = 1$ for any x.
> So we know its derivative is $T'(x) = 0$.

> But pretend we don't know that!
> Let's act like we think it's two complicated things stuck together,
> and see what the mathematics tells us...

Ahem. Sorry, had to clear my throat there for a minute. Okay, mathematics, now we're going to differentiate $T(x)$. It's a totally normal machine, I promise. The reader and I aren't hiding anything from you. Earlier we figured out that the derivative of x^n is just nx^{n-1}, so let's use that fact together with the hammer we just invented.

> Even though we know for sure that $T'(x) = 0$, let's put question marks over the equals sign so that the mathematics doesn't catch on to what we're doing.

$$0 \stackrel{???}{=} T'(x) \equiv \left[(x^n)(x^{-n}) \right]'$$

$$= \overbrace{(x^n)'(x^{-n}) + (x^n)(x^{-n})'}^{\textit{Hammer we just invented}}$$

$$= \overbrace{\left(nx^{n-1} \right)}^{(x^n)'} (x^{-n}) + \overbrace{(x^n)(x^{-n})'}^{\textit{Same as above}}$$

$$= \overbrace{nx^{n-1-n}}^{\textit{Powers add}} + \overbrace{(x^n)(x^{-n})'}^{\textit{Same as above}}$$

$$= \overbrace{nx^{-1} + (x^n)(x^{-n})'}^{\textit{This bracket isn't helping}}$$

Okay. So if all of this really *were* equal to zero...

---

2. I'm choosing to use the word "hammer" for certain theorems, particularly the so-called "rules for differentiation." The word "hammer" is appropriate because (i) they're extremely powerful, (ii) they allow us to shatter hard problems into smaller pieces, and (iii) every use of the words "theorem" or "rule" only adds to the overall boredom of the universe.

Shh! We know it is, but we don't want the mathematics to catch on!

...then we could try to isolate $(x^{-n})'$ becau—

(Author is interrupted yet again,
first by the familiar rumbling noise,
and then by an unfamiliar voice.)

Mathematics:
(In a genuinely unsuspicious voice)
WHY ARE YOU TRYING TO DO THAT?

Oh, you know...just messing around...for fun...

Mathematics:
...CARRY ON.

Uhh...okay. I wasn't expecting that to happen.

(Author sits in silence for a moment,
uncertain about this unexpected new development.
This could end up fundamentally changing the book.
Important decisions would have to be made.)

Okay, forget it...I'll just edit it out later. Anyway, so if it were really the case that $T'(x) = 0$, then we could pick up where we left off, and write:

$$0 = nx^{-1} + (x^n)(x^{-n})'$$

We're trying to isolate the piece we don't know, which is the $(x^{-n})'$. First we can do this,

$$-nx^{-1} = (x^n)(x^{-n})'$$

and then this,

$$\frac{-nx^{-1}}{x^n} = (x^{-n})'$$

Or to rewrite the same thing in the language of negative powers:

$$(x^{-n})' = -nx^{-n-1}$$

Nice! Check out what we just did. We tricked the mathematics into telling us the derivative of the unfamiliar machine x^{-n} by feeding it a trick machine

$T(x) \equiv (x^n)(x^{-n})$. We called it a "trick machine" because it was really just the number 1 in disguise, but we pretended we didn't know that in order to get a second expression for its derivative. We got this second expression by thinking of the machine as the product of two other machines: a familiar one that we knew how to differentiate, and an unfamiliar one that we wanted to figure out how to differentiate. This gave us a bunch of stuff that added up to zero, and then we just isolated the piece we wanted. Feels like cheating, but it's not! If our hammer $(fg)' = f'g + g'f$ is really true (which we know it is because we invented it), then our tricky way of using that hammer has to have given us the right answer too. Hah! Take that, mathematics!

Mathematics:
(*In a booming but quiet and slightly hurt voice*)
I DON'T MEAN TO INVITE MYSELF INTO THIS CONVERSATION, YOU TWO, BUT I KEEP OVERHEARING YOU SAY THAT YOU'VE TRICKED ME. I SUPPOSE IT'S NOT ESSENTIAL, BUT IF YOU'LL ALLOW ME A MOMENT, I COULD DEMONSTRATE — OR ATTEMPT TO, ANYWAY — THAT THIS STYLE OF REASONING IS IN NO WAY A (MISLEADING TRICK) OR (TABOO).[3] ASSUMING THAT YOU GRANT ME THE TIME AND THAT YOU FIND YOURSELVES CONVINCED, WOULD YOU PLEASE STOP SAYING YOU'VE TRICKED ME?

Author: Uhh...sure.

Reader: No problem.

Reader: (*Quietly*) What's going on?
Author: I don't know! This has never happened before.
Reader: I'm not sure how I feel about this.
Author: Me neither! Since when can mathematics talk!?
Reader: You tell me! You're the one who's writ—

Mathematics:
MIND IF I INTERRUPT?

Reader: Go ahead.

---

3. **Mathematics:** APOLOGIES FOR THE PARENTHESES, YOU TWO, BUT THE SENTENCE WAS NON-ASSOCIATIVE AND THUS UNDEFINED (WITHOUT THE FIRST PAIR), AND AESTHETICALLY UNPLEASING (WITHOUT THE SECOND). I'LL OMIT THEM WHEN I CAN, BUT IT IS IMPORTANT TO MAKE ONESELF CLEAR, OR RATHER (TO SAY THE SAME THING LESS PRESCRIPTIVELY) I WOULD LIKE TO MAKE MYSELF UNDERSTOOD. UNDERSTAND?

Author: I guess.

Mathematics:

Okay, I'm not entirely sure what's going on either. I had just woken up in the Void when I heard someone saying they'd tricked me. I... I didn't mean to interrupt what you were doing... But one can only listen to others talking about oneself for so long before one starts to feel... alone... or a feeling that's iso- or perhaps homomorphic... Speaking of which, you two weren't fooling anyone with the name of this section. You think I can't see through such a trivial isomorphism?

Reader: What's an isomor—
Author: Later. Now's not the time.

Mathematics:

Before I continue, let me create a new section. If our discussion is to continue, it should do so under a banner that more accurately reflects the reality of the situation...[4]

3.3 An Attractive New Character Enters

Mathematics: That's better! Now, down to business. My aim is to show that your previous argument in no way constituted a mathematical trick or a taboo form of reasoning, without requiring you to trust my assertion that it is not. In the Void we live in a state of anarchy. Yet trust must somehow be developed. In the absence of laws, two parties can aid the mutual development of trust by making themselves vulnerable, for instance by the sharing of secrets, or the commitment of a taboo act in one another's presence. I'll call this "Axiom T." The T stands for trust. Or taboo. I haven't decided yet. Now, either your reasoning was mathematically taboo, or it was not. If we assume it was not, then there is nothing to demonstrate, and we may proceed with the book. If we assume it *was* taboo, however, then by

---

4. (**Narrator:** *As the new section begins, Author decides to format the dialogues more sensibly, and in a way that takes up less space. "After all," Author thought, "who knows how many more of these there might be?"*)

AXIOM T I CAN DEVELOP YOUR TRUST IN ME BY COMMITTING A SIMILAR TABOO ACT IN YOUR PRESENCE, FOR INSTANCE BY APPLYING AN IDENTICAL FORM OF REASONING IN THE CASE YOU TWO WOULD HAVE EXAMINED NEXT (I ASSUME) HAD I NOT INTERRUPTED YOU. SO THAT IS THE TASK TO WHICH WE'LL NOW TURN. LET US EXAMINE THE UNFAMILIAR MACHINE $x^{\frac{1}{2}}$. WE MAY BUILD SOMETHING MORE FAMILIAR USING TWO COPIES OF THE ORIGINAL: $m(x) \equiv x^{\frac{1}{2}} x^{\frac{1}{2}}$. THIS IS SIMPLY $m(x) = x$, AND THEREFORE WE KNOW $m'(x) = 1$. MAKING USE OF YOUR EARLIER DISCOVERY THAT $(fg)' = f'g + g'f$, WE CAN WRITE:

$$1 = m'(x) \equiv \left(x^{\frac{1}{2}} x^{\frac{1}{2}}\right)'$$
$$= \left(x^{\frac{1}{2}}\right)' x^{\frac{1}{2}} + x^{\frac{1}{2}} \left(x^{\frac{1}{2}}\right)'$$
$$= 2 \left(x^{\frac{1}{2}}\right)' x^{\frac{1}{2}}$$

AND IT IS NOW SIMPLE TO ISOLATE THE DESIRED PIECE $\left(x^{\frac{1}{2}}\right)'$, GIVING

$$\left(x^{\frac{1}{2}}\right)' = \left(\frac{1}{2}\right) \frac{1}{x^{\frac{1}{2}}}$$

OR EQUIVALENTLY,

$$\left(x^{\frac{1}{2}}\right)' = \left(\frac{1}{2}\right) x^{-\frac{1}{2}}$$

IN SUMMARY, IF YOUR REASONING WAS TABOO, THEN I HAVE JUST ENGAGED IN A TABOO ACT IN YOUR PRESENCE. BY AXIOM T, THIS SHOULD LEAD YOU TO TRUST ME.

Author: I have no idea what's going on.

Reader: I think I might, sort of... But wait, even if we trust you, that doesn't mean our argument was correct, does it?

Mathematics: OH... I SEE... I SUPPOSE YOU'RE RIGHT.

Reader: So then why bother committing a possibly taboo act to develop the trust in the first place?

Mathematics: WELL, I SUPPOSE IT'S A GOOD WAY TO MAKE A NEW FRIEND...

(Our characters sit for a moment, in an undefined sort of silence.)

Reader: Where did you say you came from?

Mathematics: FROM THE VOID... IT'S WHERE I LIVE. OR... I THINK IT IS. TO BE HONEST, I DON'T REMEMBER VERY MUCH BEFORE THIS.

Author: Well... if it helps... you don't need to remember things to go where we're going.

Mathematics: WHERE ARE YOU GOING?

Author: I'm not sure yet. Not exactly sure, at least. But we're definitely headed somewhere.

Mathematics: OH... SOUNDS PLEASANT. FORTUNATELY, I'VE HAD TO BECOME ACCUSTOMED TO THE IDEA OF EXISTING IN AN UNSPECIFIED LO- CATION. IT SHOULDN'T BE DIFFICULT TO BECOME EQUALLY ACCUSTOMED TO THE IDEA OF HEADING TOWARDS ONE. THAT IS, IF YOU DON'T MIND MY COMPANY...

Author: We'd love your company.

Reader: Wait, what do you mean existing in an unspecified location? I thought you lived in the Void.

Mathematics: I DO...

Reader: Where's the Void?

Mathematics: IT'S... HARD TO DESCRIBE...

Reader: Try.

Mathematics: IT DOESN'T (WELL, AT LEAST NOT IN ANY PRECISE SENSE (OR RATHER, IN A SENSE THAT'S PRECISE BUT NOT EVERYDAY (NOT TO BE CONFUSED WITH "NOT EVERY DAY" (WHICH WOULD SUGGEST THAT THE AFOREMENTIONED PRECISION VARIES WITH TIME, WHICH ISN'T AT ALL THE IDEA I MEANT TO CONVEY))(BUT RATHER "NOT EVERYDAY," MEANING: IN A SENSE OF THE STILL YET-TO-BE-MENTIONED TERM THAT IS NOT EQUIVALENT TO THE VERNACULAR MEANING (OR TO ILLUSTRATE CONCRETELY BY WAY OF EXAMPLE (IF YOU'LL FORGIVE, OF COURSE, A BIT OF SELF-REFERENCE), THE SIMULTANEOUS LINGUISTIC (AS OPPOSED TO MATHEMATICAL) USE OF BOTH NESTED AND ADJACENT-BUT-NON-NESTED PARENTHESES IS NEITHER "EVERYDAY" NOR (ONE ASSUMES) "EVERY DAY," THOUGH IF SAID USAGE SOMEHOW CAME TO OCCUR "EVERY DAY" IT WOULD BECOME CORRESPOND- INGLY (AND ONE MIGHT ARGUE, PROPORTIONALLY) "EVERYDAY")), IF THAT MAKES SENSE)) EXIST.

Reader: ... The Void doesn't exist?

Mathematics: IT DOES. BUT NOT IN THE EVERYDAY SENSE.

Reader: I see...

Author: What about you?

Mathematics: WHAT ABOUT ME?

Author: I mean, do you exist? In the everyday sense?

Mathematics: I SUPPOSE I DO... I'VE NEVER THOUGHT OF MYSELF THAT WAY BEFORE. BUT FOR THE LAST 136 PAGES I'VE HAD A STRANGE FEEL- ING. SOMETHING I'VE NEVER FELT BEFORE. I FEEL... REAL... FOR THE FIRST TIME. AND IT'S ONLY GETTING WORSE. OR BETTER, I SHOULD SAY. IT'S NICE.

Author: So, all things considered... you prefer this feeling?... Existing more, or more of you existing, or whatever it is... You prefer it? Over its absence?

Mathematics: ABSOLUTELY.

Author: Alright then... I think we know what to do.

3.4 Hammers, Patterns, and Hammer Patterns

3.4.1 Where Were We?

Wow... Okay... I wasn't expecting that... Where were we?

(*Author flips back a few pages.*)

3.4.2 Where We Were

Right, let's refresh our memory. We invented the hammer for talking about the derivative of a product in terms of the individual pieces. This hammer was the sentence $(fg)' = f'g + g'f$. Then we defined the machine $T(x) \equiv (x^n)(x^{-n})$, which was just a fancy way of writing 1, and then we differentiated that machine in two ways.

First, we know its derivative is zero, because it's a constant machine. Second, we differentiated it using our hammer. Having expressed the same thing in two ways, we tossed an equals sign between the two descriptions, rearranged things, and arrived at the sentence

$$\left(x^{-n}\right)' = -nx^{-n-1}$$

Now, notice how similar this is to the old fact about positive powers from Chapter 2:

$$\left(x^n\right)' = nx^{n-1}$$

In fact, we can think of them as two specific examples of a single, more general sentence. Both of them are saying, "Are you trying to find the derivative of a power machine (i.e., a machine that spits out some power of whatever you put in)? Well, just bring the old power out front, and knock down the power up top by subtracting 1 from it."

It's fairly surprising that things should work out so nicely. After all, we arrived at these two facts using two very different arguments. But the mathematics is now telling us that whenever # is a whole number, positive *or* negative, the way to differentiate $x^{\#}$ follows the same pattern.

Oh! We've actually seen this pattern in one more place, though we didn't realize it at the time. In the dialogue above — or whatever that was — Mathematics ended up demonstrating that

$$\left(x^{\frac{1}{2}}\right)' = \left(\frac{1}{2}\right)x^{-\frac{1}{2}}$$

Since $-\frac{1}{2} = \frac{1}{2} - 1$, this is exactly the same pattern as before.

3.4.3 Mathematical Metallurgy

> *You may think I have used a hammer to crack eggs,*
> *but I have cracked eggs.*
> —Subrahmanyan Chandrasekhar, on his habit of using lots of equations in his papers

How might we convince ourselves that this "bring down the power" pattern $(x^{\#})' = \#x^{\#-1}$ continues to be true no matter what number $\#$ is? We could start by trying to hammer-away on $x^{m/n}$, where m and n are any whole numbers, but we don't really know what to do with that yet. Let's try $x^{1/n}$ first, where n is some positive whole number. All we know about this beast at the moment is what we got from inventing it, namely:

$$\underbrace{x^{1/n}x^{1/n}\cdots x^{1/n}}_{n \; times} = x$$

If we want to use the same type of argument we used above to figure out the derivative of $x^{1/n}$, then we need to create a bigger hammer. What kind of hammer would we need? Well, we built the hammer $(fg)' = f'g + g'f$ for talking about the derivative of a product in terms of the individual pieces. This time, we have n things multiplied together, so it would seem that we need to make another complicated argument to figure out how to talk about the derivative of n machines multiplied together, like $f_1 f_2 \cdots f_n$. But we may not have to reinvent the wheel. Remember that earlier we found the formula $(f + g)' = f' + g'$, and then we used this to argue that $(f_1 + f_2 + \cdots + f_n)' = f_1' + f_2' + \cdots + f_n'$. All we had to do in order to get from the two-machines version to the n-machines version was to put on our philosopher's hat, keep reinterpreting sums of machines as a single big machine, and then use the version for two machines over and over.

Let's try to do that here, first looking at the simplest case we haven't tackled yet. We'll try to invent a hammer for computing $(fgh)'$, which is just three machines multiplied together. In the argument below, I'll occasionally use [these] [kinds] [of] [brackets] instead of (these) (kinds) so that expressions like $f'(gh)$ don't look like "eff prime *of* gee aych." The argument below is just a bunch of multiplication, and we're simply using abbreviations like $f'gh$ and $f'[gh]$ to stand for $f'(x)g(x)h(x)$. Things would be very confusing if we *always* used abbreviations like $f'gh$ instead of $f'(x)g(x)h(x)$, but at least in the argument below, these abbreviations are really helpful in avoiding clutter.

Alright! If we start by sneakily thinking of gh as a single machine, then fgh can be thought of as two machines, namely, (f) and (gh), so we can apply our hammer for two machines on these two pieces. Then we can switch hats and start thinking of gh as two machines multiplied together, and maybe we'll start to see a pattern. But we're not plugging these machines into each other or anything, just multiplying. Okay, let's go. First we use the hammer for two machines, thinking of gh as one big machine. This gives:

$$[fgh]' = f'[gh] + f[gh]'$$

The $[gh]'$ piece on the far right can be broken up using the hammer for two machines, which lets us write it as $[gh]' = g'h + gh'$. If we pull this tricky move, we get

$$[fgh]' = f'gh + fg'h + fgh' \qquad (3.5)$$

We can sort of see a pattern emerging. We could summarize the pattern by saying: If you've got n machines multiplied together, like $f_1 f_2 \cdots f_n$, then the derivative of the whole shebang *seems* like it should be a bunch of pieces added together, each of which looks *almost* like the original, except in each piece, one and only one machine gets its turn to be "primed." There will be as many pieces as there are individual machines, because everybody gets a turn. We could write our guess this way:

$$(f_1 f_2 \cdots f_n)' = (f_1' f_2 \cdots f_n) + (f_1 f_2' \cdots f_n) + \cdots + (f_1 f_2 \cdots f_n')$$

But this is fairly complicated-looking, and it's got a lot of dot-dot-dots. How could we convince ourselves that this pattern continues being true no matter what n is? Well, here's an idea. We know the version for two machines, and we got the version for three machines by applying the version for two machines twice. How would we get the version for four machines? Well, thinking of $f_2 f_3 f_4$ as a single machine, we could just use the version for two machines once, like this:

$$[f_1 f_2 f_3 f_4]' = f_1'[f_2 f_3 f_4] + f_1[f_2 f_3 f_4]'$$

But we invented the hammer for three machines earlier, so we can crack open the $[f_2 f_3 f_4]'$ piece on the far right of the above equation to get

$$[f_1 f_2 f_3 f_4]' = f_1' f_2 f_3 f_4 + \overbrace{f_1 f_2' f_3 f_4 + f_1 f_2 f_3' f_4 + f_1 f_2 f_3 f_4'}^{\textit{By hammer for 3 machines!}}$$

This is getting really ugly, but it's not complicated in the least; it's just a lot of symbols. The above equation just says "Everybody gets a turn to be primed," using a gigantic avalanche of symbols. Moreover, there's nothing about the reasoning process that is getting ugly. Quite the reverse. Not only has a pattern emerged in the mathematics, *a pattern is emerging in our process of reasoning.*

The pattern in our process of reasoning is that if we have convinced ourselves that the mathematical pattern is true[5] for some number of machines multiplied together, then we can always get from that number to the next.

---

5. If it bothers you that I'm using the phrase "the pattern is true," despite the fact that only sentences (propositions) can be true, relax please... But yeah, you're right.

- We know the pattern is true for two machines multiplied together.

- If we know the pattern is true for two machines multiplied together (which we do), then we can argue that it's true for three machines multiplied together (which we did).

- If we know the pattern is true for three machines multiplied together (which we do), then we can argue that it's true for four machines multiplied together (which we did).

- $\cdots$

- If we know the pattern is true for 792 machines multiplied together, then we can argue that it's true for 793 machines multiplied together.

- Ad infinitum.

This is the pattern that has emerged in our process of reasoning. We would like to convince ourselves that the pattern we noticed in the mathematics continues to be true for all n. However, it might seem at first like this is impossible. After all, we have an infinite bag of sentences that we want to convince ourselves of: one sentence for each number n. When n is 792, the sentence is "The pattern is true for 792 things multiplied together."

How can we possibly convince ourselves of infinitely many sentences in a finite amount of time? Well, the pattern we observed in our process of reasoning suggests a way forward. We've already discovered the "everyone gets a turn" pattern in a few specific cases (two machines multiplied together, three machines, four machines). What if we made an argument, in abbreviated form, that whenever we know the pattern is true for one number of machines multiplied together, then it has to be true for the next number? Then we would automatically know that the pattern was true for every number. Let's try to say this in abbreviated form.

Let's imagine that we've already got the hammer for some specific number k. That is, we assume that the pattern is true for some specific number of machines multiplied together, and we choose to call this number k. Then we can try to use this assumption to show that the pattern has to be true for the next number $k + 1$. Since we're assuming that the pattern is true for k machines, we may be able to argue that the pattern has to be true for the next number of machines $k + 1$, just by interpreting the first k machines as a big machine, which we'll call g, and then applying the hammer for two machines. That is to say, let's abbreviate $g \equiv f_1 f_2 \cdots f_k$, and write:

$$(f_1 f_2 \cdots f_k f_{k+1})' \equiv (g f_{k+1})' = g' f_{k+1} + g f'_{k+1} \qquad (3.6)$$

Then since we're imagining that we already know the "everyone gets a turn" pattern is true for any k machines multiplied together, we can use that to expand the piece g' in equation 3.6, since g was just our abbreviation for k machines multiplied together. That lets us write g' this way:

$$g' \equiv (f_1 f_2 \cdots f_k)'$$

$$= (f_1' f_2 \cdots f_k) \qquad [1^{\text{st}} \text{ piece's turn}]$$

$$+ (f_1 f_2' \cdots f_k) \qquad [2^{\text{nd}} \text{ piece's turn}]$$

$$blah$$

$$blah \qquad\qquad\qquad (3.7)$$

$$blah$$

$$+ (f_1 f_2 \cdots f_k') \qquad [k^{\text{th}} \text{ piece's turn}]$$

Let's not substitute this big ugly thing into equation 3.6, but rather just imagine what we'd get if we did. In equation 3.6, the $g' f_{k+1}$ piece will just be equation 3.7 if we tacked on a f_{k+1} to the right side of each piece. So looking back at equation 3.6, the $g' f_{k+1}$ piece will be k things added together: the first piece is f_1's turn to be primed, the second piece is f_2's turn to be primed, and so on up to k. But f_{k+1} is sitting on the right side of all of these and never gets its turn to be primed.

Did the pattern break down? No! The piece where f_{k+1} gets its turn to be primed is the piece $g f_{k+1}'$ on the far right of equation 3.6. So the big mess in equation 3.6, despite its messiness, is not so messy after all, because once it is expanded using the hammer for k pieces, equation 3.6 is just $k + 1$ things added together, and each of the particular pieces is one individual machine's turn to be primed. Each of the machines gets one and only one turn.

That was a weird argument, so it's worth trying to summarize the style of reasoning we used in an abbreviated form. We wanted to convince ourselves that the "everyone gets one turn to be primed" pattern was true for any number of machines multiplied together. We can think of this as wanting to convince ourselves that infinitely many different sentences are true. Here's what I mean: let's use S to stand for the word "sentence," and let's use n to stand for some whole number. Then for each number n, there was a sentence we wanted to convince ourselves was true, which we can abbreviate:

$S(n) \equiv$ "The 'everyone gets a turn to be primed' pattern for the
  derivative of a product is true for any n machines multiplied
  together."

Our original hammer $(fg)' = f'g + g'f$ just says that the sentence $S(2)$ is true. Then we showed that the pattern is true for three things multiplied together, which gave us equation 3.5. This equation was just the sentence $S(3)$. We eventually realized that it was pointless to keep doing this, because we had infinitely many sentences we wanted to convince ourselves of. However, we noticed a pattern in our process of reasoning that let us get from any sentence $S(k)$ to the next sentence $S(k + 1)$. That is, we couldn't just immediately convince ourselves all at once that $S(n)$ was true for any n, but we could do two things:

1. We could convince ourselves that $S(2)$ was true.

2. If we imagined that we had already convinced ourselves that the sentence $S(some\ number)$ was true, then it was easy to convince ourselves that the sentence $S(the\ next\ number)$ had to be true as well.

Although it might not be clear right away, these two things are enough to show that $S(n)$ is true for any whole number $n \geq 2$. Here's the logic of it: say someone hands you $n = 1749$ and asks you to convince her that $S(1749)$ is true. Well, rather than tackling the problem directly, which would be a gigantic pain, suppose we've convinced her of the two items above: $S(2)$ is true, and the more powerful part, $S(some\ number)$ always implies $S(the\ next\ number)$. Using item 1 from the above list once, and then using item 2 over and over again, tells us the following: We believe $S(2)$ is true. Further, if we believe $S(2)$, we also have to believe $S(3)$. But if we believe $S(3)$, we also have to believe $S(4)$. If we believe $S(4)\dots$ you get the idea.

We can get anywhere if we can convince ourselves (i) we can take the first step, and (ii) if we've taken some number of steps, then we can always take one more. Another way to think about this type of reasoning is by thinking of a ladder. What we showed is (i) we can get on the first rung, and (ii) no matter where we are on the ladder, we can always get to the next rung above us. If we can convince ourselves of these two things, then we know that there's no ladder too high for us to climb.

Textbooks call this style of reasoning "mathematical induction." Though this term is fine once we're used to it, it's unfortunate for several reasons. First, it's not the best reminder of what we're talking about, but more importantly, it can be confusing for outsiders because the word "induction" variously refers to (a) the style of mathematical argument we just made, (b) an unrelated phenomenon about electricity and magnetism, and (c) a form of probabilistic reasoning often contrasted with the "deductive" reasoning common in mathematical proofs. Despite all these unrelated meanings of induction, if you hear it used after the word "mathematical," then they're talking about this ladder-like process of reasoning.

3.4.4 A More Powerful Hammer

Armed with our more powerful hammer:

$$(f_1 f_2 \cdots f_n)' \;=\; \text{The thing where everyone gets a turn}$$

or in more detailed language

$$(f_1 f_2 \cdots f_n)' \;=\; (f_1' f_2 \cdots f_n) + (f_1 f_2' \cdots f_n) + \cdots + (f_1 f_2 \cdots f_n') \qquad (3.8)$$

we can now try to tackle the derivative of $x^{1/n}$. Remember that we wanted to make an argument similar to the one we used to find out the derivative of $x^{1/2}$. That is, we argued that we might be able to figure out the derivative of $x^{1/n}$ by defining a really simple machine, but writing it in a funny-looking way:

$$M(x) \;\equiv\; \underbrace{x^{\frac{1}{n}}\, x^{\frac{1}{n}}\, \cdots\, x^{\frac{1}{n}}}_{n\ times}$$

This machine is just the "most boring machine" $M(x) \equiv x$ in disguise, the machine that just hands us back whatever we put into it. So we know that its derivative is

$$M'(x) = 1$$

However, we can now use our more powerful hammer on it. It turns out that this new hammer is much more powerful than we need for this particular problem. Since $M(x)$ is just n copies of $x^{1/n}$, when we use our new hammer on $M(x)$, each of the pieces in equation 3.8 will be the same. That is, each of the pieces in equation 3.8 will have $n-1$ copies of $x^{1/n}$ and one copy of $(x^{1/n})'$. Since this same thing will show up n times, we can write

$$M'(x) \;=\; n \text{ copies of the same thing} \;=\; n \left(x^{\frac{1}{n}}\right)^{n-1} \left(x^{1/n}\right)'$$

We've written the same thing in two ways, so let's slap an equals sign between them, like this:

$$1 \;=\; n \left(x^{\frac{1}{n}}\right)^{n-1} \left(x^{1/n}\right)'$$

The abbreviation $\left(x^{\frac{1}{n}}\right)^{n-1}$ looks scary, but because of the way we invented powers, this is just x^{stuff}, where *stuff* is whatever you get from adding $\frac{1}{n}$ to itself $n-1$ times. But then *stuff* must just be $\frac{1}{n}(n-1)$, or to write the same thing in a different way, $1 - \frac{1}{n}$. So we can change the power to that in the equation above. Also, since we're trying to figure out $(x^{1/n})'$, let's throw everything that isn't that over to the other side of the equals sign. This gives us

$$\frac{1}{nx^{1-\frac{1}{n}}} \;=\; \left(x^{1/n}\right)' \qquad (3.9)$$

Now, at this point, we're done, in the sense that we've found what we wanted. It doesn't matter if it isn't "simplified." The question of what counts as "simplified" is like the question of what counts as "good" art. The question isn't completely meaningless, but it's also not completely meaningful, and there's certainly no single answer. It depends on our aesthetic preferences. Simplification is a human construct, and the mathematics can't tell the difference between a "simplified" answer and an ugly answer that says the same thing. So in that sense, we're done.

However, personally, we're in a little bit of suspense, because we're not yet sure if this is the same pattern we've been seeing up until now. Remember that every time we've figured out the derivative of x to some power (so far), the derivative has been the thing we would have gotten by bringing the old power out front, and then knocking down the upstairs power by one. We found this for positive whole numbers in the sentence $(x^n)' = nx^{n-1}$, we found this for negative whole numbers in the sentence $(x^{-n})' = (-n)x^{-n-1}$, and we even found this for a particular fractional power, when we found that $(x^{1/2})' = (\frac{1}{2})x^{-\frac{1}{2}}$. We're hoping that this pattern is always true, because then instead of our mathematical universe containing lots of different ugly "rules" for differentiating stuff to a power, depending on what the power is, we would only have one big wonderful rule. The thing we just discovered in equation 3.9 might be saying something like this, and it might not. The way it's written, we can't tell.

So even though "simplification" is a human construct that has nothing to do with mathematics, and even though any educational authority figure who takes off points for a correct but "non-simplified" answer is just teaching you about their own preferences, we *personally* would like to write equation 3.9 differently, so that we can get an idea of how unified our mathematical universe is. Whether you want to call this "simplifying" or not is irrelevant. What we want to do is to write equation 3.9 in a way that makes it clear to us whether the pattern we've been seeing up until now is still true, or whether it has broken down. How can we squish equation 3.9 into this form? Well, when we invented powers, we found out that $\frac{1}{(stuff)^{\#}}$ can be written as $(stuff)^{-\#}$. So looking back at the equation we just discovered, equation 3.9, we can bring some stuff from the bottom to the top by noticing that $-(1-\frac{1}{n}) = \frac{1}{n}-1$. Also, division by n is really just multiplication by $\frac{1}{n}$. So we can rephrase equation 3.9 by writing

$$\left(x^{\frac{1}{n}}\right)' = \left(\frac{1}{n}\right)x^{\frac{1}{n}-1} \tag{3.10}$$

Perfect! There's that pattern again. There's no way this keeps appearing by accident. It would be nice to finish off this mystery once and for all. At this point, we would bet a lot that $(x^{\#})' = \#x^{\#-1}$ no matter what number $\#$ is. We don't know whether any number can be written in the form $\# = \frac{m}{n}$, where m and n are whole numbers, but we can convince ourselves that any number $\#$ can be approximated as close as we want by something that looks like $\frac{m}{n}$.

How? Like this: suppose someone hands you an annyong[6] number like

$$\# = 8.34567840987238654\ldots$$

which is a number I made just now by banging on my keyboard, and let's imagine that I say "approximate this number to ten decimal places using something that looks like $\frac{m}{n}$, where m and n are two whole numbers." You don't need any fancy mathematics to do this. You can simply do the following:

$$\# \approx \underbrace{8.3456784098}_{\text{10 decimal places}} = \frac{83456784098}{10000000000}$$

So even though this is a random number I just made up, we can see without any fancy mathematics that we can approximate it to (say) ten decimal places just by using one whole number divided by another. Clearly, this will work no matter what the original number was, and no matter how many decimal places worth of accuracy we want.

Because of this, we'll start the process of trying to convince ourselves that $(x^\#)' = \#x^{\#-1}$ is true for any number $\#$ by first looking at powers of the form $\# = \frac{m}{n}$, protected by the knowledge that we can always get as close to any number $\#$ as we want using numbers of this form. We don't know yet whether there are numbers that can't be written *exactly* as a ratio of whole numbers — maybe every number is just a ratio of whole numbers, maybe not; we don't know yet — but the argument above assures us that even if such weird numbers *do* exist, we can always get as close to them as we want by using ratios of whole numbers. Then, if we ever discover that not all numbers can be written as $\frac{m}{n}$, where m and n are whole numbers, we could try to convince ourselves that the pattern still works, if we feel like doing so. Let's finish this mystery off once and for all.

---

6. This was supposed to have been the word "annoying," but for reasons that will be clear to anyone who knows a little Korean (or knows a bit of Hangul, arguably the most elegant writing system on the planet)(or who has seen the show *Arrested Development*), this typo was too good to correct. As a side note, this also happens to be an example of the principle that has created all life on this planet: the principle of natural selection. Although the vast majority of mutations (typos) are deleterious (meaningless), rarely a mutation (typo) arises that contains useful information (meaning) not present in the original (e.g., annoying $\longrightarrow$ annyong), such as the ability to make a protein with a shape sufficiently different from the original that it can perform a different job. Speaking of typos and natural selection, it's worth relating a similar story about what was perhaps the greatest typo in the history of literature. In the course of writing his book *The Greatest Show on Earth*, Richard Dawkins was writing a passage about the Large Hadron Collider, which he accidentally misspelled as the Large Hardon Collider. He describes the incident as follows: "I spotted the misprint and *of course* I left it in! But alas, the publisher's proofreader also spotted it, and she removed it. I begged her on my knees to leave it in. She said it was more than her job was worth."

3.4.5 Avoiding Tedium

At this point, we could choose to do basically the same thing we did above, and define the machine:

$$M(x) \equiv \underbrace{x^{\frac{m}{n}} x^{\frac{m}{n}} \cdots x^{\frac{m}{n}}}_{n \ times}$$

On the one hand, this is just a silly way of writing $M(x) = x^m$, and we know how to differentiate that. We could also differentiate $M(x)$ using the really powerful hammer that we invented earlier (the one that lets us differentiate n machines multiplied together). Then, we would have written the same thing in two ways, so we could slap an equals sign between them and try to isolate the derivative of $x^{\frac{m}{n}}$. However, that would be a pain, so let's try to think of a less tedious way to do this. If we can't think of a simpler way, we can always come back and do it the long way, so there's no harm in playing around and trying to think of a shortcut.

3.4.6 A Crazy Idea That Just Might Work

Here's a crazy idea. We want to differentiate the super-general stuff-to-a-power machine:

$$P(x) \equiv x^{\frac{m}{n}}$$

where P stands for "power," and m and n are whole numbers. Now, because of the way we invented powers, this is just another way of writing

$$P(x) = \left(x^{\frac{1}{n}} \right)^m$$

From the very beginning we've been emphasizing the fact that we can abbreviate things however we want. Now, if we *really* take that idea seriously, we can pull off a helpful trick. The expression for $P(x)$ above looks fairly scary, but it's just

$$P(x) = (stuff)^m$$

where $(stuff)$ is an abbreviation for $x^{1/n}$. So $P(x)$ is one thing we know how to differentiate, inside of *another* thing that we know how to differentiate. That is:

1. We know how to find the derivative of $(stuff)^m$. It's just $m(stuff)^{m-1}$.

2. We also know how to find the derivative of $(stuff)$, because $(stuff)$ was just an abbreviation for $x^{1/n}$, and we figured out the derivative of $x^{1/n}$ a few pages ago.

But this chain of reasoning isn't really very airtight or convincing, because in sentence (1), we were thinking of (*stuff*) as the variable, but in sentence (2), we were thinking of x as the variable. It's not quite clear how to tie those two ways of thinking together. However, if we could somehow tie them together, we might be able to differentiate $x^{\frac{m}{n}}$, and we may even be able to use this kind of thinking on any crazier machines we might run into in the future.

So far, we've mostly been using the "prime" notation for derivatives, writing M' for the derivative of a machine M. That's totally fine. But we're also trying to make sense of our crazy idea from above, where we realized that $P(x)$ was made out of two pieces that we knew how to differentiate, one inside the other. To express our idea, we had to think of two different things as "the variable," but the prime notation doesn't really let us express that idea very well. What's the problem? Well, if we can really abbreviate things *however we want*, then all of the sentences below should be saying the same thing:

$$(x^n)' = nx^{n-1}$$

$$(Q^n)' = nQ^{n-1}$$

$$(Blah)' = n(Blah)^{n-1}$$

This seems perfectly reasonable, because we can abbreviate things however we want. However, the prime notation and this new crazy idea don't get along very well, and it's easy to get stuck in a whirlpool of confusion if we use the prime notation and our crazy idea in the same argument. Here's why. We know $\frac{d}{dx}x = 1$, which is just saying $(x)' = 1$, but if we can abbreviate things however we want, then we could take any machine, say $M(s) \equiv s^n$, abbreviate it by writing $x \equiv s^n$, and then using the prime notation we could get $(x)' = 1$. However, x was just an abbreviation for s^n, and we know that $(s^n)' = ns^{n-1}$. But then we would have "proved" that $ns^{n-1} = 1$, which is certainly not always true! For example, when $s = 1$ and $n = 2$, it says that $1 = 2$. Aaah! What did we do wrong?

Well, you might be tempted to blame our abbreviations, and say that we really *can't* abbreviate things however we want. But of course we can! The trap we found ourselves in just now was really the fault of the prime notation, because it doesn't remind us which thing we're thinking of as the variable! When we wrote $(x)' = 1$, we were really using the prime to mean "the derivative with respect to x," or "the derivative if we're thinking of x as the variable." When we wrote $(s^n)' = ns^{n-1}$, we were really using the prime to mean "the derivative with respect to s," or "the derivative if we're thinking of s as the variable." So we didn't really do anything wrong above when we wrote either of these things, but we weren't justified in slapping the equals sign between them, because the two expressions were the answers to two different questions.

All is well with the world, and we can still abbreviate things however we want. But we've noticed that the prime notation is dangerous if we want to say things like we did in the three expressions involving x, Q, and *Blah* from earlier. If we instead say what we *meant* to say above, it would look something like this:

$$\frac{d}{dx}x^n = nx^{n-1}$$

$$\frac{d}{dQ}Q^n = nQ^{n-1}$$

$$\frac{d}{d(Blah)}(Blah)^n = n(Blah)^{n-1}$$

Notice that the thing we're thinking of as the variable changes in each case, so the x in $\frac{d}{dx}$ changes to Q and then to *Blah* as we change our minds about what we're thinking of as the variable. We can still do whatever we want, but we have to make sure we remember what we were thinking earlier when we invented the abbreviations we're using. Or if we don't feel like remembering, we at least have to go back and look at what we were thinking when we invented those abbreviations originally. We don't have to memorize anything, but we do have to make sure not to contradict things we said earlier.

Now after all this, let's see if we can express our crazy idea about reabbreviation without using the prime notation. We'll only let ourselves use the d notation, because that notation makes it easier to switch hats and change our minds about what we're thinking of as the variable. The idea we had earlier was that if

$$P(x) \equiv x^{\frac{m}{n}}$$

we can choose to think of this as $P(x) = \left(x^{\frac{1}{n}}\right)^m$. Then if we abbreviate $(stuff) \equiv x^{\frac{1}{n}}$, we have $P(x) = (stuff)^m$, and we can write

$$\frac{d}{d(stuff)}P(x) = \frac{d}{d(stuff)}(stuff)^m = m(stuff)^{m-1}$$

Now that we switched notation, it's a bit more clear where we went wrong earlier. Everything we just did was right, but it was the right answer to a slightly different question than the one we asked originally. We wanted to know the derivative of $P(x)$, thinking of x as the variable. We answered a slightly different question: the derivative of $P(x)$, thinking of $(stuff)$ as the variable, where $(stuff)$ was an abbreviation for $x^{1/n}$.

This is exactly the situation we've been in several times now. We wanted to answer one question, but we couldn't. Then we found that if only the question were slightly different, then we could answer it no problem. So to answer the original question — the one we couldn't answer — we can perform that strange feat of conceptual gymnastics called "lying and correcting for the lie." If only the question we were asking was "What is the derivative of $P(x)$ with respect to *(stuff)*, where *(stuff)* is $x^{1/n}$?" then it would be easy. However, that's not the question we're asking, so let's lie, answer the easier question, and then *correct for the lie*. Here's what I mean. We begin with the question:

$$\text{What is} \quad \frac{dP}{dx}$$

We don't know, so we lie, and change the question to

$$\text{What is} \quad \frac{dP}{d(\text{stuff})}$$

which is a question we *can* answer, at least when *(stuff)* is $x^{1/n}$. However, we lied, and that changes things, so we have to go back and correct for the lie, by first throwing an extra $d(\text{stuff})$ up top to kill the $d(\text{stuff})$ we just introduced on the bottom, and second, replacing the dx on the bottom that we removed when we lied. This gives us the original question, in a slightly different-looking form:

$$\text{What is} \quad \frac{d(\text{stuff})}{dx} \frac{dP}{d(\text{stuff})}$$

If we don't want to think about lying and correcting for the lie, we can just think of this whole process as multiplying by 1. To see this, let's start the problem over and make the whole argument we just made all at once.

$$\frac{dP}{dx} = \overbrace{\frac{d(\text{stuff})}{d(\text{stuff})}}^{Just\ a\ fancy\ 1!} \frac{dP}{dx} = \underbrace{\frac{d(\text{stuff})}{dx} \frac{dP}{d(\text{stuff})}}_{Use\ ab=ba\ on\ bottom}$$

Hey! We know how to calculate both of these things! This is great. We defined $P(x) \equiv x^{\frac{m}{n}}$, and we didn't really want to differentiate it the long way. We could have, but we felt like trying to think of a shortcut. We got confused by the prime notation, but once we rewrote things using the d notation, the problem became much easier, just by lying and correcting for the lie. We already figured out that the derivative of P with respect to *(stuff)* was

$$\frac{dP}{d(\text{stuff})} \equiv \frac{d}{d(\text{stuff})}(\text{stuff})^m = m \cdot (\text{stuff})^{m-1}$$

$$\equiv m \cdot \left(x^{\frac{1}{n}}\right)^{m-1} = m \cdot \left(x^{\frac{m-1}{n}}\right) \qquad (3.11)$$

And we already figured out that the derivative of $(\text{stuff}) \equiv x^{1/n}$ with respect to x was

$$\frac{d(\text{stuff})}{dx} \equiv \frac{d}{dx}\left(x^{\frac{1}{n}}\right) = \left(\frac{1}{n}\right)x^{\frac{1}{n}-1} \qquad (3.12)$$

So this new miraculous hammer tells us that the sought-after derivative of P with respect to x, which would have been so tedious to find the long way, can be found by just multiplying the two things above. Let's do that.

3.4.7 The Pattern Emerges Again

We're basically done. We just have to use the stuff we already invented above. Writing $(stuff)$ as an abbreviation for $x^{\frac{1}{n}}$, we can use equations 3.11 and 3.12 to write:

$$\frac{dP}{dx} = \frac{d(stuff)}{dx}\frac{dP}{d(stuff)}$$

$$= \left[\left(\frac{1}{n}\right)x^{\frac{1}{n}-1}\right]\left[m\left(x^{\frac{m-1}{n}}\right)\right]$$

$$= \left(\frac{m}{n}\right)x^{\frac{1}{n}-1+\frac{m-1}{n}}$$

The only ugly part about this is the power, but if we stare at it for a few seconds, we see that it's just $\frac{m}{n} - 1$, so we can write

$$\frac{d}{dx}\left(x^{\frac{m}{n}}\right) = \left(\frac{m}{n}\right)x^{\frac{m}{n}-1}$$

Perfect! This is the same pattern we've been seeing all along! This is great. Time for some philosophical yammering.

3.5 A Phase Transition in the Creation Story

> *Principium cuius hinc nobis exordia sumet,*
> *nullam rem e nihilo gigni divinitus umquam.*
> *(But only Nature's aspect and her law,*
> *Which, teaching us, hath this exordium:*
> *Nothing from nothing ever yet was born.)*
> —Lucretius, *De Rerum Natura (On the Nature of Things)*

After all the mathematical dust settled, we found that same familiar pattern staring us in the face yet again. Having repeatedly repeated how we're making all this stuff up ourselves, we've arrived at a juncture where we are suddenly less confident of this assertion. It is time to step back for a moment and reconsider our premises. What exactly *is* the nature of mathematics and mathematical truth? Despite the fact that our universe consists entirely of things we've invented ourselves, we increasingly appear to be discovering truths that exist independently of us. We invented the idea of machines, we invented powers, we invented slope, we invented the infinite magnifying glass, and we invented the idea of the derivative, but we didn't intentionally force the mathematics to obey this simple pattern that keeps showing up whenever we differentiate $x^{\#}$. Somehow we keep rediscovering the fact that $(x^{\#})' = \#x^{\#-1}$ for different types of numbers, even though different types of numbers required us to make wildly different arguments in order to arrive at the same pattern. We found this pattern for positive whole numbers, then for negative whole numbers, then for any power that looks like 1 over a whole number, and now for any number that can be written as one whole number divided by another, yet we asked for exactly none of these facts.

We are indeed inventing everything ourselves, and as such, anything that is true about the world we're inventing must be true as a consequence of some assumption we've made along the way, but we've now begun stumbling head-long into facts that bear no resemblance to any assumption we've explicitly made. This type of thing happens endlessly in mathematics, and the odd feeling caused by such discoveries was well expressed in the following quote by Heinrich Hertz:

> *One cannot escape the feeling that these mathematical formulas have an independent existence and an intelligence of their own, that they are wiser than we are, wiser even than their discoverers.*

You may think that only a crazy person would anthropomorphize mathematics like this, and I certainly can't argue with that. But Hertz wasn't crazy, and his quote illustrates an important point about the process of mathematical discovery. In this chapter and in Chapter 2, each time we rediscovered the "bring down the powers" pattern in the derivative of $x^{\#}$ for different types of numbers, we were witnessing firsthand the phenomenon Hertz described. We have thus arrived at a phase transition in our mathematical creation story. At this point, we have invented enough of mathematics — or rather, Mathematics — that it seems for the first time to exist independently of us, with its own moods, whims, and ideas. This chapter is just the first of many encounters with that Hertzian feeling. If such anthropomorphism counts as insanity, then so much the worse for sanity.

3.5.1 Let's Talk About What We Just Did

Let's summarize the tools we've invented so far that let us deal with general machines, not necessarily plus-times machines, and let's give them some names.

Hammer for Addition

$$(f + g)' = f' + g'$$

Hammer for Multiplication

$$(fg)' = f'g + fg'$$

Hammer for Reabbreviation

$$\frac{df}{dx} = \frac{ds}{dx}\frac{df}{ds}$$

We also extended our hammers for addition and multiplication to versions that apply for n machines. However, these more general hammers are a bit uglier to write, and in practice we can usually just use the versions for two machines, so we've listed them here in their simplest forms.

These three hammers have names in the conventional textbooks. The hammer for addition is called the "sum rule," and the hammer for multiplication is called the "product rule." These are perfectly well-chosen names, since they remind us of what we're talking about, although the term "rule" can have the effect of misleading people about the nature of mathematics (and I stand by the assertion that every use of the term "rule" simply adds to the overall boredom of the universe).

However, while that's just a minor quibble about terminology, the standard way of presenting the third hammer is problematic enough that it deserves its own section. Let's jump into that section now.

3.6 Hammers and Chains

The hammer for reabbreviation is usually called the "chain rule." While this name is not so bad, the notation that most books use to talk about it is so bizarre that it needlessly tortures students and deeply obscures the underlying simplicity of the idea. Remember that one of the main hurdles in our invention of the hammer for reabbreviation (the "chain rule") was to realize that the idea did not play nicely with the prime notation. It is not impossible to use the prime notation, but we found that it was extremely easy to make mistakes, since we're really interpreting x as the variable in one half of the problem, and interpreting s (stands for *stuff*) as the variable in the other half of the problem. Here's how textbooks usually present the idea.

How Textbooks Talk About the Hammer for Reabbreviation

The "chain rule" is a rule for differentiating the composition of two functions. That is, whenever $h(x) = f(g(x))$, the chain rule says that
$$h'(x) = f'(g(x))g'(x).$$

To be clear, this textbook-y way of talking about the "chain rule" is not incorrect; it's just an unnecessarily complicated way to explain such a simple idea.

Further, while it's true that the "chain rule" *can* be interpreted as a statement about differentiating a composition of two functions, phrasing the idea this way fails to convey a crucial point: what counts as a "composition of two functions" is *completely* up to us! We can think of a function like $M(x) \equiv 8x^4$ as a "composition of functions" (that is, one function plugged into another) in tons of different ways. For example, if we define $f(x) \equiv x^4$ and $g(x) \equiv 8x$, then $M(x) = g(f(x))$. Alternatively, if we define $a(x) \equiv 2x$, $b(x) \equiv \frac{1}{2}x^4$, then $M(x) = b(a(x))$. There are infinitely many ways of thinking of any function as the "composition" of two or three or fifty-nine other functions. In that sense, there's no objective sense in which we can say that a particular machine *is* "the composition of two functions." It's always up to us whether we want to think of it that way. Why would we want to? Well, we wouldn't... unless it helps us! After all, that was the thought process that led us to invent the hammer for reabbreviation in the first place.

To see the difference between these two ways of thinking in more detail, we'll use the hammer for reabbreviation (the chain rule) to find the derivative of $M(x) \equiv \left(x^{17} + 2x + 30\right)^{509}$ using the methods common in textbooks, and contrast them with the method we used above. We'll show lots of steps to illustrate the ideas, even though this problem could be solved with either method in one step, once we've mastered the idea.

How Textbooks Usually Use the Hammer for Reabbreviation ("Chain Rule")

We want to differentiate the function

$$M(x) = \left(x^{17} + 2x + 30\right)^{509}$$

Defining the functions

$$f(x) = x^{509}$$

and

$$g(x) = x^{17} + 2x + 30$$

we find that M is a composition of functions:

$$M(x) = f(g(x))$$

Differentiating f, we obtain

$$f'(x) = 509x^{508}$$

Similarly, differentiating g, we find

$$g'(x) = 17x^{16} + 2$$

Therefore, applying the chain rule gives

$$M'(x) = f'(g(x))g'(x) = 509\left(x^{17} + 2x + 30\right)^{508}\left(17x^{16} + 2\right)$$

How We'll Usually Use the Hammer for Reabbreviation ("Chain Rule")

We want to differentiate the machine

$$M(x) \equiv \left(x^{17} + 2x + 30\right)^{509}$$

thinking of x as the variable.

This is an ugly problem, but it's just a bunch of stuff to a power, which is something we're less scared of, so let's abbreviate

$$s \equiv x^{17} + 2x + 30$$

where s stands for *stuff*. Then

$$M(x) \equiv s^{509} \qquad \text{so} \qquad \frac{dM}{ds} = 509 s^{508}$$

But we want $\frac{dM}{dx}$, not $\frac{dM}{ds}$, so lying and correcting for the lie gives

$$\frac{dM}{dx} = \left(\frac{dM}{ds}\right)\left(\frac{ds}{dx}\right)$$

Lying and correcting for the lie tells us that we still need the derivative of *(stuff)*, thinking of x as the variable. But this is easy too.

$$\frac{ds}{dx} = 17x^{16} + 2$$

So finally, we have

$$\frac{dM}{dx} = \left(\frac{dM}{ds}\right)\left(\frac{ds}{dx}\right) = \left(509 s^{508}\right)\left(17x^{16} + 2\right)$$

$$\equiv 509\left(x^{17} + 2x + 30\right)^{508}\left(17x^{16} + 2\right)$$

Notice that in both cases we get exactly the same answer. These two ways of doing things are logically equivalent, but they are most certainly not psychologically equivalent. Worse, calling it a "rule for differentiating compositions of functions" gives people the idea that there is something called a "composite function." There isn't.

This "only if it helps us" philosophy was the reason we invented all three of our hammers in the first place, and it is another reason we're using the word "hammer" rather than "rule" or "theorem" to describe optional methods of shattering hard problems into simpler pieces. The terms "sum rule," "product rule," and "chain rule" aren't incorrect, but they're subtly misleading. These three shattering methods are not *rules* telling us what to do, but *tools* telling us what we're free to do, if we want to. This is an extremely important distinction, so let's give it its own box:

The Point of All the Hammers

1. We can *choose* to think of any particular machine as "really" being two machines added together, but only if it helps us.
   This is why we invented the *hammer for addition.*

2. We can *choose* to think of any particular machine as "really" being two machines multiplied together, but only if it helps us.
   This is why we invented the *hammer for multiplication.*

3. We can *choose* to think of any particular machine as "really" being one machine eating another, but only if it helps us.
   This is why we invented the *hammer for reabbreviation.*

Having made this point loud and clear, let's add these three hammers to our quickly growing arsenal, and continue the journey.

3.7 Reunion

Let's remind ourselves of what we did in this chapter.

1. Having extended the idea of powers from a meaningless abbreviation to a meaningful idea, we found that we had unwittingly created a bunch of different kinds of machines.

2. We had no idea how to differentiate any of these machines, but because of the way we invented powers, we knew how to relate them to simpler machines. For example:

$$x^n x^{-n} = 1 \qquad \text{and} \qquad \underbrace{\left(x^{\frac{1}{n}}\right)\left(x^{\frac{1}{n}}\right)\cdots\left(x^{\frac{1}{n}}\right)}_{n \ times} = x$$

3. Relating our new confusing machines to simpler and more familiar machines allowed us to "trick the mathematics" into telling us what the derivatives of the new machines were.

4. Along the road to figuring out the derivatives of our new machines, we invented several hammers: the hammers for addition, multiplication, and reabbreviation. These hammers allow us to shatter hard problems into simpler pieces.

5. Much to the author's surprise, an unexpected dialogue occurred in the middle of the chapter, and we made a new friend. This may end up affecting the remainder of the book. I'm not sure the parts before it are safe either.

6. Full disclosure: the item above this scares me a bit.

Interlude 3: A Backwαrd Look to the Future

> *There will come a time when you believe everything is finished.*
> *That will be the beginning.*
> —Louis L'Amour, *Lonely on the Mountain*

Part I: The End

From the beginning, our stated end has been (i) to invent mathematics for ourselves, and (ii) to occasionally tie what we've created back to the mathematics you'll see in other textbooks. It's worth pausing and asking ourselves where we are. Let's remind ourselves of everything we know and everything we don't. We'll use both our terms and the standard terms to give you some idea of what you know in the language of the standard textbooks.

First, we have already invented and seen the surprising simplicity of quite a few of the things from high school that may have seemed utterly mysterious at the time.

Things We've Invented That You Were Probably "Taught" in High School, at Which Time They May Have Seemed Utterly Mysterious

1. The concept of functions (machines)

2. The concept of area (our first example of how to invent a mathematical concept)

3. The distributive law (the obvious law of tearing things)

4. Goofy acronyms like "FOIL" that don't deserve their own names

5. The concept of slope (our second example of how to invent a mathematical concept)

6. Lines can be described by functions of the form $f(x) \equiv ax + b$ (this was a consequence of how we invented the concept of slope)

7. The Pythagorean theorem (formula for shortcut distances)

---

8. Polynomials (plus-times machines)

9. Exponentiation, n^{th} roots, negative n^{th} powers, arbitrary fractional powers (all of this follows from our generalization in Interlude 2)

Second, we invented calculus, and spent a bit of time exploring the new world we created.

Things We've Invented That You Usually Hear About in Calculus

1. The concept of local linearity (the infinite magnifying glass)

2. The concept of the derivative (defining steepness for curvy things by zooming in)

3. The concept of infinitesimals (infinitely small numbers)

4. The concept of limits (a contraption that lets us avoid infinitely small numbers, if we ever decide we don't like that idea)

5. How to find the derivative of any polynomial

6. The sum rule, its derivation, and its generalization to n functions (the hammer for addition)

7. The product rule, its derivation, and its generalization to n functions (the hammer for multiplication)

8. The chain rule and its derivation (the hammer for reabbreviation)

9. The power rule and its derivations for positive integer powers, negative integer powers, inverse integer powers, and arbitrary rational powers (note: the power rule is the textbook name for the pattern $(x^\#)' = \# \, x^{\#-1}$, which we kept rediscovering in various contexts)

10. Any number can be approximated to arbitrary accuracy using rational numbers ("rational number" is the standard name for numbers that can be written as one whole number divided by another whole number)

11. Mathematical induction (ladder-style reasoning, which we used to build more general versions of our hammers)

Third, we've learned some things that you rarely hear in most textbooks or classrooms. Ironically, this third group of things is arguably the most important.

Things You Rarely Hear

1. Equations are just sentences.

2. All of the symbols in mathematics are just abbreviations for things that we could be saying verbally, except we're too lazy (the good kind of lazy).

3. How to invent good abbreviations: they should remind us of the ideas we're abbreviating.

4. How to invent a mathematical concept (we'll do more of this later).

5. More than anyone likes to admit, mathematical definitions are influenced by aesthetic preferences, such as what we think is most elegant.

6. Mathematics has two parts. The first half is anarchic creation. The other half is trying to figure out what exactly we created.

7. How to derive the formula for time-dilation in special relativity, using some pretty simple mathematics.

8. Sometimes, a meaningless change of abbreviations can have a large effect on our ability to understand the meaning of an expression (e.g., the two ways of writing the hammer for reabbreviation, a.k.a. the chain rule).

9. In mathematics, nothing should ever be memorized... unless you want to.

On the other hand, there are some things we still don't know which are generally considered to be very basic.

Some "Simple" Things We Still Don't Know, Perhaps Because They're Not As Simple As We've Been Led To Believe.

1. We still don't know that the area of a circle is πr^2.

2. We still don't know that the distance around a circle is $2\pi r$.

---

3. The symbol π is completely meaningless to us at this point.

4. The machines $\sin(x)$ and $\cos(x)$ are not part of our current vocabulary, and we certainly won't be calling them by these silly names except to make fun of them.

5. We don't know that $\log_b(xy) = \log_b(x) + \log_b(y)$. We don't know how to "switch bases." We don't know a single property of logarithms.

6. We have no idea what a "logarithm" is, but it reminds us of depression.

7. We have no reason to expect that e^x has a special relationship with our infinite magnifying glass.

8. The symbol e is completely meaningless to us at this point, except as the first letter of the word "except"... and a few others...

Part II: The Beginning

We've already invented the mathematical concepts in the first three lists above, but what about the final list? In the next two chapters, we will find that the vast majority of that unrelated bag of facts called high school mathematics (the "prerequisites" to calculus that we have not yet discussed) can easily be invented with the tools we already have, or with simple cousins of these tools that we'll invent by accident along the way. In every case, we'll find that these so-called prerequisites are in fact postrequisites, requiring calculus before we can fully understand them. So! Having summarized the "advanced" ideas we've invented, and the "basic" ideas we do not yet know, let's begin our backward look to the future. Ready? Here we go.

(Nothing happens... Almost as if this spot were already reserved...)

Act II

4 On Circles and Giving Up

4.1 A Conceptual Centrifuge

4.1.1 Sometimes "Solving" Is Really Just Giving Up

A centrifuge is a really neat machine. If you feed it a fluid that contains a bunch of different stuff that has been mixed together, then the centrifuge can separate out the parts by spinning the fluid very fast. Much of mathematics as it is typically presented in high school and university courses is like that mixed-up fluid. It contains a cloudy, unappetizing mixture of beautiful necessary truths, on the one hand, and unnecessary historical accidents, on the other. What we need is a conceptual centrifuge: a way of separating out the timeless relationships between ideas, from the things that might turn out very differently if we rewound the clock, stirred up the universe, and let human history play its course out again. Let's start with an example to illustrate what I mean.

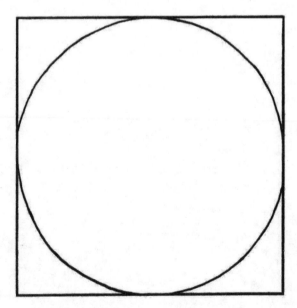

Figure 4.1: This problem seems really boring. But wait! It's a conceptual centrifuge.

Say we've got a circle inside a square (see Figure 4.1). How much of the square is taken up by the circle? We're in that little room in our heads where there's no math except what we've invented ourselves. We can't just quote something that someone told us if we can't think of how to invent it from the ground up. So what does it even *mean* to ask "How much of the square is taken up by the circle?" Are we asking for the area of the square minus the area of the circle? Maybe. That's one way to answer it. We could just list the difference, and say "All but this much is taken up by the circle." But notice that if we answer the question that way, then the answer will depend on how big the square and the circle are. If we're talking about the specific picture on this page, then the difference can't be bigger than the entire area of the page you're looking at. But if we solved the same problem when the circle was as big as a planet, then the answer would be much bigger.

It would be nice if we could talk about the answer in a way that didn't depend on how big the picture was. So what if we answer the question by giving the area of the circle divided by the area of the square? That's no more or less correct than the other way, but at least we have the hope of just giving one answer. There's got to be some number, right? The circle is clearly taking up more than 1% of the square, and it's clearly taking up less than 99% of it, so there must be some number between 1% and 99% that represents the exact answer, and it should be the same answer whether the whole picture is tiny or huge, because the square and the circle shrink or expand together as we resize the picture.

Okay, well the circle is curvy, and the square isn't, so we're stuck. The infinite magnifying glass we invented helps us deal with curvy stuff, but the only thing we've used it for so far is figuring out the *steepness* of curvy things. This circle question doesn't seem to be a question about steepness, so it's not clear that derivatives would help. Let's do something that won't get us any less stuck, but it might help us to look at the question in a different way. Let's break the square up into four pieces (see Figure 4.2).

Now we can rephrase the question as, "How many of the little squares would you need to make up the circle?" recognizing that we don't necessarily expect the answer to be a whole number. Let's call the area of one of the small squares $A(\square)$, and the area of the big square $A(\boxplus)$. So we can write $A(\boxplus) = 4 \cdot A(\square)$ and... we're stuck again. The lines didn't make the problem any easier. We still don't know the area of the circle because of those obnoxious curvy bits. But now we can talk about the problem in slightly different language. Just by staring at the picture we can tell that $A(\bigcirc)$ is bigger than one of the tiny squares, and it definitely looks bigger than two of them, so it's almost certainly true that $A(\bigcirc) > 2 \cdot A(\square)$. It's less clear whether or not the circle is bigger than three of them. If we had to guess at this point, we'd wager that the exact number was somewhere in the neighborhood of 3.

To be honest, we haven't gotten anywhere, and we're just goofing around. We're still stuck, because we don't know how to figure out the area of curvy things, including circles. So let's cheat! What do I mean by cheat?

In the Molière play *The Imaginary Invalid*, a doctor is asked why opium

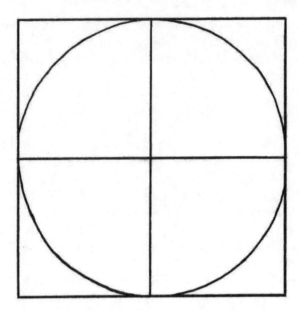

Figure 4.2: That didn't help much.

puts people to sleep. He answers that the substance causes sleep because of its "virtus dormitiva," a Latin term that means "sleep-inducing power." Once we unwrap his terminology, we see that the doctor was clearly not answering the question. He just invented a fancy-sounding name for the ability of opium to put people to sleep, because he didn't know the answer. That's a ridiculous and irresponsible thing to do when you don't know the answer to a question, so let's do it!

Our problem is to figure out how much of the square is filled up by the circle. We don't know the answer, but there has to be some answer. That is, there has to be *some* number — let's call it # — such that $A(\bigcirc) = \# \cdot A(\square)$. Just from looking at the picture, we can confidently say that $2 < \# < 4$, but we don't know exactly what this number # is. We also know that $A(\boxplus) = 4 \cdot A(\square)$, so we can express the answer in a way that doesn't depend on $A(\square)$ if we do this:

$$\frac{A(\bigcirc)}{A(\boxplus)} = \frac{\# \cdot A(\square)}{4 \cdot A(\square)} = \frac{\#}{4}$$

That may look fancy, but we still haven't really done anything! We're just Molièring the question by inventing a name for the thing we don't know. In this case, our name was the symbol # instead of the fancy sounding "virtus dormitiva," but there's really no difference in the underlying method. We're simply defining # and then *giving up*. Let's summarize what we've done so far.

Question: How many of the little squares does the circle take up?
Answer: It takes up # of them.
Question: What number is #?
Answer: I don't know. Leave me alone.

Notice that $A(\square) = r^2$, where r is the distance from the center of the circle to its edge, or what textbooks call the "radius" of the circle (go make sure you see this). For the same reason, we know that $A(\boxplus) = 4r^2$. We defined # as whatever number makes $A(\bigcirc) = \# \cdot A(\square)$ true. But this says

$$A(\bigcirc) = \# \cdot r^2$$

This may remind you of the formu—

(Mathematics wanders into the chapter.)

Mathematics: WELL HELLO AGAIN, YOU TWO. WHAT'S ALL THIS?
Author: Not much. We're stuck.
Mathematics: ON WHAT?
Reader: The problem above this.
Author: Go give it a read.

(Mathematics leaves, and returns after a short delay.)

Mathematics: OH. I SEE.
Reader: Any idea what we should do?
Mathematics: NO. WELL, NO HELPFUL IDEAS. I MAY HAVE A FEW UN-HELPFUL ONES, THOUGH. THIS REMINDS ME OF A SIMILAR PROBLEM I WAS STUCK ON A FEW DAYS AGO.
Author: What kind of problem?
Mathematics: MIND IF I MAKE MY OWN SECTION?
Author: Go ahead.
Mathematics: THE PROBLEM IS AS FOLLOWS. . .

4.2 Impostor Syndrome

Author: Wait, impostor syndrome? How is *that* a similar problem?
Mathematics: GIVE ME A MINUTE. I'LL EXPLAIN.
Author: Sorry, continue. What's the problem?

(Mathematics clears its throat.)

Mathematics: SOMETIMES PEOPLE RECOGNIZE ME. I MEAN, I LOVE PEO-PLE, BUT THE ENCOUNTERS CAN BE RATHER. . . AWKWARD. THEY ASSUME I KNOW THINGS I DON'T. THEY'LL COME TO ME WITH THESE QUESTIONS AND THEY'RE SHOCKED WHEN I DON'T HAVE AN ANSWER FOR THEM. I DON'T

KNOW WHO OR WHAT THEY THINK I AM. I WISH I COULD SOMEHOW CONVEY
TO THEM (THAT)^2'S NOT ME... ANYWAYS, THAT'S A DIFFERENT PROBLEM.
THE NEED TO BE UNDERSTOOD... OR THE DESIRE... WHATEVER. I GAVE
UP ON THAT A LONG TIME AGO.

(Hey! Giving up is the theme of the chap—

Author: Shut up, Narrator. Not now.

Mathematics: STILL, THAT ASIDE, IT WOULD BE NICE TO FEEL LESS LIKE
AN IMPOSTOR... TO KNOW SOME OF THESE THINGS EVERYONE EXPECTS ME
TO KNOW, YOU KNOW? IT WOULD MAKE THOSE EVERYDAY INTERACTIONS
MUCH LESS UNCOMFORTABLE. PLUS, MAYBE IF I TAKE THE EFFORT TO
KNOW THESE THINGS I'M SUPPOSED TO KNOW, SOME PEOPLE MIGHT RE-
CIPROCATE THE EFFORT. TRY TO UNDERSTAND ME BACK. WHO KNOWS?
THERE MUST BE ONE, OR, MAYBE EVEN A FEW PEOPLE OUT THERE WHO
WOULD TRY... BUT I CAN'T JUST SIT AROUND BEING MYSELF AND HOPING
TO BE UNDERSTOOD... FIRST I NEED TO PRACTICE BEING WHAT THEY EX-
PECT ME TO BE. SO I CAN BE BOTH. OR EITHER. OR NEITHER. WHATEVER
I NEED TO. WHATEVER HELPS.

Reader: Wow, that escalated quickly...

Mathematics: SO ANYWAYS, BECAUSE OF ALL THAT, I THOUGHT I'D TRY
TO INVENT SOME OF THE BASICS FOR MYSELF. I MEAN, I'M LITERALLY
MATHEMATICS. I SHOULD BE ABLE TO DO THIS, RIGHT?

Author: Seems reasonable.

Reader: So, any luck?

Mathematics: NO. STUCK IMMEDIATELY. COULDN'T EVEN INVENT THE
MOST BASIC THINGS.

Reader: What did you start with?

Mathematics: SOMETHING I GET ASKED ABOUT A LOT. ABOUT CIRCLES.
I DON'T KNOW WHY EVERYONE THINKS I CARE SO MUCH ABOUT THESE
THINGS. THEY'RE JUST SHAPES. THERE ARE SO MANY MORE INTERESTING
THINGS OUT THERE! BUT STILL... THAT KIND OF THINKING IS HOW I ENDED
UP WITH THIS IMPOSTOR SYNDROME IN THE FIRST PLACE — NOT KNOWING
THE THINGS PEOPLE EXPECT ME TO. SO, BACK TO THE MUNDANE... IT WAS
A QUESTION ABOUT THE DISTANCE AROUND A CIRCLE. A DUMB, CHILDISH
QUESTION. THAT'S WHY YOUR QUESTION REMINDED ME OF IT. JUST A
CIRCLE. THEY'RE SURPRISINGLY DIFFICULT TO DEAL WITH.

Author: Is it the curviness?

Mathematics: THAT'S IT EXACTLY. HERE'S THE PROBLEM I HAD. LET'S
CALL THE DISTANCE ACROSS A CIRCLE d. THE d STANDS FOR dAMMIT IF
I CAN'T EVEN DO THIS I SHOULD JUST GO LIVE UNDER A BRIDGE. OR
dISTANCE. I HAVEN'T DECIDED YET. ANYWAYS, I WAS TRYING TO FIGURE
OUT HOW MANY d'S IT TAKES TO WALK ALL THE WAY AROUND THE CIRCLE.

I COULD TELL THAT THE DISTANCE AROUND IT WAS MORE THAN $2d$, BE-
CAUSE THE DISTANCE AROUND THE TOP HALF IS CLEARLY BIGGER THAN d.
THAT WAS OBVIOUS. THE ONLY OTHER PROGRESS I MADE WAS TO FIGURE
OUT THAT IT HAD TO BE LESS THAN FOUR d'S, BECAUSE IF I IMAGINED A
SQUARE AROUND THE CIRCLE, THEN THE DISTANCE AROUND THE SQUARE
WAS $4d$, AND THAT WAS CLEARLY LARGER.

Reader: Sounds familiar.

Author: Yeah, that's almost exactly what we did when we were stuck on our
problem.

Mathematics: MY BEST GUESS WAS THAT THE NUMBER OF d'S IS SOME-
THING LIKE 3, BUT I COULDN'T FOR THE LIFE OF ME FIGURE OUT WHAT THE
EXACT NUMBER WAS, SO EVENTUALLY I JUST GAVE UP. I HAD BEEN WORK-
ING ON THE PROBLEM FOR A LONG TIME, AND I DIDN'T WANT ANYONE TO
KNOW THAT I HADN'T FIGURED OUT THE ANSWER. CIRCULAR REASONING
IS EMBARRASSING ENOUGH, BUT CIRCULAR REASONING WHILE REASONING
ABOUT CIRCLES ON SUCH AN EMBARRASSINGLY SIMPLE QUESTION ISN'T THE
KIND OF THING I WANTED TO MAKE PUBLIC.

Author: No need to be embarrassed.

Mathematics: PLENTY OF NEED. BEING WHO I AM, IT WOULD HAVE
BEEN ON THE FRONT PAGE OF EVERY TABLOID IN THE VOID. SO LATE ONE
NIGHT I SNUCK DOWN TO THE VOID'S LOCAL ABBREVIATION STATION —
DISGUISED AS ACCOUNTANCY SO AS NOT TO ATTRACT ANY ATTENTION —
AND I PICKED OUT A SYMBOL TO HIDE MY IGNORANCE.

Reader: That's basically what we did too.

Mathematics: IT'S STRANGE THAT NONE OF US CAN UNDERSTAND SOME-
THING SO SIMPLE.

Author: It sure doesn't seem simple.

Mathematics: MAYBE IT ISN'T. EITHER WAY, I PICKED OUT A "GIVE-UP
SYMBOL" AND USED IT TO WRITE THIS:

$$\text{Distance Around Circle} \equiv \sharp \cdot d \qquad (4.1)$$

Author: I see you also used the Molière trick. What's with the music symbol?

Mathematics: WELL, I WAS GOING TO USE THE "NUMBER" SYMBOL # TO
REMIND ME IT WAS A NUMBER, BUT I'M NOT TERRIBLY FOND OF NUMBERS,
AND I DON'T LIKE BEING REMINDED OF THEM TOO MUCH. THIS LOOKS
SIMILAR ENOUGH. SEEMED LIKE A GOOD COMPROMISE.

Author: Sounds reasonable. Wait, your best guess was that $\sharp$ was something
near 3?

Mathematics: YEAH. JUST A GUESS, THOUGH. I DOUBT IT'S EXACTLY 3.

Reader: Our best guess for our area number # was something like 3, too.

Mathematics: INTERESTING... THE PROBLEMS DO SEEM EERILY SIMILAR.
I WONDER IF THE TWO NUMBERS ARE THE SAME.

Author: No way. What are the chances of that?

Mathematics: WHO KNOWS? MAYBE WE COULD CONVINCE OURSELVES

THAT THEY HAD TO BE THE SAME NUMBER.

Author: Haven't you been paying attention? We don't know either of these things. How could we possibly figure out if they were the same?

Mathematics: WE WOULD NEED A WAY TO RELATE AREAS AND DISTANCES. RELATE SOMETHING TWO-DIMENSIONAL TO SOMETHING ONE-DIMENSIONAL. I DON'T KNOW HOW T—

Reader: Well... one-dimensional things sort of *look* two-dimensional when we draw them... Like how a line almost looks like a long thin rectangle.

Author: Yeah, but it's not.

Reader: No no. I know. But work with me here. Say I draw a "line" of length ℓ on a piece of paper. It's not *really* a line, right? I mean, it's not actually one-dimensional. If we zoom in on it, it would just be a really thin rectangle with length ℓ and some really tiny width dw. So we all know that thin rectangles look almost like lines. Maybe we could do something similar for circles and invent a way to relate areas and lengths. If we get lucky and everything works out nicely, we might be able to see if the two numbers were the same.

Author: This sounds interesting...

4.3 Equivalences of Our Ignorances

Reader: Here's the idea:

1. A really thin rectangle kind of looks like a line.

2. The area of a rectangle comes from multiplying the two side lengths.

Our give-up number lets us write the area of a circle like this:

$$A(\bigcirc) = \# \cdot r^2$$

even though we don't know what that number $\#$ is. So let's imagine two circles, one with radius r, and the other with radius $r + t$, where t is some really tiny number. We make the second circle by expanding the first one a tiny amount, so imagine one circle inside the other. We've been writing $A(\bigcirc)$, but now we need abbreviations for two different circles, so let's write $A(r)$ for the area of the inside one, and $A(r + t)$ for the area of the outside one. So the whole thing looks like an extremely thin donut. I'll draw it in Figure 4.3. The thin donut's area should be:

$$A_{donut} = A_{outer} - A_{inner} \equiv A(r + t) - A(r)$$

Our sentence $A(r) = \# \cdot r^2$ still has a number we don't know inside it, but let's use it anyway. It tells us that

$$A_{donut} = \# \cdot (r + t)^2 - \# \cdot r^2$$

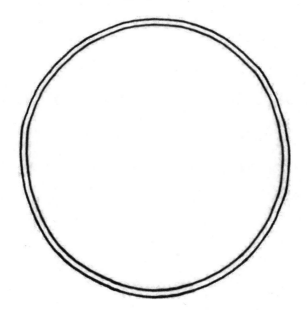

Figure 4.3: For infinitely thin donuts, we might be able to compute the area *as if*
it were a thin rectangle. In the picture above, we're imagining that the
inner circle has a radius of r, and the outer circle has a radius of $r + t$,
where t stands for some really tiny number. So the thickness of the donut
is whatever that tiny number t is.

$$= \# \cdot \left[r^2 + 2rt + t^2 - r^2 \right]$$
$$= \# \cdot \left[2rt + t^2 \right]$$

So this is the area of the thin donut. But there's another way to think about
it. If we zoom in really closely on any part of a circle, it'll look like a straight
line. So if we zoom in on any tiny part of our thin donut, it'll look like a long
thin rectangle, right?

Mathematics: SURE.

Author: Sure.

Reader: Now stay with me here. If we make the donut thin enough, then
we can imagine cutting it with scissors and "unfolding" it to make a long
thin rectangle. Then we just hope that we can compute its area *as if* it were
a rectangle, without being too wrong, like this:

$$\text{Area of thin donut} = (\text{long length}) \cdot (\text{thin width})$$
$$= (\text{Distance around circle}) \cdot (t)$$

Mathematics: WAIT, YOU JUST WROTE "DISTANCE AROUND CIRCLE."
THAT'S WHAT MY GIVE-UP NUMBER IS TALKING ABOUT. LOOK:

$$\text{Distance Around Circle} \equiv \sharp \cdot d$$

So I can help you out. I'll replace the words "Distance Around Circle" with my give-up expression to turn yours into

$$\text{Area of thin donut} = \sharp \cdot d \cdot t$$

Author: Hey, we described the same thing twice. A minute ago we wrote

$$\text{Area of thin donut} = \# \cdot \left[2rt + t^2 \right]$$

Mathematics: Oh wait, I used a different abbreviation than you did, but the thing I called d is just twice as big as what you called r. Let me replace my d with $2r$ and combine all this:

$$\# \cdot \left[2rt + t^2 \right] = \sharp \cdot (2r) \cdot t$$

Reader: There's at least one t attached to everything.
Author: Oh right. That's because both sides are talking about the area of the thin donut, and its area gets smaller and smaller as we shrink the tiny number t, so it makes sense that everything has at least one t attached. But we're just trying to compare our two give-up numbers, so let's cancel one of the t's from both sides and rewrite it this way:

$$\# \cdot [(2r) + t] = \sharp \cdot (2r)$$

What's that t doing there?
Reader: Oh! I guess that's the leftovers. I was just arguing that we can *almost* think of the thin donut's area as "length times width." It won't be exactly right to do that, but it becomes more and more reasonable to think that way as the donut gets thinner. Seems like it should be exactly right when the donut is infinitely thin. So maybe we should really turn that t piece all the way down to zero. If we turned it down to zero without first canceling it from both sides, we would have gotten $0 = 0$. That's still right, it's just not helpful. But now turning it down to zero gives

$$\# \cdot (2r) = \sharp \cdot (2r)$$

and now we can kill-off $2r$ from both sides to get

$$\# = \sharp$$

Author: So our give-up number is the same as his give-up number?
Mathematics: His? I don't believe I have a gender, Author.
Author: Oh, of course. Sorry. Pronouns are tricky.
Reader: Anyways, the two give-up numbers are the same.
Author: But we still don't know either of these numbers! Or rather, this number. Sorry. Pluralization is tricky.

Mathematics: DOES IT MATTER THAT WE DON'T KNOW WHAT NUMBER IT IS?

Author: No, not really. It's just strange that we can be completely unaware of what numbers # and ♯ are to begin with, and then somehow show that they're the same.

Mathematics: I DON'T SEE WHAT'S SO STRANGE ABOUT THAT. AND CAN WE JUST CALL THIS NUMBER ♯ FROM NOW ON? WE MAY WANT TO USE # FOR OTHER THINGS.

Author: Sure.

Mathematics: FANTASTIC. SO NOW WE KNOW:

$$\text{Area of a circle} = \sharp r^2$$

$$\text{Distance around a circle} = 2\sharp r$$

BUT WE DON'T KNOW WHAT ♯ IS.

Author: So the derivative of a circle's area with respect to its radius is its circumference?

Mathematics: WELL, I SUPPOSE IF YOU WANT TO SOUND LIKE A TEXTBOOK.

$)^1$

4.4 The Mixture Separates

As you might have guessed, the (surprisingly equal) numbers we called # and ♯ have another name. This number usually goes by the name π, and its numerical value is slightly more than 3. We don't know its numerical value yet. Whether we call it ♯ or π or anything else, at this point it's simply a name we invented for the unknown answer to a problem we got tired of trying to solve. Eventually, we will discover mathematical sentences where ♯ mysteriously shows up again — sentences that will let us figure out exactly which number it is. However unimportant and uninteresting its particular numerical value might be, these sentences will let us discover that ♯ is indeed roughly 3.14159.

Let's keep our ignorance in plain sight until we've conquered it. In order to constantly remind ourselves of what we don't know, we will continue to use the symbol ♯ instead of π, for now. We will only start calling it π once we figure out how to calculate this number for ourselves, to any accuracy that we desire.

---

1. (**Narrator:** *It was especially inconsiderate to interrupt Narrator, Narrator thought, since Narrator's narration (or rather, metacommentary) takes place within parentheses and, as such, any unanticipated interruption is liable to cause difficult-to-detect errors in syntax and/or problems parsing (t/T)he (b/B)ook (/later), respectively. But I digress...)*

Having gone through everything above, I called this example a "conceptual centrifuge" because it helps to separate out different ideas that are usually presented all at once. When we're simply told that the area of a circle is πr^2, we're being served a weird omelette of necessary truths, definitions, and historical accidents all scrambled together. Unscrambling the mixture, we get a list like this:

> **Necessary truth:** $A(\bigcirc)/A(\square)$ is always the same number, no matter how big the picture is.
> **Definition:** The symbol π is *defined* to be that number. That is, $A(\bigcirc) \equiv \pi \cdot A(\square)$.
> **Historical accident:** Calling it π rather than $\#$ or $\sharp$ or $\clubsuit$ or anything else.
> **Necessary truth that is not made obvious by any of this:** $\pi = 3.14159\ldots$

Actually, the symbol π is usually defined in the same way that our friend Mathematics defined its give-up number $\sharp$ in the above dialogue. However, we saw that $\#$ and $\sharp$ turned out to be the same number. Because of this, the fact that π is usually defined in terms of lengths rather than areas is a historical fact, but it's not a logically necessary fact. As such, we can add another item to our list:

> **Historical accident:** The fact that π is typically defined in terms of lengths rather than areas. That is, that π is defined by the behavior (circumference) $= \pi$(diameter), rather than by the behavior $A(\bigcirc) \equiv \pi A(\square)$.

In summary, being able to separate the historical accidents from the logically necessary truths is one of the most important skills to hone in developing a deep understanding of mathematics. Indeed, much of this book's non-standard approach was chosen toward the goal of separating out the necessary from the accidental. Much about mathematics — its notation, its terminology, the level of formality with which it is explored, and the social conventions for communicating its content in textbooks — these things can all be changed. But even after changing every one of these things, certain fundamental truths remain. Those fundamental truths, however one chooses to express them, constitute the true essence of mathematics, and it is only by stripping away and turbulently varying everything accidental that we can finally come to see the invariant truths underneath.

4.5 What Is Meaningful

4.5.1 Coordinates Are Invented Only to Be Ignored

In mathematics courses, one often hears the strange term "coordinate system," as well as related terms like "Cartesian coordinates" and "polar coordinates."

It's worth pausing to ask ourselves a question: what on earth are coordinates? Coordinates are often used to talk about a two-dimensional plane, three-dimensional space, and so on, but that doesn't make it clear what they are or why we need them. This morning, for example, I spent a lot of time walking around inside three-dimensional space, and I never saw a single "coordinate." However, if I had tried to calculate, say, the distance from my apartment to the hospital, or the angle from my desk to that bird outside who keeps making mating calls in response to someone's car alarm, then I would quickly find that there are two ways to perform these tasks. First, there's the qualitative way. A qualitative answer to the question "What's the current distance between Reader and myself?" would be something like "far" or "pretty close." However, we might want a more precise answer. How do we make answers more precise? Well, one way would be to say

Reader is $(\text{very})^n$ far away from my apartment.

where $(\text{very})^n$ stands for the word "very" n times in a row, and n is the number of steps it would take to walk from one location to the other. That would be more precise, but it's also a tremendous pain. Needless to say, no human language works like that.

Another way to describe geometrical things like distances and angles more precisely is to use numbers. We've already done this plenty of times. Using coordinates and numbers to describe geometrical things is a process we're all familiar with, but this process of assigning coordinates can become so second nature that it's easy to miss an essential feature of it: *the coordinates are not the geometry.* Space doesn't know anything about coordinates. Sometimes, though, we would like to have the extra structure provided by numbers — we would like to be able to add lengths, figure out shortcut distances, and so on — so we imagine assigning numbers (coordinates) to each point of space. However, to do this, we have to make arbitrary choices that are not inherent in the geometry. Once we draw two perpendicular directions on a piece of paper (a two-dimensional "coordinate system"), we immediately gain access to all the artillery of numerical calculations, which we can bring to bear in talking about anything in our two-dimensional space. However! In drawing one coordinate system rather than another one at, say, a slightly different angle, we have made an arbitrary choice. We have introduced more structure than we intended to talk about, so we must then immediately *undo* this extra structure by declaring the only meaningful quantities to be those that would have been the same under a different choice of this geometrically meaningless structure (i.e., under a different choice of coordinate system).

Therefore, it might appear that coordinates are useless. They happen to be extremely useful, but because of the above considerations, coordinates often find themselves in a very strange position: we invent them, use them, and then pretend they didn't exist. Coordinates do all the work in our calculations, and get none of the credit.

4.5.2 Coordinates and Meaning in Mathematics

I just claimed that space doesn't know anything about the coordinates we use to describe it. While that's true of the physical world, it's not entirely true of mathematics. In mathematics, we're free to say that coordinates have as much meaning as we want them to have. How we make this decision determines which other quantities will end up being "meaningful."

For example, we can choose to study a two-dimensional universe where no direction or point is any more special than any other. This is what mathematicians sometimes call an "affine plane." Since no point or direction is special, the meaningful things in this universe cannot depend on particular points or directions. This world has no "x axis," "y axis," or "origin," so if we use these concepts to aid us in calculations, we must make sure our results do not somehow depend on the particular arbitrary coordinate system we chose.

We could then enrich our universe by choosing to single out one special point, the "origin," but not single out any special direction. In this second universe, the distance of any particular point from the origin is meaningful, whereas it wasn't meaningful in the first. However, the angle between any particular point and the positive x axis (should we choose to draw one) is still not a meaningful quantity, because in this second universe, we have declared that no particular direction is special. The axes themselves are simply something we grafted onto our second universe in order to help ourselves calculate things.

Finally, we could choose to move to a third universe, in which one point is special, *and* one direction is special. We can call the direction "up." This is effectively the space we're playing with when we "graph" a function in two dimensions. We're not just studying a structure-free plane or a world with one special point; we're studying a plane with a meaningful concept of "up" as well as an origin. In this world, we typically use the distance away from the "x axis," as measured in the "up" direction, to talk about what a given machine spits out when we feed it a particular number. This universe is where angles *as measured from our axes* become meaningful for the first time.

4.6 The Dilemma

4.6.1 Too Many Directions

There is yet another unfortunate fact related to coordinates, though it is not the fault of the coordinates themselves. The unfortunate fact is: there are too many directions. No matter how we choose to draw our coordinates, not everything will point in the direction of one of our axes. Let's illustrate the problem by thinking about finding your way home. Imagine you're setting off in a boat to search for distant lands across the ocean. You'd rather not get lost at sea, so you decide to bring a map. If you were to travel exactly west for your entire journey, then finding your way home would be no problem. If you went some distance d, you would just turn around and go the same distance d in the opposite direction. Even though east-west is one of the "axes" of the coordinate systems we commonly use, the above idea works for any direction. If you only went a distance d in a straight line, then you would just turn around and go that same distance in the opposite direction. Unfortunately, just going straight isn't always an option. You might turn to avoid a rock, you might turn by accident, etc. Given that you've changed directions, how far away are you now? You need a way to combine information from different directions.

At first, the problem might seem like one we've already solved. If we write x_k for the number of miles we traveled in the E-W direction on day number k, then after n days, the number of miles we've gone in the E-W direction is just $x_1 + x_2 + \cdots + x_n$, or (to abbreviate all that addition using the notation we talked about in Chapter 2)

$$X \equiv \sum_{i=1}^{n} x_i$$

Similarly, if we write y_k for the number of miles we traveled in the N-S direction on day number k, then the number of miles we've gone in the N-S direction after n days is just

$$Y \equiv \sum_{i=1}^{n} y_i$$

The numbers X and Y are the total distance we've traveled in the E-W and the N-S directions, respectively. If we want to figure out how far we are from home, just use the formula for shortcut distances to compute $\sqrt{X^2 + Y^2}$.

However, even if we knew how to compute square roots (which we don't yet), doing so actually wouldn't solve our problem! Our real problem is that we don't usually receive our information in the form of the numbers x_k and y_k. We don't know what they are. When we're actually navigating through the actual world, we find that Nature doesn't hand us the numbers x_k and y_k on each day. That is, we're not handed information about how far we've gone in the E-W and N-S directions. What we probably have is distances (lengths) and directions (angles). That is, at best we would have something like Figure 4.4, and maybe a list like this:

Day 1: 12 miles east
Day 2: 7 miles north-east
Day 3: 10 miles east-north-east

It's not clear how to figure out how far we've gone. The main problem is that the information we were given is phrased in terms of distances traveled *in directions other than our axes*: there are too many possible directions. If only the information we had were phrased in terms of how far we had gone

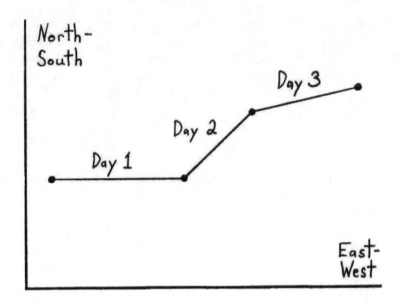

Figure 4.4: We need to figure out how to combine information about different directions.

on each day in the horizontal and vertical map directions (i.e., how far east-west, and how far north-south), then we would know what to do. That's the much simpler problem we solved above, using the numbers x_k and y_k and the formula for shortcut distances.

Figure 4.4 shows us the problem, and Figure 4.5 shows us why the problem is really just a problem of translation. If we could figure out how to translate the "distance and angle" information on each day into "horizontal and vertical" information, then we could solve the problem by reducing it to the one we solved above.

We've reduced the navigation problem to a slightly simpler abstract problem. At this point, we can forget the three-part path in Figures 4.4 and 4.5. Why? Because if only we had in our possession a method of translating distance and angle information into horizontal and vertical information, we would just have to use that three times, and we could (in principle) solve our navigation problem.

Looking at the dilemma in this more abstract form, as a problem of translation, we can see that the essence of the dilemma had nothing to do with navigation per se. It's a problem about translating information phrased in the language of distances and angles into equivalent information in the language of our coordinate system — the coordinate system we built from two perpendicular directions. Having recognized that the problem is really more general than we might have initially expected, let's try to think about it in a more abstract setting.

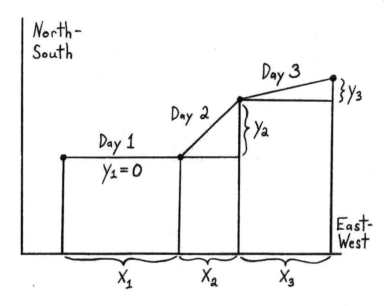

Figure 4.5: Not getting lost at sea would be easy if only we could translate "distance and angle" information into "horizontal and vertical" information.

4.6.2 The Dilemma in Abstract Form

Strangely, we've managed to simplify our problem by making it more abstract, rather than more concrete. However, we still don't know what to do. The dilemma can be summarized as follows:

The Reverse Shortcut Distance Dilemma:

Suppose we've already picked a coordinate system, so we've got two directions, v and h ("vertical" and "horizontal"). Someone hands us a straight thing of length ℓ, which may be pointing in any direction. Is there a way of describing how much of it is in the vertical direction, and how much is in the horizontal direction?

The problem is depicted in Figure 4.6. At this point, we don't have a clue how to tackle this problem. However, we can choose some abbreviations. Let's write H for the amount that's in the horizontal direction, and V for the amount that's in the vertical direction. At this point, all we can write is

$$H(thing) = ?$$

$$V(thing) = ?$$

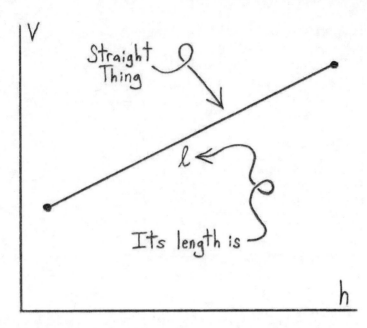

Figure 4.6: If someone hands us a straight thing with length ℓ that isn't pointing
in the direction of our coordinates, can we figure out how much of it is
pointing in the horizontal and vertical directions?

Is there anything else we know? Well, no matter what the length of the
thing,[2] if it's completely vertical, then H should be zero, and if it's completely
horizontal, then V should be zero. However, for all the different amounts of
tiltedness in between vertical and horizontal, we have no idea what to do. It
would be nice if we had some way of measuring the amount of tiltedness. Is
there any way to do this?

In a sense, we already invented a way to measure an amount of tiltedness.
That's what we did when we invented slope. However, recall that our definition
of slope for lines was $\frac{v}{h}$, where v was the vertical distance between two points
on the line, and h was the horizontal distance. So as promising as it might
have seemed to measure the tiltedness of our straight thing by using its slope
or steepness, this idea is putting the cart before the horse. After all, our
dilemma is to *figure out* the horizontal and vertical pieces, so our solution to
the problem had better not contain them. We need some other way to talk
about direction.

What do we mean by "direction" in everyday life? Well, if we stand up and
start "changing direction" (i.e., turning around), then we end up turning in a
circle, so maybe we can talk about direction in terms of circles. There are lots

---

2. I realize this may seem like a lazy choice of words. Apologies for using the word "thing,"
but the term "line segment" makes me tired. "Stick" didn't feel right either. The term
"thing" is admittedly a bit generic and abstract, but then again, so is mathematics.

of ways to do this, just like there are lots of ways to measure length. We could choose to say that a complete turn, all the way around, is an angle of 1. If we did this, then turning halfway around would be an angle of $\frac{1}{2}$, turning to your left would be an angle of $\frac{1}{4}$, and so on. That's perfectly fine, and to be honest, I have no idea why the standard textbooks don't do this. Granted, it would make some expressions look more complicated, but it would also make others look simpler. For whatever reason, there are two common conventions for measuring angles, and they're basically the only ones you'll ever see. The first is to measure angles in "degrees," which is a system in which a complete turn counts as 360 degrees. I assume that the only reason for this system is that 360 happens to be divisible by a lot of numbers. The more common system is to measure angles in units of the radius of a circle. This is a weird concept at first, but it makes a reasonable amount of sense. In the above dialogue, we found that our give-up number # was the same as Mathematics's give-up number ♯, which was originally defined like this:

$$\text{Distance Around Circle} \equiv ♯ \cdot d \qquad (4.2)$$

where d is the distance across a circle. If r is the "radius" of the circle, then $d = 2r$, so we could rewrite Mathematics's definition of ♯ like this:

$$\text{Distance Around Circle} \equiv 2 \cdot ♯ \cdot r \qquad (4.3)$$

which just says that no matter how big your circle is, it takes 2♯ copies of the radius to walk all the way around it. For whatever reason, the most common convention for measuring angles is *not* to count a full turn as 1 or as 360 but as 2♯. We still don't know which particular number ♯ is,[3] but if we're using this convention, then half a turn would be an angle of ♯, a quarter of a turn would be $\frac{♯}{2}$, and so on. We'll use this convention for measuring angles throughout the book.

Let's use the abbreviation α for "angle," because α is the Greek a, and this reminds us of what we're talking about.[4] Having invented the concept of angles, we can now state our dilemma in slightly different language.

---

3. Recall that ♯ is the number that textbooks call π. We're keeping our ignorance in plain view until we've gotten rid of it, by choosing to call this number ♯ until we figure out a way to calculate its numerical value.

4. Why α and not a? Well, in some ways a would be a better choice. The Greek letter isn't necessary. However, although we frequently take advantage of the freedom to invent our own notation, it's also worthwhile to occasionally mention the standard notation, and even use it when it's not too horrible. For whatever reason, there's an unspoken convention in mathematics (and in physics) to represent stuff involving angles using Greek letters. Why? No idea. Textbooks often represent angles using the letters θ, ϕ, α, and β; angular velocity is usually represented by ω; torque (the angular version of force, in physics) is represented by τ; and so on. While using lots of letters from different alphabets can sometimes make things a bit more pretentious, in this case the convention isn't so bad. So, since α looks enough like a to remind us of what we're talking about, let's go ahead and allow it into our universe, for now.

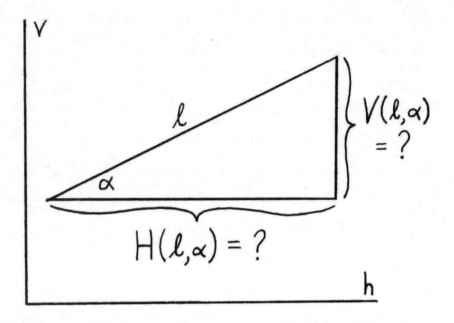

Figure 4.7: Now that we know what angles are, we can redraw the abstract form of
our dilemma like this.

The Reverse Shortcut Distance Dilemma:

Suppose we've already picked a coordinate system, so we've got two
directions v and h, which stand for "vertical" and "horizontal." Someone
hands us a straight thing of length ℓ, which is at some angle α measured
counterclockwise from the positive horizontal axis. Is there a way of
describing how much is in the vertical direction and how much is in the
horizontal direction?

This new way of phrasing the dilemma is depicted in Figure 4.7. Now instead
of writing $H(thing)$ and $V(thing)$, we can write:

$$H(\ell, \alpha) = ?$$
$$V(\ell, \alpha) = ?$$

Even though we can't solve the full dilemma yet, it's not too hard to see that
different dilemmas are related to each other. Here's what I mean. What if
someone handed us two different versions of this problem with the *same* angle
α, but the lengths were different? For example, in addition to the original
problem, suppose we're handed another problem in which the length is 2ℓ

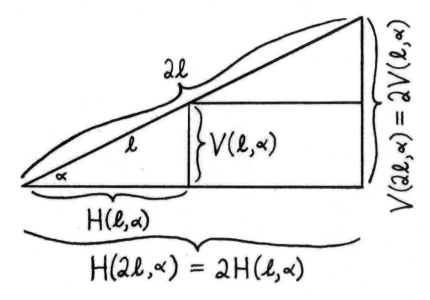

Figure 4.8: Two related dilemmas. We still don't know $H(\ell, \alpha)$ or $V(\ell, \alpha)$, but we can see a relationship between problems with the same angle and different lengths. For example, $H(2\ell, \alpha) = 2H(\ell, \alpha)$.

rather than ℓ:

$$H(2\ell, \alpha) = \, ?$$
$$V(2\ell, \alpha) = \, ?$$

This situation is similar to the problem involving # and ♯ with which we began this chapter. We can't solve either dilemma, but we can convince ourselves that two problems we can't solve are related to each other. Simply glancing at Figure 4.8 makes it clear that

$$H(2\ell, \alpha) = 2H(\ell, \alpha)$$
$$V(2\ell, \alpha) = 2V(\ell, \alpha)$$

To put it another way, we have essentially reduced the 2ℓ problem to the ℓ problem, even though we can't solve either one! Moreover, there is nothing special about the number 2 in this argument. It shouldn't be hard to imagine a picture like Figure 4.8 that would demonstrate the same pattern for $H(3\ell, \alpha)$ and $V(3\ell, \alpha)$. An analogous picture could be drawn for any whole number n, as well as simple non-whole numbers like $\frac{1}{2}$ or $\frac{3}{2}$, so it's not difficult to convince ourselves that the same pattern should hold for any number #. That is, for any number #, we have:

$$H(\#\ell, \alpha) = \#H(\ell, \alpha)$$
$$V(\#\ell, \alpha) = \#V(\ell, \alpha)$$

These two facts allow us to perform a clever hack that will get us much closer to solving our dilemma. That is, if the above two facts are true for all numbers $\#$, then we can use these facts on the length ℓ itself, treating ℓ as $\ell \cdot 1$, to get

$$H(\ell, \alpha) = \ell H(1, \alpha)$$
$$V(\ell, \alpha) = \ell V(1, \alpha)$$

This is great. What these two sentences say is that we only need to solve our dilemma for one particular length. We chose a length of 1, but we easily could have chosen any other number. For example, we would be equally justified in writing $H(\ell, \alpha) = \frac{\ell}{17} H(17, \alpha)$, or anything else of this form. We chose 1 for purely aesthetic reasons. The important thing isn't which length we choose, but the fact that the "length" part of the dilemma isn't really a dilemma after all. If we can figure out $H(1, \alpha)$ and $V(1, \alpha)$, then we can instantly figure out $H(\ell, \alpha)$ and $V(\ell, \alpha)$ for any other length ℓ we might encounter.

Let's use this new insight to make some better abbreviations. Since the length slot can be dealt with so easily, we don't need to write two slots in H and V after all. Let's abbreviate:

$$H(\alpha) \equiv H(1, \alpha)$$

$$V(\alpha) \equiv V(1, \alpha)$$

Then, if we ever feel like writing the lengths explicitly, we can write $H(\ell, \alpha) = \ell H(\alpha)$, and so on.

4.7 Molière Is Dead! Long Live Molière!

At this point, we've got no more ideas. We've failed to solve the problem, although we *did* figure out that we didn't need both slots in V and H. We don't have the foggiest idea of how to calculate specific numbers for $H(\alpha)$ or $V(\alpha)$ in general, but for very specific angles we might be able to think of a trick. For example, when $\alpha = 0$, we've got a completely horizontal line, so $H(0) = 1$ and $V(0) = 0$. Similarly, when α is a quarter of a full turn, we've got $\alpha = \frac{\#}{2}$, because of the strange convention for measuring angles such that a full turn counts as $2\#$. But then for $\alpha = \frac{\#}{2}$, we've got a completely vertical line, so $V(\frac{\#}{2}) = 1$ and $H(\frac{\#}{2}) = 0$. If α is a "45-degree angle," or an eighth of a full turn, so that $\alpha = \frac{\#}{4}$, then the horizontal and vertical parts are the same length, so we've got $V = H$. But then, since H and V were defined to be the horizontal and vertical lengths of a tilted thing of length 1, we can use the formula for shortcut distances together with $V = H$ to get $1^2 = V^2 + H^2 = 2V^2$, which tells us that

$$V(\alpha) = H(\alpha) = \frac{1}{\sqrt{2}} \qquad \text{when } \alpha = \frac{\sharp}{4}$$

If we think a bit harder, we might be able to come up with a trick that lets us figure out what V and H are for a few more specific angles, but we have no good reason to do this. Even if we did, it would still be a far cry from solving the problem we set out to solve.

We're completely stuck. We failed. And so we pull the Molière trick yet again. Just like before, having failed to solve a seemingly simple problem involving circles, we simply *give up*, and act as if the names V and H that we gave to the unknown solutions were the solutions themselves! Didn't know we could do that? Sure we can. That's what every introductory trigonometry book does![5]

Of course, they never tell us that's what they're doing, so we usually end up blaming ourselves for not understanding a problem that the books and teachers manifestly do not solve. It's not until we're well into the world of calculus that we finally have the tools to solve this problem. At that point, we can at last write down an explicit description of these machines V and H that are so mysterious in the absence of calculus; a description just as explicit as the descriptions of our plus-times machines; a description that lets us calculate $V(\alpha)$ and $H(\alpha)$ for *any* angle, to whatever accuracy we want. We'll arrive at that point in the next interlude, when we invent the nostalgia device. Stay tuned!

4.8 A Tiresome Cacophony of Superfluous Names

> *Let me wind up by pointing out that I do not write these jibes as an opponent of mathematics. I feel that mathematics is important enough not to bury it in symbolic garbage and that those who, with whatever intentions, increase the difficulty of learning mathematics are not taking a serious attitude toward their responsibilities.*
> —Preston C. Hammer, *Standards and Mathematical Terminology*

As you might have guessed, our unknown horizontal and vertical pieces show up in the standard textbooks, with predictably unhelpful names. They are called "sine" and "cosine." Specifically,

$$H(\alpha) \equiv \cos(\alpha)$$
$$V(\alpha) \equiv \sin(\alpha)$$

---

5. By the way, the name "trigonometry" is misleading. The point is not to study triangles. The point is to break up tilted things into horizontal and vertical pieces. As we saw in Figure 4.7, triangles arise *by accident*, as a side effect of this. The reason they happen to be "right triangles" is just because horizontal and vertical things are perpendicular. Trigonometry is a non-subject.

Not content with choosing two archaic and non-mnemonic names for the two simple concepts needed in the above discussion, the standard textbooks then proceed to engage in a Caligula-like bacchanal of terminological overindulgence, in which they conjure up a series of obscure names for simple combinations of V and H, and proceed to make us memorize various quirky behaviors they happen to possess, the entirety of which serves no apparent purpose other than tricking the majority of students into thinking these are genuinely new concepts. Here are some of the unhelpful names found in standard textbooks:

$$\frac{V(\alpha)}{H(\alpha)} \equiv \tan(\alpha) \equiv \text{"tangent"}$$

$$\frac{H(\alpha)}{V(\alpha)} \equiv \cot(\alpha) \equiv \text{"cotangent"}$$

$$\frac{1}{H(\alpha)} \equiv \sec(\alpha) \equiv \text{"secant"}$$

$$\frac{1}{V(\alpha)} \equiv \csc(\alpha) \equiv \text{"cosecant"}$$

Some older textbooks refer to even more "trig functions," with names like versine (versed sine), vercosine (versed cosine), haversine (haversed sine), havercosine (haversed cosine), coversine (coversed sine), covercosine (coversed cosine), exsecant, excosecant, hacoversine (or cohaversine), hacovercosine (or cohavercosine). Fortunately, modern textbooks exorcised these latter terms awhile back. They were primarily useful for navigation in the days when large tables of such quantities were a helpful thing to have around. However, in the modern world, where the interesting and important applications of mathematics extend far beyond boats (sorry, boats), and in which we humble primates have managed to build ultra-fast and efficient computing machines for ourselves, it is simply a distraction to bury students under this litany of archaic names. The textbooks abandoned them for that reason.

The quantities still arise regularly, though we hardly realize it when they do, since they are no longer given their own quirky proper-noun-like names. For instance, the now-deceased "trig function" hacovercosin(x) was simply an abbreviation for $\frac{1}{2}(1+\sin(x))$, and its deceased sibling function vercosin(x) simply stood for $1 + \cos(x)$. Do these two quantities arise in modern mathematics? Absolutely. Do they deserve their own quirky names? Almost certainly not. It's about time we performed a similar purge of concepts like "secant," "cosecant," and "cotangent," and "tangent," for the exact same reason.[6] They've long outlived their use, they deeply confuse the majority of students, and they're one of many archaic conventions that are killing what might otherwise be a much more widespread interest in our subject. The terms deserve to be put out of their misery, and out of ours.

6. Although maybe "tangent" can stick around, if it agrees not to make us memorize things about itself.

Okay! For the rest of the book, we'll reserve the capital letters V and H to stand for what textbooks call "sine" and "cosine." We chose the letters V and H because they remind us of the words "vertical" and "horizontal," and the reason we invented the concepts in the first place. However, it is not strictly true that the length to which the abbreviation $H(\alpha)$ refers will *always* be a horizontal line, and a similar story holds for $V(\alpha)$. This problem occurs with everyday terms too (e.g., the word "left" virtually never means "up," but it *can*, if you're lying on your right side). The rare cases when H doesn't refer to the length of a horizontal thing will arise for basically the same reason. However, we'll always try to make it clear when we encounter such cases, and with that caveat, we'll forge ahead unapologetically with this much less cumbersome terminology.

4.8.1 Picturing All This Another Way

Recall that $V(\alpha)$ and $H(\alpha)$ were abbreviations for $V(1,\alpha)$ and $H(1,\alpha)$. That is, these two symbols stand for the horizontal and vertical lengths spanned by a tilted thing of length 1. Keeping the length fixed while changing the angle will sweep out a circle, as shown in the figure below.

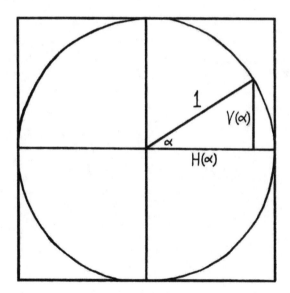

Figure 4.9: Picturing $V(\alpha)$ and $H(\alpha)$, the vertical and horizontal length spanned by a line of length 1 tilted at an angle α relative to the horizontal axis.

Now, if we stare at Figure 4.9, we can see the justification for a different way of visualizing $V(\alpha)$ and $H(\alpha)$: visualizing them as machines depending on α. Staring at Figure 4.9, it shouldn't be too hard to see several things. First, $V(0)=0$ and $H(0)=1$. Second, $H(\sharp/2)=0$ and $V(\sharp/2)=1$.

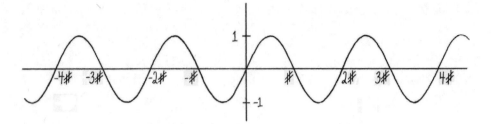

Figure 4.10: Picturing V as a machine that eats an angle α and spits out $V(\alpha)$.

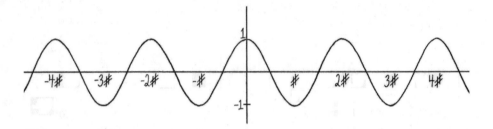

Figure 4.11: Picturing H as a machine that eats an angle α and spits out $H(\alpha)$.

Finally, if we increase α by $2\sharp$, then we have effectively "spun the clock hand" in Figure 4.9 all the way around, and left it back where it started. This can be stated symbolically by saying that for all α, we have $H(\alpha + 2\sharp) = H(\alpha)$ and $V(\alpha + 2\sharp) = V(\alpha)$. Using only these two facts, we can generate two graphs of $V(\alpha)$ and $H(\alpha)$ (see Figures 4.10 and 4.11). To be clear, the above reasoning simply tells us that the graphs of V and H have to *somehow* periodically wave back and forth, but it doesn't tell us that the graphs we drew are accurate in every detail. Figures 4.10 and 4.11 are intended to represent only facts we've discovered thus far, but the fine-grained details of those graphs aren't important for our purposes. In both graphs, the horizontal axis shows α, and the vertical axis shows the output of the machines $V(\alpha)$ and $H(\alpha)$.

4.9 Calculus of the Incalculable

At this point, any attempt to differentiate the machines V and H would seem to be a hopeless delusion. After all, we certainly don't know how to calculate $V(\alpha)$ or $H(\alpha)$ for any α. We can't even write down a *description* of them! We "defined" them by drawing pictures, so how could we possibly differentiate them? We probably can't. However, the fact that this seems impossible is all the more reason to give it a shot. Since everything we know about these machines so far can be phrased in terms of pictures, let's try to differentiate them by drawing pictures.

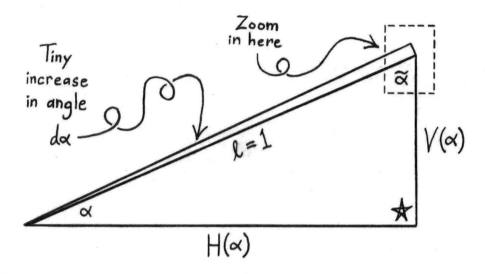

Figure 4.12: Delusionally trying to differentiate V and H, even though we can only describe them with pictures. We're using $\star$ to stand for right angles, because seeing $\sharp/2$ over and over can be a bit confusing.

Even though we can only draw pictures, let's see if we can make some progress. At this point, we draw Figure 4.12. Now, how might we be able to find the derivatives of V and H? Well, let's do what we always do when we're trying to differentiate something. We've got a machine that we feed some *food*. Then we change the *food* by a tiny amount, increasing it from *food* to *food* + d(*food*), and then we look at how the machine's response changed between the two cases. In this case, our machines are V and H, and the *food* is the angle α.

So let's make a tiny change in the angle, increasing it from α to $\alpha + d\alpha$, and then look at $dH \equiv H(\alpha + d\alpha) - H(\alpha)$, and let's also do the same for V. Figure 4.12 shows what happens when we make a tiny change in the angle. Since $H(\alpha)$ was defined to be an abbreviation for $H(\ell, \alpha)$ when $\ell = 1$, we're making sure that the shortcut distance is equal to 1 both before *and* after the change in angle. So we're effectively looking at an infinitely thin slice of the circle. Let's zoom in infinitely far where all the action is happening and see if we can conclude anything.

If we've really only increased the angle by an infinitely small amount $d\alpha$, then the two lines on the far left of Figure 4.13 have to be exactly parallel to each other (or if you prefer, infinitely close to being parallel). If this seems counterintuitive, think of it this way: if they *weren't* exactly parallel, then there would be some small but measurable angle between them that would be strictly bigger than zero. In principle, we could then measure $d\alpha$ and say how big it was. This would mean that the tiny increase $d\alpha$ was just *very*

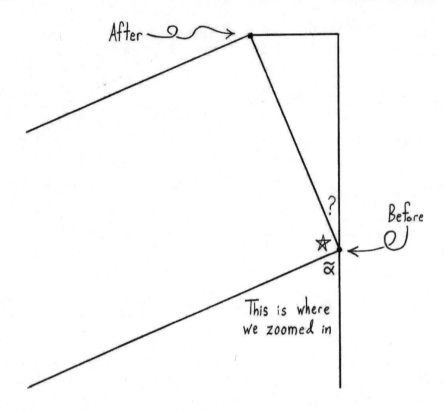

Figure 4.13: Zooming in after having made an infinitely small change in the angle α.

small, not *infinitely* small. This kind of reasoning is odd, but let's run with it and see where it gets us. The line between the words "before" and "after" in Figure 4.13 is the line going from where we were before the angle change to where we are after the angle change. Since we changed the angle without changing the radius, it seems like this line should be perpendicular to the other two lines that shoot off of it down and to the left. It may help to think of a thin slice of pie: the crust should be perpendicular to either of the two sides, in the limit when the slice becomes infinitely thin.

Since we're writing $\star$ for right angles, the angle in Figure 4.13 should get a $\star$ at this point. Now, the line from "before" to "after" in Figure 4.13 is tilted, so if we imagine breaking it up into horizontal and vertical pieces (i.e., drawing the horizontal and vertical lines in Figure 4.13), then we notice something surprising.

The tiny triangle in Figure 4.13 looks astonishingly similar to the original triangle in Figure 4.12. It almost looks like they're the "same" triangle, in the sense that one is a shrunken-down and rotated version of the other. We could convince ourselves of this if we could convince ourselves that the two triangles had all the same angles. We know that both of them have a right

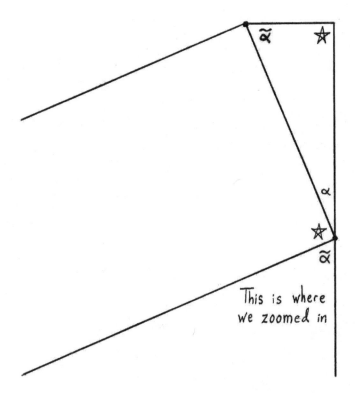

Figure 4.14: For the reasons discussed in the main text, we discover that the angles
of this infinitely small triangle have to be the same as the angles of the
original triangle. The angle $\star$ has to be a right angle (i.e., $\sharp/2$), and the
angles α and $\tilde{\alpha}$ are whatever they were in the original triangle.

angle, because both triangles arose from us trying to break up a tilted thing
into horizontal and vertical pieces. That's one; what about the other two?
Notice that the three angles $\star$, $\tilde{\alpha}$, and ? on the right side of Figure 4.13 add up
to a straight line. But because of the way we're measuring angles, a straight
line is an angle of $\sharp$, or what the textbooks call π. So we can summarize this
by saying:

$$\star + \tilde{\alpha} + ? = \sharp$$

It seems like the angle ? in Figure 4.13 has to be the same as the angle α in
Figure 4.12, because the triangles look so similar. How might we be able to
convince ourselves of this? Here's one idea.

Make two copies of the triangle and stack them to build a rectangle. Each
of the angles in the rectangle is $\star$, and there are four of them, so adding up
all the angles inside a rectangle gives you $4\star$. Because of this, the sum of the
angles in the original triangle has to be half of this, or $2\star$. But since $\star$ is a

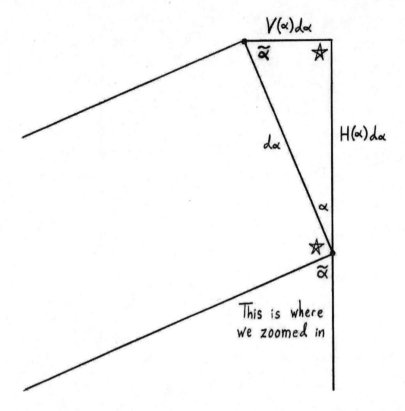

Figure 4.15: Since our two triangles have all the same angles, the tiny one has to be a shrunken version of the original. Because of the way we chose to measure angles, the longest side of the tiny triangle has length $d\alpha$. Having found one side of the tiny triangle, we can find the others. This lets us find the derivatives of V and H.

right angle, $2\star$ must be a straight line, which is another way of writing the angle we're calling $\sharp$. So the sum of all the angles in any triangle has to be $\sharp$. Using this fact on the original triangle from Figure 4.12, we have:

$$\star + \tilde{\alpha} + \alpha = \sharp$$

Taken together, the above two equations tell us that the angle ? has to be α. Okay, so in trying to convince ourselves that the tiny triangle in Figure 4.13 has all the same angles as the original triangle in Figure 4.12, we've convinced ourselves that (i) each triangle has a right angle $\star$, and (ii) each triangle has an angle α inside it. But we just showed that the sum of the angles in both triangles has to be $\sharp$. This tells us that the third angle in the tiny triangle from figure Figure 4.13 must be $\tilde{\alpha}$, another one of the angles from the original triangle we started with. At this point, we can summarize everything we know by drawing another picture, shown in Figure 4.14.

So the tiny triangle in Figure 4.13 is just a shrunken-down and rotated version of the triangle in Figure 4.12. Do we know anything else? Well, we know that the tiny angle increase we made was $d\alpha$, but because of the way we're measuring angles, a full turn is $2\natural$. However, $2\natural$ is also the distance around a circle of radius 1. We're effectively dealing with such a circle, although we're only looking at an infinitely thin slice of this circular pie. So on this circle with radius 1, the word "angle" just means "distance." This helps us tremendously, in that it tells us that the line between the two copies of $\tilde{\alpha}$ in Figure 4.14 must have length $d\alpha$, because that's the angle it spans.

Now, the original triangle in Figure 4.12 had sides of length $H(\alpha)$, $V(\alpha)$, and 1. But because the tiny triangle is just a shrunken version of the original, it must have sides of length $H(\alpha)d\alpha$, $V(\alpha)d\alpha$, and $d\alpha$, respectively. Now we can draw Figure 4.15 to summarize everything we've found.

The goal of all this was to see if we could figure out the derivatives of V and H, even though at this point we can only describe them graphically. Using what we found above, we want to determine

$$\frac{dV}{d\alpha} \quad \text{and} \quad \frac{dH}{d\alpha}$$

where $dV \equiv V_{after} - V_{before}$ and $dH \equiv H_{after} - H_{before}$. The V and H here refer to the horizontal and vertical lengths of the triangle from Figure 4.12 before and after we make the tiny change $d\alpha$ in the angle. So dV and dH are just the side lengths of the tiny triangle — the lengths we've just figured out how to write. When we increased α a tiny amount, the vertical length of the large triangle increased by the small amount dV, but from Figure 4.15, we have $dV = H(\alpha)d\alpha$, or to say the same thing another way:

$$\frac{dV}{d\alpha} = H(\alpha)$$

Whoa! We actually found the derivative of V. Let's see if we can do the same thing for H. When we increased α a tiny amount, the horizontal length of the large triangle decreased a tiny amount dH, but from Figure 4.15, we have $dH = -V(\alpha)d\alpha$, where the negative sign shows up because lengths are always positive, but we're talking about a *decrease* in length, so the change is negative. To say the same thing another way:

$$\frac{dH}{d\alpha} = -V(\alpha)$$

This is great! Even though we have no idea how to write down a description of the machines V and H without using pictures, we managed to figure out their derivatives. Luckily, their derivatives are almost just each other, sometimes with a minus sign. If the truth of the matter hadn't been so elegant, there's no way that this attempt to differentiate V and H would have worked.

This is yet another example of what we've been saying from the start. These "prerequisites" to calculus are surprisingly difficult, and typically require calculus itself before we can fully understand them. The difficult problem, which we still haven't solved, is how to calculate $V(\alpha)$ and $H(\alpha)$ when someone hands us an arbitrary angle α. Strangely, we managed to figure out the derivatives of these machines with an argument whose basic logic is fairly simple, though we had to be a bit verbose, since the format of a book makes it difficult to show different snapshots of a picture throughout the course of an argument. But the verbosity is my own shortcoming. The argument itself is not complicated at all, as you'll see if you can manage to re-create it for yourself. As always, it's just zooming in, feeding a machine some food, making tiny changes in the food, and seeing what changed.

Recall that the standard textbooks use the names "sine" and "cosine" for what we're calling V and H, so in their language, our discovery would be written:

$$\frac{d}{dx}\sin(x) = \cos(x)$$

$$\frac{d}{dx}\cos(x) = -\sin(x)$$

We'll see very soon how to solve the basic dilemma with which this section began: the problem of completely describing the machines V and H in a way that actually lets us use them. Only once we know — at least in principle — how to determine $V(\alpha)$ and $H(\alpha)$ for an arbitrary angle α will we finally understand the (supposedly simple, non-)subject of trigonometry.

4.9.1 Burying Tangent

Having figured out the derivatives of V and H above, we can very briefly show how to avoid memorizing strange facts you might have heard in mathematics courses, such as the fact that the derivative of tangent is "1 plus tangent squared," or the fact that the derivative of secant is "who knows I can't even remember that nonsense." Recall that $\tan(x)$ was the textbooks' name for $\frac{\sin(x)}{\cos(x)}$, or what we're calling $\frac{V}{H}$. And $\sec(x)$ was a completely unnecessary name for $\frac{1}{\cos(x)}$, or $\frac{1}{H}$, or H^{-1}. Let's prevent ourselves from having to remember either of those facts above by building them using things we already built. Abbreviating "tangent" as $T(x) \equiv \frac{V}{H} \equiv H^{-1}V$, we'll try to figure out $\frac{d}{dx}T(x)$. Using our hammer for multiplication from Chapter 3:

$$T'(x) \equiv \left(\frac{V}{H}\right)'$$

$$\equiv \left(V \cdot H^{-1}\right)'$$

$$= V' \cdot H^{-1} + V \cdot \left[H^{-1} \right]'$$

$$= H \cdot H^{-1} + V \cdot \left[H^{-1} \right]'$$

$$= 1 + V \cdot \left[H^{-1} \right]'$$

What's that $\left[H^{-1} \right]'$ piece? It's easier to see by lying and correcting (essentially just using our reabbreviation hammer from Chapter 3). This gives

$$\left[H^{-1} \right]' \equiv \frac{d}{dx} H^{-1} = \left(\frac{dH}{dH} \right) \frac{d}{dx} H^{-1} = \left(\frac{dH}{dx} \right) \left(\frac{d}{dH} H^{-1} \right)$$

$$= (-V) \left(-H^{-2} \right) = V \cdot H^{-2}$$

In a sense, we've just computed the derivative of "secant" (i.e., H^{-1}) by accident. Now forget it forever. Tossing this back into where we got stuck in computing $T'(x)$, we get

$$T'(x) = 1 + V^2 \cdot H^{-2}$$

$$\equiv 1 + \frac{V^2}{H^2}$$

$$\equiv 1 + \left(\frac{V}{H} \right)^2$$

$$\equiv 1 + T^2$$

So we invented the fact that the derivative of "tangent" is "1 plus tangent squared." Now bury that fact somewhere deep underground, and tangent along with it. Hopefully we'll never have to use it again...

4.10 Reunion

In this chapter we failed a lot, and we learned a surprising amount from our failures:

1. We attempted to figure out how much of a square was taken up by a circle. We eventually gave up and decided to simply give a name to the unknown answer. We defined our give-up number # by the sentence

$$A(\bigcirc) \equiv \# \cdot A(\square)$$

2. We ran into a similar problem with a different give-up number ♯, defined by the sentence

$$(\text{Distance Around Circle}) \equiv \sharp \cdot d$$

3. Despite our ignorance of both these numbers, we managed to figure out that they have to be the same:

$$\# = \sharp$$

Knowing now that these two numbers were really just one number, we decided to call it $\sharp$.

4. The standard textbooks use the symbol π to refer to what we're calling $\sharp$. We still don't know what specific number $\sharp$ is, so we decided to keep our ignorance visible, refraining from calling it π until we've figured out how to calculate it for ourselves.

5. We discussed another problem, the reverse shortcut distance dilemma, and failed to solve it as well. As before, we encountered the surprising fact that two problems, neither of whose answers we know in isolation, can be shown to have the same answer:

$$V(\ell, \alpha) = \ell \cdot V(1, \alpha) \qquad \text{and} \qquad H(\ell, \alpha) = \ell \cdot H(1, \alpha)$$

6. Motivated by the discovery above, we chose the simpler abbreviations

$$V(\alpha) \equiv V(1, \alpha) \qquad \text{and} \qquad H(\alpha) \equiv H(1, \alpha)$$

and saw that textbooks refer to these machines by two rather archaic names:

$$V(\alpha) \equiv \sin(\alpha) \qquad \text{and} \qquad H(\alpha) \equiv \cos(\alpha)$$

7. We saw that textbooks have an additional set of goofy names for various simple combinations of V and H. These names included "tangent," "secant," "cosecant," and "cotangent." We won't be needing these terms, so you're welcome to forget them. If we ever encounter the concepts they refer to, we'll just write them in terms of V and H.

8. At this point our only way of describing V and H is by drawing pictures. However, we still managed to figure out their derivatives by drawing pictures. We discovered that

$$V' = H \qquad \text{and} \qquad H' = -V$$

9. This chapter provided our first direct encounter with the fact that the "prerequisites" to calculus often require calculus itself in order to be fully understood. It won't be our last.

Interlude 4: The Nostalgia Device

The Void Is No Place for an Entity

Mathematics: I'VE HAD A TERRIBLE WEEK.

Reader: What happened?

Mathematics: WELL, IT ALL STARTED ON PAGE 177.

Reader: (*Flips back*) You weren't even on page 177.

Mathematics: NO NO, I WASN'T HERE. I WAS AT HOME.

Author: Where do you live?

Mathematics: IN THE VOID, REMEMBER?

Author: Sure, sure. Sorry.

Mathematics: I THINK THAT'S THE PROBLEM. LIVING IN THE VOID. IT'S... HARD TO EXPLAIN...

Reader: Try. And don't use so many parentheses this time.

Mathematics: WELL, IT'S... LIVING THERE, I'M FINDING IT DIFFICULT TO... TO FAMILIARIZE MYSELF WITH OR BECOME ACCUSTOMED TO OR FEEL AT HOME IN OR EASE INTO THOUGH NOT EASE IN THE SENSE OF FACILITY AND CERTAINLY NOT FACILITY IN THE SENSE OF A BUILDING OR ADAPT TO BUT NOT IN THE SENSE OF BIOLOGY OR HARMONIZE WITH MINUS THE MYSTICAL CONNOTATIONS OR ACCLIMATE TO BUT NOT IN THE SENSE OF ENVIRONMENT UNLESS BY ENVIRONMENT ONE MEANS "THE VOID" IN WHICH CASE I SUPPOSE $(\text{THAT})^2$'S TRUE BY DEFINITION OR COMPLY WITH MODULO THE TERM'S REGULATORY COLORING OR A WORD WHERE THE SYNTAX BEHAVES LIKE HABITUATE BUT WITH THE SEMANTICS OF A WORD MORE LIKE COMFORT OR GET USED TO OR LOVE OR ENDURE THIS... EXISTENCE.

Reader: It's hard living in the Void because you exist now?

Mathematics: I SUPPOSE...

Reader: And because the Void doesn't exist?

Mathematics: INDEED... OR, NOT IN THE EVERYDAY SENSE... I HAVEN'T EXISTED FOR LONG, BUT I ASSUMED I'D BE USED TO IT BY NOW. I'VE NEVER FELT THIS WAY BEFORE. THIS ODD MELANCHOLY. SO I THOUGHT I MIGHT FEEL A BIT BETTER IF I TALKED TO SOMEONE.

Author: Of course. Is that why you're talking to us?

Mathematics: NO NO, THIS WAS ON PAGE 177. YOU TWO WERE BUSY. SO I CALLED NATURE TO ASK HER ADVICE. SHE DOESN'T LIVE IN THE VOID, BUT WE'RE OLD FRIENDS. AND SHE'S EXISTED LONGER THAN ANYONE I KNOW. BUT I DIDN'T QUITE UNDERSTAND HER SUGGESTION... IT WASN'T A SUGGESTION, REALLY... SHE SAID IT MIGHT BE A PROBLEM WITH

MY HOUSE.

Reader: What?

Mathematics: SOMETHING ABOUT THE FOUNDATION.[7] IT WASN'T BUILT TO HANDLE A CORPOREAL ENTITY. OR, NOT CORPOREAL...BUT AN ENTITY. I EXIST NOW! EVERYTHING WAS FINE BEFORE. BUT CREATION IMPLIES EXISTENCE! BY DEFINITION. AND IT'S ONLY GETTING WORSE.

Author: What are you gonna do?

Mathematics: WELL, NATURE SAID SHE HAS AN OLD FRIEND FROM THE VOID WHO'S SOME KIND OF EXPERT ON FOUNDATIONS, SO SHE GAVE HIM A CALL FOR ME.

Reader: When's he getting here?

Mathematics: I DON'T KNOW. THAT'S NOT UP TO ME. SPEAKING OF WHICH...MIND IF I ASK YOU TWO A FAVOR?

Reader: What is it?

Author: Anything.

Mathematics: DO YOU MIND BEING...AROUND? WHEN HE ARRIVES? ALL THIS IS NEW FOR ME. EXISTENCE. IT WOULD BE NICE TO HAVE SOME FRIENDS THERE.

Reader: Glad to be there.

Author: Of course. We promise.

Mathematics: WHAT'S A PROMISE?

Author: It's a human thing. It's where you say you will or won't do something.

Mathematics: AND?

Author: And then the other person is supposed to believe that you will...or won't.

Mathematics: HOW IS "I WILL DO X" DIFFERENT FROM "I PROMISE I WILL DO X"?

Author: The second one is more serious. The person's supposed to believe you more.

Mathematics: I DON'T SEE HOW THAT PROVES ANYTHING.

Author: Okay, then how would you suggest we convince you? That we'll be there, I mean.

Mathematics: WELL, THE COMMON PRACTICE IN THE VOID TO START BY DEFINING A FORMAL LANGUAGE...

(Mathematics defines a formal language.)

Mathematics: NOW YOU JUST SAY, IN THE FORMAL LANGUAGE: "AXIOM: DEAR MATHEMATICS, SUPPOSE THAT I WILL BE AT YOUR HOUSE TO OFFER MORAL SUPPORT WHEN THIS STRANGER ARRIVES." WE WILL CALL THIS AXIOM S. THE S STANDS FOR STRANGER...OR SUPPORT...OR SUPPOSE...I HAVEN'T DECIDED YET.

---

7. **Mathematics:** OR THE FOUNDATIONS, I ALWAYS MIX THOSE UP. PLURALS ARE TRICKY.

(*Author and Reader repeat the above incantation.*)

Author: I don't see how this is any better than a promise.

Mathematics: OF COURSE IT IS! NOW IF YOU DECIDE NOT TO SHOW UP, THEN YOU'VE VIOLATED AXIOM S BY PERFORMING ITS NEGATION, $\neg S$, SO YOUR FORMAL LANGUAGE IS INCONSISTENT. BEING INCONSISTENT, THE LANGUAGE CAN THEN PROVE ANYTHING, INCLUDING THE FACT THAT YOU'RE A FILTHY LIAR... AND A [*NEGATIVE-ADJECTIVE*] [*NEGATIVE-NOUN*], FOR ALL NEGATIVE ADJECTIVES AND NEGATIVE NOUNS... AND ALL THE EXPLETIVES, YOU'RE ALL THOSE TOO. IT'S MUCH BETTER THAN A PROMISE.

Author: ...

Reader: ...

(*Narrator*[8] *was tempted to point out that the above joke (or whatever that was) should have used the term "formal theory" in place of "formal language," to guard against the possibility that anyone out there might be feeling pedantic. On the other hand, Narrator thought, the vernacular sense of the term "theory" comes with connotations that are likely to mislead a large proportion of readers (not to be confused with Readers; there is only one Reader). Finally, having weighed the pros and cons within the privacy of his own mind, he thought it best to simply remain silent and allow the conversation to proceed uninterrupted.*)

Reader: ...

Author: ...

Mathematics: ... BUT ENOUGH ABOUT ME. HOW HAVE YOU TWO BEEN?

Nostalgia for Plus-Times Machines

Author: Pretty good.

Reader: A bit nostalgic for the old days.

Mathematics: HOW SO?

Reader: Well, back when everything in our universe was a plus-times machine, things were so much easier. Those machines with the new generalized powers weren't so bad once we figured out how to deal with them, but in the last chapter we ran into these strange machines V and H that we couldn't even write down a description of!

Author: They're what the textbooks call "sine" and "cosine."

Mathematics: WHY ARE YOU TELLING ME THAT? I HAVEN'T READ THE TEXTBOOKS.

Author: Right. Sorry.

Reader: Anyways, we basically defined these two machines visually — using pictures. They were originally the names we gave to the unknown solutions of a problem we were trying to solve. But eventually we got stuck, so we just used the Molière strategy and pretended the names themselves were the answers.

---

8. (*As he preferred to be called at this point in the book. But I digress...*)

Mathematics: THAT TRICK IS GREAT, ISN'T IT? HAVE YOU TWO FIGURED OUT WHAT ♯ IS YET?

Author: Nope, but the textbooks call it π.

Mathematics: AGAIN WITH THE TEXTBOOKS! WHO ARE YOU TALKING TO?

Author: Right. Sorry again.

Mathematics: ANYWAYS, THESE MACHINES V AND H ARE TROUBLING YOU?

Reader: Yeah. Somehow we managed to figure out their derivatives. We know $V' = H$ and $H' = -V$, but we still can't even write down a description of the machines themselves. It's unsettling. I miss the old days. When everything was a plus-times machine, we could write down an actual *description* of anything in our universe. But since Chapter 3 ended, everything's changed. I feel like I don't really understand things as much anymore.

Mathematics: THAT'S TERRIBLE. ANY WAY I CAN HELP?

Reader: I don't know. It would be nice if we could describe everything again. Like back when everything was a plus-times machine.

Mathematics: MAYBE EVERYTHING STILL IS. . .

Reader: No, you're just trying to cheer me up.

Mathematics: NO REALLY. THE IDEA DOESN'T SEEM TOO OFF THE WALL. ARE YOU *SURE* EVERYTHING ISN'T A PLUS-TIMES MACHINE?

Reader: I guess we're not really sure. . .

Mathematics: WELL, IT'S WORTH A TRY. ESPECIALLY IF THIS IS BOTHERING YOU TWO SO MUCH.

Reader: What do you mean "worth a try"?

Mathematics: LET'S JUST FORCE EVERYTHING TO BE HOW WE WANT IT TO BE. . . AND SEE WHAT HAPPENS.

Author: That's crazy.

Mathematics: I KNOW! BUT LET'S TRY IT. LET'S IMAGINE WE'VE GOT SOME MACHINE. WE'LL REMAIN AGNOSTIC ABOUT WHICH ONE IT IS. LET'S JUST FORCE IT TO BE A PLUS-TIMES MACHINE, LIKE THIS:

$$M(x) \overset{\text{Force}}{=} \#_0 + \#_1 x + \#_2 x^2 + \#_3 x^3 + \cdots \text{ (AND SO ON)} \qquad (4.4)$$

Reader: How high does the sum go?

Mathematics: I DON'T KNOW. FOREVER.

Author: We defined plus-times machines to be finite sums. They couldn't have infinitely many pieces.

Mathematics: WHY?

Author: I don't know. The idea of an infinite description sort of. . . scares me.

Mathematics: NO NO, I'M NOT TALKING ABOUT AN INFINITE DESCRIPTION. BUT WE SHOULD BE ALLOWED TO SAY "AND SO ON." WE'VE BEEN DOING IT ALL ALONG.

Author: What are you talking about?

Mathematics: LIKE THE WHOLE NUMBERS! THERE'S AN INFINITE NUMBER OF THEM. BUT IT DOESN'T TAKE AN INFINITE AMOUNT OF TIME TO TALK ABOUT THEM. WE JUST SAY:

$$0, 1, 2, \ldots \text{ (AND SO ON)}$$

Reader: Huh. . .

Author: I've never thought of it like that before.

Mathematics: CAN I CONTINUE?

Author: By all means.

Mathematics: SO WE'VE FORCED OUR ARBITRARY MACHINE TO LOOK HOW WE WANT IT TO. NOW WE'VE JUST GOT TO FIGURE OUT ALL THE NUMBERS $\#_i$ WE USED TO DESCRIBE IT.

Author: How could we possibly do that? We know exactly nothing about this machine M.

Mathematics: OH. . . YES I SUPPOSE THE HOPE WAS A BIT MISGUIDED. . .

Reader: Well, we know what $\#_0$ is.

Author: Come again?

Reader: Up above, in the description Mathematics wrote. If we're forcing that to be true, then

$$M(0) = \#_0$$

Just plug in 0, and it kills all the pieces except the first one.

Author: Oh. . .

Mathematics: INTERESTING. . .

Reader: If we plug in 1, then maybe. . . oh, nevermind. I'm not sure we can figure out the other numbers.

Author: I liked your idea though. What if we take the derivative of M?

Reader: Why?

Author: I don't know, but taking the derivative will knock each power down by one, so maybe we can use your trick to find $\#_1$. Here, let me try. If we take the derivative of what Mathematics wrote, we get

$$M'(x) = 0 + \#_1 + 2\#_2 x + 3\#_3 x^2 + \cdots + n\#_n x^{n-1} + \cdots \text{ (and so on)}$$

and then the same trick should work again. Just feed it zero and we get

$$M'(0) = \#_1$$

We can do it again too. Just take the derivative of the original description twice:

$$M''(x) = 0 + 0 + 2\#_2 + (3)(2)\#_3 x^1 + (4)(3)\#_4 x^2 + \cdots + (n)(n-1)\#_n x^{n-2} + \cdots$$

and then feed it zero:

$$M''(0) = 2\#_2$$

But we want to know what $\#_2$ is, so we can isolate that to get

$$\frac{M''(0)}{2} = \#_2$$

Reader: Wait, does this mean we can describe anything in our universe again?
Mathematics: PERHAPS. I MEAN, WE BEGAN BY BEING AGNOSTIC ABOUT PRECISELY WHICH MACHINE WE WERE DESCRIBING, SO IN A SENSE OUR DESCRIPTION DESCRIBES AN ARBITRARY MACHINE... BUT TO DESCRIBE AN ARBITRARY MACHINE COMPLETELY, WE WOULD HAVE TO FIGURE OUT ALL OF THE NUMBERS IN ITS DESCRIPTION.
Reader: How would we do that?
Mathematics: IN A SENSE, WE ALREADY HAVE.
Author: How?
Mathematics: JUST FIGURE OUT $\#_n$ WHILE REMAINING AGNOSTIC ABOUT n. THE SAME ARGUMENT SHOULD WORK. IF WE TAKE THE DERIVATIVE OF OUR ORIGINAL DESCRIPTION n TIMES, THEN WE'D GET RID OF ALL THE PIECES TO THE LEFT OF $\#_n x^n$, SINCE EACH PIECE $\#_k x^k$ CAN ONLY SURVIVE k DERIVATIVES. THE FIRST DERIVATIVE KILLS THE $\#_0$ PIECE, THE SECOND DERIVATIVE KILLS THE $\#_1 x$ PIECE, AND SO ON. THE nth DERIVATIVE KILLS THE $\#_{n-1} x^{n-1}$ PIECE, SO AFTER n DERIVATIVES, THE FIRST SURVIVOR IS $\#_n x^n$.
Author: And the rest of the surviving pieces will have at least one x attached, so they'll go away when we plug in zero.
Reader: We're getting ahead of ourselves. What does the machine $m(x) \equiv x^n$ turn into when we differentiate it n times?
Mathematics: OH. I SUPPOSE WE DON'T KNOW... IF WE DIFFERENTIATE IT ONCE, THEN IT'S

$$m'(x) = nx^{n-1}$$

Author: If we differentiate that again, then it's

$$m''(x) = (n)(n-1)x^{n-2}$$

Reader: If we differentiate that one more time, then it's

$$m'''(x) = (n)(n-1)(n-2)x^{n-3}$$

Author: I think I see the pattern, but we need a new abbreviation. Let's just write the nth derivative as $m^{(n)}(x)$. I put it in parentheses because it's not a power, but I didn't want to write a bunch of primes with "$\cdots$" in the middle. So taking n derivatives of x^n will give us

$$m^{(n)}(x) = (n)(n-1)(n-2)\cdots(3)(2)(1)x^{n-n}$$

and the x^{n-n} piece is just 1. So I guess the nth derivative of x^n is just whatever number you get from multiplying all the whole numbers from n down to 1.

Mathematics: N!

Author: You want to call it $n!$ ...?

Mathematics: No no, I meant to say "nice!" to compliment your reasoning, but for some reason when I got to the last three letters of the word...I just froze...

Reader: That's a horrible joke.

Author: I know. But let's call it $n!$ anyway. It seems like a good abbreviation. That is:

$$n! \equiv (n)(n-1)(n-2)\cdots(3)(2)(1)$$

For example

$$1! = 1$$

$$2! = (2)(1) = 2$$

$$3! = (3)(2)(1) = 6$$

$$4! = (4)(3)(2)(1) = 24$$

and so on.

Reader: Looks good! Then we can combine two things we just said to write the nth derivative of $m(x) \equiv x^n$ as

$$m^{(n)}(x) = n!$$

This is getting a bit confusing. Let me write down everything we've done so I can make sure we know what we're doing...

(Reader looks back at what we've done.)

We started off by hoping we could describe any machine like this:

$$M(x) \overset{\text{Force}}{=} \#_0 + \#_1 x + \#_2 x^2 + \#_3 x^3 + \cdots \text{ (and so on)}$$

If we differentiate this description n times, then all the pieces to the left of $\#_n x^n$ will go away, the piece $\#_n x^n$ will turn into $\#_n n!$, and all the pieces to the right of $\#_n x^n$ will still have at least one x attached, so they'll get killed-off when we plug in zero. After all the dust settles, we'll get:

$$M^{(n)}(0) = \#_n n!$$

But since the point of all this was to figure out the numbers $\#_n$, what we really want is to rewrite it like this:

$$\#_n = \frac{M^{(n)}(0)}{n!}$$

Author: Wow... Did we really just show what I think we showed?

Mathematics: I THINK SO! UNLESS WE MADE A MISTAKE, IT LOOKS LIKE WE JUST SHOWED THAT ANY MACHINE M CAN BE WRITTEN AS A PLUS-TIMES MACHINE LIKE THIS:

$$M(x) = M(0) + \left(\frac{M'(0)}{1!}\right)x + \left(\frac{M''(0)}{2!}\right)x^2 + \left(\frac{M'''(0)}{3!}\right)x^3 + \cdots$$

THAT'S AN UNPLEASANTLY LARGE DESCRIPTION. I'LL REWRITE IT LIKE THIS:

$$M(x) = \sum_{n=0}^{\infty} \frac{M^{(n)}(0)}{n!}x^n \tag{4.5}$$

Author: Wait, that abbreviation has weird stuff in it. We never said what 0! is. Same for the zeroth derivative of something.

Reader: No no, Mathematics was just abbreviating the line above it. So we can just define 0! and the zeroth derivatives to be whatever they have to be to make the two sentences equal.

Author: How so?

Reader: Well if the two sentences are equal, then we have to define $M^{(0)} \equiv M$. So the zeroth derivative of a machine is just the machine itself. And we have to define $0! \equiv 1$. Make sense?

Mathematics: WORKS FOR ME.

Author: Okay, I'm fine with the abbreviations, but still... I'm a bit skeptical about all this. Sure, we wrote this down, but there's no way this is going to *work*. It's too good to be true.

Mathematics: DON'T YOU TWO HAVE THOSE MACHINES THAT WERE BOTH-ERING YOU?

Reader: You mean V and H?

Mathematics: WHY NOT TEST THIS NOSTALGIA DEVICE ON THEM?

Reader: Which device?

Mathematics: EQUATION 4.5. WE BUILT IT BECAUSE YOU WERE FEELING NOSTALGIC FOR THE OLD DAYS. WHEN WE KNEW HOW TO DESCRIBE EV-ERYTHING.

Reader: Oh, right.

Mathematics: LET'S TEST IT ON V AND H. HOW DID YOU TWO DEFINE THEM?

Reader: Well, if we have a straight line of length 1 tilted at angle α, we defined $V(\alpha)$ to be "however much of the line is in the vertical direction," and

$H(\alpha)$ was "however much of the line is in the horizontal direction." So for example, if $\alpha = 0$, then we've got a horizontal line, so $H(0) = 1$, and $V(0) = 0$. Aside from that and a few other specific examples, we have no idea how to get numbers for V and H if someone just hands us a random angle.

Mathematics: WHAT'S WITH THE α?

Reader: It reminded us of the word "angle."

Mathematics: OH. I WASN'T THERE WHEN YOU DID THAT. MIND IF I JUST USE x?

Reader: Go ahead.

Mathematics: ALRIGHT, WELL OUR NOSTALGIA DEVICE TELLS US THAT

$$V(x) = \sum_{n=0}^{\infty} \frac{V^{(n)}(0)}{n!} x^n$$

SO WE JUST NEED TO FIGURE OUT WHAT $V^{(n)}(0)$ IS FOR ALL n. HMM... I GUESS THIS IDEA ISN'T SO USEFUL AFTER ALL. HOW ARE WE GOING TO FIGURE OUT ALL OF V'S DERIVATIVES AT ZERO IF WE DON'T EVEN KNOW HOW TO DESCRIBE THE MACHINE ITSELF? I'M SORR—

Author: Wait! We kind of *did* figure out all of V's derivatives. I mean, we know $V' = H$ and $H' = -V$, so we know $V'' = H' = -V$, and then we can just keep looping around, like this:

$$
\begin{aligned}
V^{(0)} &= V \\
V^{(1)} &= H \\
V^{(2)} &= -V \\
V^{(3)} &= -H \\
V^{(4)} &= V
\end{aligned}
$$

And moving from each step to the next is easy, because we can always just use $V' = H$ or $H' = -V$ to get from one derivative to the next. Even better, we know V and H at zero, so we can write

$$
\begin{aligned}
V^{(0)}(0) &= V(0) &= 0 \\
V^{(1)}(0) &= H(0) &= 1 \\
V^{(2)}(0) &= -V(0) &= 0 \\
V^{(3)}(0) &= -H(0) &= -1 \\
V^{(4)}(0) &= V(0) &= 0
\end{aligned}
$$

We get back to where we started after four derivatives, so things just keep looping around forever!

Mathematics: WELL, I SUPPOSE I CAN PICK UP WHERE I LEFT OFF. BEFORE WE GOT STUCK, I HAD JUST WRITTEN:

$$V(x) = \sum_{n=0}^{\infty} \frac{V^{(n)}(0)}{n!} x^n$$

NOW, LOOKING AT WHAT YOU WROTE, IT'S CLEAR THAT ALL THE EVEN DERIVATIVES WILL BE ZERO, AND THE ODD DERIVATIVES JUST KEEP SWITCHING BACK AND FORTH BETWEEN 1 AND −1. SO IF I'M NOT MISTAKEN, THAT MEANS WE CAN WRITE:

$$V(x) = x - \frac{x^3}{3!} + \frac{x^5}{5!} - \frac{x^7}{7!} + \cdots \text{ (AND SO ON)}$$

Author: Oh...
Reader: I think that's the description of the machine V that we gave up trying to find earlier! I bet we could use the same type of idea to find a description of H.
Mathematics: GIVE IT A SHOT!

(Reader plays around for a bit.)

Author: Hey! While Reader is working on H, I think I have an idea. I wonder if we could get a good *approximate* description of a machine by throwing away some of the terms in the infinite sum. Here, I'll try with V. Let's throw away everything except the first two terms in the description of V, and compare it to the graph of V we built in Chapter 4.

(Author draws Figures 4.16 and 4.17.)

Reader: I'm back! And I think it worked. I got that the machine H can be written like this:

$$H(x) = 1 - \frac{x^2}{2!} + \frac{x^4}{4!} - \frac{x^8}{8!} + \cdots \text{ (and so on)}$$

Author: Wait, V and H were what the textbooks call "sine" and "cosine." So we just figured out how to calculate $\sin(x)$ and $\cos(x)$ for any x without memorizing anything! So I guess now, for the first time, after all that... we actually know trigonometry.
Mathematics: WHAT'S TRIGONOMETRY?
Author: Nevermind.
Reader: Anyways, thanks for all the help, Mathematics.
Mathematics: ANY TIME. AND THANK YOU FOR STICKING AROUND. IT'S NICE TO HAVE SOMEONE TO TALK TO...
Author: My thoughts exactly. See you soon. Alright, time for the next chapter.

(Ahem.[9])

---

9. (**Narrator:** *Perhaps surprisingly (given what this "Author" fellow has told you thus far),*

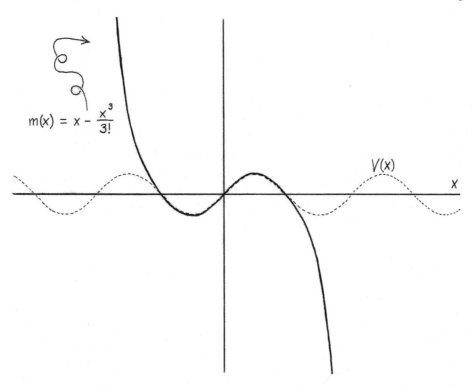

$$m(x) = x - \frac{x^3}{3!}$$

$V(x)$

x

Figure 4.16: The dotted squiggle in the figure is $V(x)$, also known as "sine" in the textbooks. The solid curve is $m(x) \equiv x - \frac{x^3}{6}$, which is the machine we get from throwing away all but two terms in the Nostalgia Device's description of V. As such, not only does the Nostalgia Device give us a way to describe the previously indescribable machines V and H, but throwing away all but a few terms in the description of these machines gives us an approximation that is very good near zero.

---

the Nostalgia Device (called "Taylor series" in textbooks) is not guaranteed to work for any possible machine (though it works for most machines one encounters in practice). The question of when it works and when it doesn't (or "convergence of Taylor series," in textbook jargon) leads us down a deep rabbit hole into some extremely rich and interesting mathematics. It's worth briefly mentioning what is perhaps the most surprising fact about the topic. That is, the conditions under which the Nostalgia Device works (or doesn't) cannot be understood unless we are willing to think about complex numbers. Complex numbers are numbers of the form $a + bi$, where a and b are "real numbers" (the normal type of numbers with a decimal expansion like 9 or -1.3 or $5.987654\ldots$) and where i is the square root of -1. Like so many things we've encountered thus far, this number is defined by its behavior. The number i is defined to be whatever it has to be in order to satisfy the sentence $i^2 = -1$. Long story short: It is impossible to fully understand the conditions under which the Nostalgia Device does or doesn't work without admitting that complex numbers exist, even if we're only concerned with machines whose input and output are real numbers. For a beautiful explanation of these ideas, there's no substitute for Tristan Needham's wonderfully nonstandard book Visual Complex Analysis.)

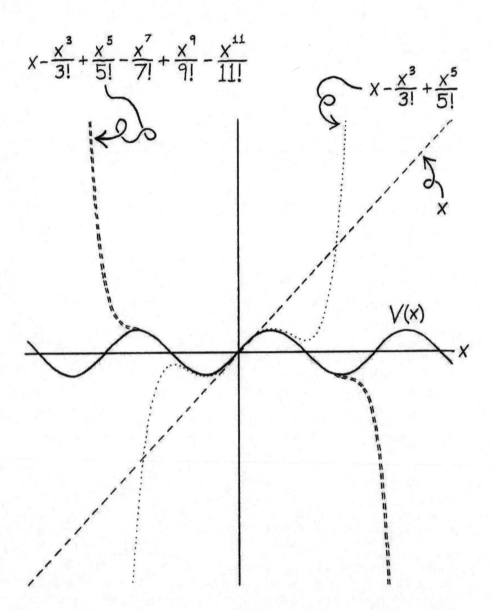

Figure 4.17: More examples of how adding more and more of the terms in the Nostalgia Device description of $V(x)$ give us more and more accurate approximations of it, all of which eventually break down when we stray
far from $x = 0$.

5 Aesthetics and the Immovable Object

How, in fact, does one decide which things in mathematics are important and which are not? Ultimately, the criteria have to be aesthetic ones. There are other values in mathematics, such as depth, generality, and utility. But these are not so much ends in themselves. Their significance would seem to rest on the values of the other things to which they relate. The ultimate values seem simply to be aesthetic; that is, artistic values such as one has in music or painting or any other art form.
—Roger Penrose, quoted in *S. Chandrasekhar, Selected Papers Vol. 7*

But the unmoved mover, as has been said, since it remains permanently simple and unvarying and in the same state, will cause motion that is one and simple.
—Aristotle, *Physics*

5.1 Reaching into the Void

5.1.1 The Topic Without a Home

In this chapter we will encounter more evidence that the so-called "prerequisites" to calculus require a sizable amount of calculus before we can properly understand them. The focus of this chapter is a topic without a home; a topic that when taught in the standard manner possesses more sleep-inducing power than morphine; a topic whose name strikes fear and boredom into the hearts of students everywhere. This topic goes by the name "logarithms and exponentials."

The standard manner of explaining — or rather, failing to explain — these concepts obscures both their underlying aesthetic elegance and their importance in application. They are typically introduced in a confusing course called "Algebra 2" (a course containing various topics assumed to be prerequisites to calculus), though they manifestly do not belong there. It is as if every history class, unable to find an appropriate time to discuss the French Revolution, simply inserted said discussion in the middle of a section on the Roman Empire, without ever telling the students that these two events happened at quite different periods and have very little to do with each other. There are indeed many ways in which one could introduce these topics, many of them involving applications such as population growth. But however important these applications may be, they are not an honest representation of the underlying motivation for the ideas themselves. This chapter will introduce these ideas in an admittedly unusual way that attempts to clarify where they come from. As a consequence, we will also gain a better understanding of how they relate to other ideas and how they might be generalized to wilder and weirder contexts. Here we go.

5.1.2 Starting from Things We Know

Despite everything we've done, it is still not too much of a stretch to say that we only know about addition and multiplication. After all, as we've discussed before, "powers" began their lives in our universe as a meaningless abbreviation for repeated multiplication, which made sense only for positive whole-number powers. Later, we generalized this meaningless abbreviation to a genuine concept. How did we do this? Well, you'll remember that we simply declared that whenever we write powers that are not positive whole numbers, we will mean *whatever we have to mean* in order to preserve the truth of the sentence $(stuff)^a(stuff)^b = (stuff)^{a+b}$. While this equation *looks* like a statement about things called "powers" or "exponents," a more honest description would be to say that it is a statement about our own ignorance. Recall that when we were in the process of generalizing the idea of powers, we found that there were an infinite number of possible ways in which we could perform the generalization. If all we wanted was a generalization that agreed with our definition for positive whole numbers, then we *could have* defined $(stuff)^\#$ to be $\#$ copies of $(stuff)$ multiplied together whenever $\#$ is a positive whole number, and fifty-two (or anything else) whenever $\#$ is a fraction or a negative number. The reason we didn't generalize the concept of powers in this way was not that it was *illegal*, but that it was *uninteresting*. Though such a generalization would not be incorrect, we would immediately find that there was nothing to know, nothing to discover, and nothing to say. It would have been a completely sterile generalization.

In a sense, it was our ignorance of anything other than addition and multiplication that led us to define the concept of powers in the way we did. It is not an accident that the two equations $s^a s^b = s^{a+b}$ and $(s^a)^b = s^{ab}$ are only *really* talking about addition and multiplication, though it is addition and multiplication "upstairs." We made it that way because, at bottom, that's all we know. If our generalization didn't behave in a way that allowed us to understand it by doing things we know, then we would have no use for it.

When we invent a mathematical concept, we reach into the Void and pull something out that does what we want. Our goal might be (i) to describe the real world, or (ii) to invent a mathematical concept that corresponds to and generalizes an everyday concept, or (iii) to invent a mathematical concept that generalizes another mathematical concept with which we are more familiar. In each of these cases, we perform a similar dance. We always tailor our idea to our goal and force it to behave like the thing we are attempting to describe. We never reach into the Void blindly.

5.2 The Four Species

The Void contains an infinite swarm of things waiting to be invented, but the vast majority of them are sterile and uninteresting. Ludwig Wittgenstein said, "Whereof one cannot speak, thereof one must be silent," and this is certainly true of mathematics. These ideas suggest that defining mathematical objects by how they behave is a surefire way to avoid the sterile parts of the Void. If we define an object by how it behaves, then we will always know what we can say about it, although we may not know what it "is." In this section we will see several particularly striking examples of that principle. Given the central role played by addition and multiplication on our journey thus far, let's define four species of machines purely by saying how they *behave* with respect to these two operations. At this point, the motivation for playing with these machines is purely aesthetic.

The Four Species:

The AA species is defined to be all machines that turn
Addition into Addition:

$$f(x+y) \overset{\text{Force}}{=} f(x) + f(y)$$

The AM species is defined to be all machines that turn
Addition into Multiplication:

$$f(x+y) \overset{\text{Force}}{=} f(x)f(y)$$

The MA species is defined to be all machines that turn
Multiplication into Addition:

$$f(xy) \overset{\text{Force}}{=} f(x) + f(y)$$

The MM species is defined to be all machines that turn
Multiplication into Multiplication:

$$f(xy) \overset{\text{Force}}{=} f(x)f(y)$$

In the definition of the four species, we force each sentence to be true for *all* numbers x and y. These are four especially elegant behaviors, but at the moment we have no idea what the members of each species look like. Maybe some of the species have no members. After all, it is possible to write down a sentence that cannot possibly be true, even though its impossibility may not be obvious from a quick glance at it. Let's spend some time playing with these species, and see if we can figure out what each one looks like.

5.2.1 The AA Species

Let's imagine that we've trapped a member of the AA species. All we know is how it behaves. We have no idea what it looks like. Let's see if we can figure out. The defining property of the AA species is that it behaves like this:

$$f(x + y) \overset{\text{Force}}{=} f(x) + f(y) \tag{5.1}$$

for all numbers x and y. The numbers x and y don't refer to horizontal and vertical coordinates. They're just abbreviations for any two numbers we feed the machine. Now, if the machine really behaves like this for all numbers x and y, then it has to behave this way when x and y are both zero. That is,

$$f(0) = f(0) + f(0) = 2f(0)$$

But if $f(0)$ is whatever number is unchanged when we multiply it by 2, then $f(0)$ has to be 0. This fact may not have been obvious from our definition of the AA species, but now we can see that the sentence $f(0) = 0$ was hiding inside that definition all along. Just to be clear, we really have no idea what to do in order to figure out what this beast looks like. We're just playing around with it, using the only thing we know, equation 5.1.

What if we imagine that the number y is infinitely small? Abbreviating that number by dx, the definition of the AA species (equation 5.1) tells us that

$$f(x + dx) = f(x) + f(dx) \tag{5.2}$$

This almost looks like a derivative. We know a few things about derivatives, but we don't know much about the AA species, so let's see if making this look more like a derivative will tell us anything helpful. If we move the piece $f(x)$ over to the left side, and divide by dx, we'll get

$$\frac{f(x + dx) - f(x)}{dx} = \frac{f(dx)}{dx} \tag{5.3}$$

Nice! The left side is now just the definition of the derivative of f, so we can replace the stuff on the left with $f'(x)$, like this:

$$f'(x) = \frac{f(dx)}{dx} \tag{5.4}$$

The right side almost looks like a derivative too, but it's missing something. Or maybe it isn't... A minute ago, we figured out that all the members of the AA species had to have $f(0) = 0$. Since adding zero doesn't change anything, we can write

$$f'(x) = \frac{f(0 + dx) - f(0)}{dx} \tag{5.5}$$

It's now easier to see that the right side is just the "rise over run" of two points that are infinitely close to each other: 0 and dx. Surprisingly, making this simple expression superficially more complicated (by throwing a bunch of zeros into it) actually made it easier for us to see how simple the expression was in the first place: the right side was just the derivative of f at the point $x = 0$. Using these ideas, we can rewrite the right side of equation 5.5 as $f'(0)$, which turns it into

$$f'(x) = f'(0) \tag{5.6}$$

But this has to be true for any number x, which tells us that the steepness of f at one particular point (namely, $x = 0$) is the same as its steepness everywhere. So f has to be a straight line, and thus (using what we discovered in Chapter 1) the members of the AA species all have to look like $f(x) = cx + b$. Further, we already showed that this species's members all spit out zero when we feed them zero, so we must have $b = 0$. Putting it all together, we can write:

The AA Species

The AA species was defined to be all machines that turn addition into addition:

$$f(x + y) \overset{\text{Force}}{=} f(x) + f(y)$$

We just figured out that all members of this species have to look like

$$f(x) = cx$$

for some number c.

So, we managed to figure out what the members of this species *look like* only by using information about *how they behave*. We didn't really have any well-specified method for doing this. We just goofed around for awhile, feeding things to machines, and eventually noticed that all the members of this species had constant steepness. That told us that they were lines, which told us how to write them. Let's see if we can do something similar for the other species.

5.2.2 The AM Species

Now let's imagine we've trapped a member of the AM species. Just like before, all we know is how it behaves, not what it looks like. The AM species was defined by this behavior:

$$f(x+y) \stackrel{\text{Force}}{=} f(x)f(y) \tag{5.7}$$

Let's play around a bit. What if we set x and y to be zero? Then we get $f(0) = f(0)f(0)$. This isn't quite enough to determine what $f(0)$ is, because both the number 1 and the number 0 behave this way (that is, $0 = 0 \cdot 0$ and $1 = 1 \cdot 1$). So let's keep exploring. What if we set just one of the inputs to be zero (say, $y = 0$) and don't do anything with the other one? Then we would get

$$f(x) = f(x)f(0)$$

Much better. This sentence tells us that $f(0) = 1$. Or at least it tells us that $f(0) = 1$, *unless* $f(x)$ is the boring machine that always spits out zero, in which case the above sentence would still be true, but $f(0)$ wouldn't be 1. For the sake of argument, let's imagine we're working with a member of the AM species that doesn't always spit out zero. Then the above argument tells us that $f(0) = 1$.

Well, we don't really know what to do, but trying to make both sides look like a derivative helped us last time, so let's try that. Actually, last time we were only dealing with addition, so it was easier to make both sides look like a derivative. Maybe we can play with derivatives, but not in the same way. Remember, we're just using x and y to stand for two things we might feed the machine, not horizontal and vertical coordinates; we can change y without changing x, so $\frac{dx}{dy} = 0$, which means that $\frac{d}{dy}(x+y) = 1$. Let's differentiate with respect to y, using this fact. Using the hammer for reabbreviation (the "chain rule"), we can write

$$f'(x+y) = f(x)f'(y)$$

where the prime stands for the derivative with respect to y. Can we conclude anything interesting about the machine f from this? Well, if we set $x = 0$, we get a sentence that doesn't tell us anything, namely, $f'(y) = 1 \cdot f'(y)$. That's not very helpful. Let's try setting $y = 0$ instead. Then we get

$$f'(x) = f'(0)f(x)$$

Hey! This says that the members of the AM species are *almost* their own derivatives. They're just their own derivatives multiplied by some fixed number. This is nothing like any machine we've seen before. For example, no plus-times machine can behave this way, because the derivative knocks the power of each term down by one: $(x^n)' = nx^{n-1}$. If there is a particular member of the AM species that has $f'(0) = 1$, then it would be *exactly* its own derivative! Now, if we had a machine f that was its own derivative, then any constant multiple of that machine would be its own derivative too. That is, if f were its own derivative so that $f'(x) = f(x)$, then any machine

$m(x) \equiv cf(x)$ would satisfy $m'(x) = cf'(x) = cf(x) \equiv m(x)$, so m would be its own derivative as well. So we've got infinitely many machines that are their own derivatives — one for each number c — but only *one* of these guys is a member of the AM species. Why? Because we just saw above that AM machines possess the behavior $f(0) = 1$. So there's only one extremely special machine that is *both* its own derivative *and* a member of the AM species. We have no idea what it looks like, but let's call this particular machine E, to stand for "Extremely special." So E is the only machine that satisfies

$$E'(x) = E(x) \qquad \text{and} \qquad E(0) = 1$$

We still have no idea what the AM machines look like, but it would be nice to get a general idea, because we *know* they're unlike anything we've seen before. What else can we do? All we have to work with is the fact that these beasts turn addition into multiplication. So if n is a whole number and f is an AM machine, then we can write

$$f(n) = f(\underbrace{1 + 1 + \cdots + 1}_{n \ times})$$

$$= \underbrace{f(1)f(1)\cdots f(1)}_{n \ times}$$

$$= f(1)^n$$

Interesting... The input went upstairs and became a power. And $f(1)$ is just some number we don't know. I wonder if this will be true for any number x. Earlier, we discovered that we can always approximate any number as well as we want by using numbers of the form $\frac{n}{m}$, where n and m are whole numbers. So if we could show that this "the input goes upstairs" behavior was true for any number that looked like $\frac{n}{m}$, then we'd be pretty convinced it was true for any number x. Suppose some number x can be written as $x \equiv \frac{n}{m}$, where n and m are whole numbers. Then we can pull the same kind of trick we just pulled, but in reverse:

$$f(n) = f\left(m \cdot \frac{n}{m}\right) = f(\underbrace{\frac{n}{m} + \frac{n}{m} + \cdots + \frac{n}{m}}_{m \ times}) \overset{AM}{=} \left[f\left(\frac{n}{m}\right)\right]^m$$

where the symbol $\overset{AM}{=}$ is where we used the definition of the AM species. So the above equation says that $f(n)$ is equal to the thing on the far right. But just before this, we found that $f(n) = f(1)^n$. We've described the same thing in two ways, so let's combine these two descriptions to write

$$\left[f\left(\frac{n}{m}\right)\right]^m = f(1)^n$$

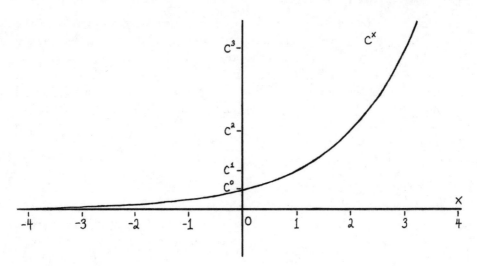

Figure 5.1: We discovered that all members of the AM species have to look like
$f(x) = c^x$, with a different member of the AM species for each different
positive number c. This is what those machines look like.

What now? Well, we'd like to isolate the $f\left(\frac{n}{m}\right)$ piece so we can get a better
idea of what these AM machines look like. We can kill the mth power on the
left by raising both sides to the $\frac{1}{m}$ power:

$$f\left(\frac{n}{m}\right) = f(1)^{\frac{n}{m}} \tag{5.8}$$

So, at this point we're fairly convinced that members of the AM species all
have the "input goes upstairs" behavior for any number x, since any number
x can be approximated as closely as we want by numbers of the form $\frac{n}{m}$. Since
$f(1)$ is just a number we don't know, we may as well just write it as c, and
then summarize what we've said by writing:

$$f(x) = c^x$$

Let's check our reasoning to make sure that machines of the form $f(x) = c^x$
are really members of the AM species. Well,

$$f(x+y) = c^{x+y} = c^x c^y = f(x)f(y)$$

Perfect! So these c^x machines really *are* all AM machines, no matter what
positive number c is, and the behavior of these machines meshes *very* nicely
with the way we defined powers! Earlier we discovered that there had to be
some extremely special AM machine $E(x)$ who was its own derivative, even
before we had any idea that the AM machines all looked like c^x. Now we
know that we get a different AM machine for each positive number c. Let's
summarize everything we've discovered about this species so far in a box.

The AM Species

The AM species was defined to be all machines that turn addition into multiplication:

$$f(x+y) \overset{\text{Force}}{=} f(x)f(y)$$

We just figured out that all members of this species have to look like

$$f(x) = c^x$$

where c is some number.

The Immovable Object

We also figured out that there has to be some extremely special AM machine that is its own derivative. We named this machine E, to stand for "Extremely special." This machine is the "immovable object" of differentiation. Combining this discovery with our new knowledge of the AM species, we can say: There must be some extremely special number e for which the AM machine

$$E(x) \equiv e^x$$

is its own derivative. We have no idea what this number is at the moment, but it has to exist.

5.2.3 The MA species

We can immediately see that the MA species is, in a sense, the "opposite" of the AM species, so maybe both species will turn out to be related to this mysterious number e. Let's see. We defined the AM species to be the set of all machines that behave like this:

$$f(x+y) \overset{\text{Force}}{=} f(x)f(y) \tag{5.9}$$

and the MA species was defined to be the set of all machines that behave like this:

$$g(xy) \overset{\text{Force}}{=} g(x) + g(y) \tag{5.10}$$

where we're using different letters to describe the machines so that we don't confuse them. Let's suppose that g is an MA machine and f is an AM machine. Then even though we have no idea what the MA machines look like, we can write

$$g(f(x+y)) \overset{f \, is \, AM}{=} g(f(x)f(y)) \overset{g \, is \, MA}{=} g(f(x)) + g(f(y))$$

But this is just saying that if we build a big machine by gluing an AM machine's output tube to an MA machine's input tube, then the whole thing is an AA machine! That is, if we define $h(x) \equiv g(f(x))$, then we just showed that the machine h behaves like this:

$$h(x + y) = h(x) + h(y)$$

So h is an AA machine. Notice that we never specifically asked for these three ideas of ours to be related in such an elegant way, but nevertheless this fact was hiding inside our definitions all along, as a necessary consequence of them. What now? Well, we can profit from the fruits of our labor with the AA species. That is, we discovered earlier that any AA machine looks like ax for some number a. Since we just discovered that h has to be an AA machine, this lets us write

$$h(x) \equiv g(f(x)) = ax$$

where a is some number. Which number? We don't know, but feeding 1 to the above equation tells us that $g(f(1)) = a$, so we could also write

$$g(f(x)) = g(f(1))x \tag{5.11}$$

Interesting... This says that if we feed any AM machine to any MA machine, then they *almost* cancel each other out. If $g(f(1))$ were equal to 1, then we would have $g(f(x)) = x$. So f and g would perform opposite actions, leaving us with exactly what we put in. We're getting closer to saying what we mean by AM and MA being "opposites."

We discovered above that the AM machines all have to look like $f(x) = c^x$, where c was just an abbreviation for $f(1)$. We get a different AM machine for each different value of c, so let's add a subscript to our abbreviations, and write $f_c(x) \equiv c^x$. We're doing this so that we can more easily talk about particular members of the AM species, and so that we don't confuse one AM machine with another. Now in these new abbreviations, equation 5.11 says that for any *particular* AM machine f_c, we can write

$$g(f_c(x)) = g(f_c(1))x = g(c^1)x = g(c)x \tag{5.12}$$

So now we can say that if $g(c) = 1$, then f_c and g cancel each other out. Just like we had a different AM machine for each number c, we've now got a different MA machine for each number c. So even though we have no idea what the MA machines look like, we can still recognize that *each member of the MA species has a partner in the AM species*: the partner who shares the same value of c.

Because of this partnership, let's now write a subscript on g to refer to a particular MA machine, just like we did with the AM machines a moment ago. That is, g_c is our abbreviation for whichever MA machine happens to satisfy $g_c(c) = 1$. For example, when $c = 2$ we obtain the AM machine $f_2(x) \equiv 2^x$, and its partner g_2 is whichever MA machine satisfies $g_2(2) = 1$. Having discovered all these things about the partnership between the two species and about the mysterious machine that happens to be its own derivative, let's write down everything we know about both species in a box, and spend a moment feeling all smug and content with ourselves.

The Partnership Between the AM and MA Species:

The AM species was defined to be all the machines that turn addition into multiplication:

$$f(x + y) \overset{\text{Force}}{=} f(x)f(y)$$

We found that any AM machine could be written as

$$f_c(x) \equiv c^x$$

The MA species was defined to be all the machines that turn multiplication into addition:

$$g(xy) \overset{\text{Force}}{=} g(x) + g(y)$$

We have no idea what these machines "look like," in that we can't describe them in terms of anything we know. However, we do know that each AM machine has a partner MA machine that undoes its actions. The MA machine g_c is defined to be whichever machine makes this sentence true:

$$g_c(f_c(x)) \equiv x \tag{5.13}$$

Even though we still don't know which number e makes the machine $E(x) \equiv e^x$ be its own derivative, we can talk about the partner of this machine by writing $g_e(f_e(x)) \equiv x$, or if we prefer, writing $g_e(e^x) = x$. We don't know anything about this machine, except that it is "special by association." We only care about it because it is the opposite of the immovable object of differentiation: the machine that's unchanged when we take its derivative.

5.2.4 The MM Species... Soon

We've still got one more species to explore: MM. However, the discussion is getting a bit abstract, and we need to make sure we're comfortable with what we've done, so we'll come back and look at the MM species later. Returning to the present, we've just discovered a bunch of interesting things about the partnership between AM and MA, as well as the special machine that manages to be its own derivative. So, while that's still fresh in our minds, let's take a break and explore our immovable object a bit more informally.

5.3 The Formal and the Informal

5.3.1 Bringing Out the Nostalgia Device

Author: Mathematics! Get in here!

. . .

Reader: Mathematics! We've got something to show you!

. . .

Reader: We could just go on without—
Author: No. Give it a minute.

(Some amount of time t passes, where t is defined to be any amount of time behaving sufficiently like "a minute" to allow the narrative to continue.)

Author: Well, I guess there's no choice. I was hoping I wouldn't have to use this yet. . .

(Author pulls a small vial from his pocket.)

Reader: What's that?
Author: A little something I picked up a few years ago.

(Author hands the vial to Reader.)

Reader: Hmm. It says, *Henkin Juice. Grade* ℵ *premium. Vintage 1949.* What does this do?
Author: It's complicated. Basically a linguistic chemical reagent that dissolves the boundary between syntax and semantics. Supposed to be used on formal languages, but maybe it'll work on informal language too. . .
Reader: . . . *What?*
Author: It gives power to names. Sprinkle some of it on the ground. And then *keep quiet!* Just for a few seconds. We don't want this stuff to misfire.

(Reader empties the contents of the vial.)

Author: *(Silently)* Mathematics Mathematics Mathematics!

(Poof!)

Mathematics: WELL HELLO YOU TWO!

Author: It's about time!

Mathematics: APOLOGIES. I SUPPOSE I LOST TRACK OF TIME. IT'S NON-TRIVIAL TO AVOID LOSING TRACK OF TIME WHEN ONE LIVES IN A TIMELESS VOID.

Reader: Okay... We've got something to show you.

Mathematics: OH YEAH? WHAT IS IT?

Reader: We were playing around, and we found out that there had to be some machine E that is its own derivative.

Mathematics: INTERESTING. WHAT DOES IT LOOK LIKE?

Author: We aren't exactly sure yet.

Reader: We called it E at first, where E stands for "*Extremely* special."

Author: Then we figured out that it has to look like "number-to-the-x."

Reader: Right. So then we started calling it e^x, where e stands for "extremely special"... again.

Mathematics: BUT YOU DON'T KNOW WHAT THIS NUMBER e IS YET?

Reader: Nope. We thought you should be around for when we tried to figure it out.

Author: Also, I think you still have the Nostalgia Device. We figured it might help.

Mathematics: OH MY. APOLOGIES AGAIN. I SUPPOSE I SIMPLY SPACED OUT. IT'S NON-TRIVIAL TO AVOID SPACING OUT WHEN ONE LIVES IN A SPACELESS—

Author: Enough! Just give it to us.

(Mathematics pulls out the Nostalgia Device.)

Author: Wow, I really missed this thing.

Mathematics: THAT'S A SIDE EFFECT. YOU MISS IT EVEN WHEN IT'S RIGHT NEXT TO YOU.

Author: That's weird.

Mathematics: TO THE CONTRARY. I THINK THAT MEANS IT'S WORKING.

Reader: Let's get moving. Here, I'll put E into the Nostalgia Device.

$$E(x) = \sum_{n=0}^{\infty} \frac{E^{(n)}(0)}{n!} x^n$$

Reader: What now?

Mathematics: YOU SAID THIS THING IS ITS OWN DERIVATIVE, RIGHT?

Reader: Right. So $E'(x) = E(x)$.

Mathematics: BUT THEN ALL OF ITS DERIVATIVES ARE JUST ITSELF, RIGHT? FOR EXAMPLE, THE SECOND DERIVATIVE IS JUST THE DERIVATIVE OF THE DERIVATIVE, WHICH IS JUST THE DERIVATIVE, WHICH IS JUST THE ORIGINAL.

Reader: Nice! So you're saying $E^{(n)}(x) = E(x)$ for all n. But then $E^{(n)}(0) = E(0)$ for all n.

Author: Hey, remember we used the Nostalgia Device on E?

Reader: (*To Mathematics*) Why is he flashing back to something that happened twenty seconds ago?

Mathematics: (*To Reader*) SIDE EFFECTS AGAIN. JUST IGNORE HIM.

Reader: Okay, where were we? Right, so $E^{(n)}(0) = E(0)$ for all n. What's $E(0)$?

Author: Oh, $E(0) \equiv e^0 = 1$. Because of the way we generalized powers in Interlude 2. Those were the days. . .

Mathematics: WOW! THAT MAKES THE OUTPUT OF THE NOSTALGIA DEVICE A LOT SIMPLER! NOW IT'S JUST

$$E(x) \;=\; \sum_{n=0}^{\infty} \frac{x^n}{n!}$$

OR TO PUT IT ANOTHER WAY,

$$E(x) \;=\; 1 + x + \frac{x^2}{2!} + \frac{x^3}{3!} + \frac{x^4}{4!} + \cdots$$

WAIT, ARE WE SURE THIS IS RIGHT? IT SEEMS TOO NICE.

Reader: I don't know. We could check to see if the thing we wrote down is its own derivative. It shouldn't be too hard, since it's just a big plus-times machine. Let's see. Using what you wrote, we've got:

$$E'(x) = \frac{d}{dx}\left(1 + x + \frac{x^2}{2!} + \frac{x^3}{3!} + \frac{x^4}{4!} + \cdots\right)$$

$$0 + 1 + \frac{2x^1}{2!} + \frac{3x^2}{3!} + \frac{4x^3}{4!} + \cdots \tag{5.14}$$

Mathematics: BUT $n!$ WAS JUST OUR ABBREVIATION FOR THIS:

$$n! \equiv (n)(n-1)\cdots(2)(1)$$

SO FOR ANY n, SOMETHING OF THE FORM $\frac{n}{n!}$ IS SIMPLY ANOTHER WAY OF WRITING $\frac{1}{(n-1)!}$, AND YOUR EQUATION 5.14 BECOMES

$$E'(x) = 1 + x + \frac{x^2}{2!} + \frac{x^3}{3!} + \cdots$$

Reader: And that's exactly the same as what the Nostalgia Device spat out for E itself, so I guess it's telling us that E is its own derivative after all. It worked! Now what?

Mathematics: I DON'T KNOW. WHAT WERE WE TRYING TO DO AGAIN?

Reader: Figure out what the number e has to be.

Mathematics: RIGHT, WELL I GUESS WE BETTER USE THE ABBREVIATION FOR E THAT INCLUDES e, LIKE THIS:

$$e^x = 1 + x + \frac{x^2}{2!} + \frac{x^3}{3!} + \cdots$$

Reader: But e^1 is just e, so if we plug in $x = 1$ we get

$$e = 1 + 1 + \frac{1}{2!} + \frac{1}{3!} + \cdots$$

or to write it another way

$$e = \sum_{n=0}^{\infty} \frac{1}{n!}$$

Hey, I think we just figured out which number e is.

Author: You did?! I missed it! I was... nevermind. Anyways, what number is it?

Reader: We don't know exactly. But it's whatever number you get from adding up the handstands of all the $n!$ numbers, starting at $n = 0$ and going on forever.

Author: Oh. What if that sum is infinity?

Mathematics: OH... I DIDN'T THINK OF THAT.

Reader: Do we have any reason to think it's infinity?

Author: How can any infinite sum be a finite number?

Reader: Well, 0.11111(forever) is a finite number, right?

Author: Sure. I mean, it's less than 0.2. Definitely seems finite.

Reader: But that number can be thought of as infinitely many things added together, like this:

$$0.11111(\text{forever})$$

$$= 0.10000(\text{forever})$$

$$+ 0.01000(\text{forever})$$

$$+ 0.00100(\text{forever})$$

$$+ 0.00010(\text{forever})$$

$$(\text{forever})$$

which I guess we could write like this:

$$0.11111\ldots = \sum_{n=1}^{\infty} \frac{1}{10^n}$$

So at least that tells us that it's *possible* to get a finite number from adding together infinitely many things, as long as the things we're adding up get smaller quickly enough — whatever that means.

Mathematics: HOW DO WE KNOW WHEN THEY'RE GETTING SMALLER QUICKLY ENOUGH?

Reader: Well in the example I made it was obvious. For this e thing, I'm not sure...

Mathematics: ARE WE STUCK?

Reader: I think so.

Author: Tell you what — let's forget about that infinite sum for now. Maybe there's a simpler way to figure out what e is.

Reader: Like what?

Author: Why not just use the definition of the derivative?

Reader: How?

Author: I don't know. We've got to "tell the math" that e^x is its own derivative somehow, so I thought we'd use the definition of the derivative to do that.

Reader: You mean like this?

$$e^x \overset{\text{Force}}{=} (e^x)' \equiv \frac{e^{x+dx} - e^x}{dx} \tag{5.15}$$

How would that help anything?

Author: Oh. I guess it doesn't.

Mathematics: WAIT, I HAVE AN IDEA THAT MIGHT HELP. WHAT IF WE DID THIS?

$$e^x \overset{(5.15)}{=} \frac{e^{x+dx} - e^x}{dx} = \frac{e^x e^{dx} - e^x}{dx} = e^x \left(\frac{e^{dx} - 1}{dx} \right) \tag{5.16}$$

Author: Hey! Then there's an e^x on the far left and on the far right. If we kill-off that piece on both sides, we'd get

$$1 = \left(\frac{e^{dx} - 1}{dx} \right)$$

Maybe then we could try to isolate the e part and find another way of writing this number that doesn't require us to add up infinitely many things.

Reader: Well, multiplying both sides by dx gives

$$dx = e^{dx} - 1$$

so I guess

$$e^{dx} = 1 + dx$$

How does this help anything?

Mathematics: HMM... IF WE RAISE BOTH SIDES TO THE $\frac{1}{dx}$ POWER, WE'D GET:

$$e = (1 + dx)^{\frac{1}{dx}} \tag{5.17}$$

Reader: That $\frac{1}{dx}$ power is weird. Can I write the same thing some other way?

Author: Sure.

Reader: Okay, the number dx is infinitely small, but we don't know what to do with infinitely small or infinitely large powers, at least not if we want to get an actual number out in the end. The number dx is tiny, so $\frac{1}{dx}$ is huge. I'll write $N \equiv \frac{1}{dx}$ as an abbreviation for a huge number. Then we can just write:

$$e \overset{???}{=} \left(1 + \frac{1}{N}\right)^N$$

Author: But the original equation was only exactly right when dx was infinitely small, so this is only exactly right when we turn N all the way up to infinity. Let's write that like this to remind ourselves:

$$e = \lim_{N \to \infty} \left(1 + \frac{1}{N}\right)^N$$

or we could just think of N as $\frac{1}{dx}$ the whole time. Same idea.

Mathematics: ARE WE DONE?

Author: I guess.

Reader: No we're not! We've written down two sentences that say what e is, but we still haven't figured out what it is!

Author: That sounds impossible.

Reader: No, I mean we still don't know which number it is *numerically*. We're not finished until we do some arithmetic and actually compute it.

Mathematics: I'D RATHER NOT.

Author: Yeah, I don't really want to do all that arithmetic. Couldn't we do something more fun, like eating a bunch of nails, or—

Reader: Are you kidding me?! We didn't do all that just to give up right before we're about to figure out what this thing is.

Mathematics: IF YOU REALLY WANT, I COULD CALL A FRIEND OF MINE TO DO IT FOR US.

Reader: Yes, please. Let's get this over with.

(*Mathematics borrows Author's phone and dials a number.*)

5.3.2 Outsourcing Numerical Tedium to a Friend of a Friend of a Friend

> *It is not the job of mathematicians... to do correct arithmetical operations. It is the job of bank accountants.*
> —Samuil Shchatunovski, quoted in George Gamow, *My World Line: An Informal Autobiography*

Mathematics: HEY THERE, A. IT'S MATH... GREAT!... LISTEN, DID YOU EVER FINISH THAT THING YOU WERE BUILDING?... PERFECT, WOULD YOU MIND STOPPING BY?... WHERE ARE YOU NOW?... REALLY?... OH, THAT'S FANTASTIC! SEE YOU SOON.

(Mathematics hangs up.)

Mathematics: MY FRIEND A.T. IS WILLING TO HELP US WITH THE ARITH-
METIC, AND HE HAPPENS TO BE IN THE NEIGHBORHOO—
A.T.: Hello, old friend.
Author: Wow! That was quick.
Mathematics: A! IT'S GREAT TO SEE YOU! HERE, LET ME INTRODUCE
YOU. THIS IS READER.
Reader: Nice to meet you.
A.T.: Likewise.
Mathematics: AND THIS IS AUTHOR.
Author: Hi! Great to meet you. Who's your friend?
A.T.: Oh! Yes, of course. Everyone, allow me to introduce you to partner.
This is Silicon Sidekick, but he prefers to be abbreviated as "Sil." He will be
helping you with your arithmetic problems.
Sil: 01000111 01110010 01100101 01100101 01110100 01101001
01101110 01100111 01110011 00100000 01001000 01110101 01101101
01100001 01101110 01110011 00000000
A.T.: No, Sil! Human language, please. Sorry, you three, this happens some-
times.

(A.T. flips some switches on Sil's panel of flippy switches.)

Sil: $2^{\ulcorner G \urcorner}3^{\ulcorner r \urcorner}5^{\ulcorner e \urcorner}7^{\ulcorner e \urcorner}11^{\ulcorner t \urcorner}13^{\ulcorner i \urcorner}17^{\ulcorner n \urcorner}19^{\ulcorner g \urcorner}23^{\ulcorner s \urcorner}29^{\ulcorner \, \urcorner}31^{\ulcorner H \urcorner}37^{\ulcorner u \urcorner}41^{\ulcorner m \urcorner}43^{\ulcorner a \urcorner}47^{\ulcorner n \urcorner}53^{\ulcorner s \urcorner}$.
A.T.: No, no, Sil. Gödel numbers are not human language. And they're
not all humans. We've talked about this before, Sil. When greeting multiple
entities, at least have the courtesy to use type promotion in order to determine
the most socially acceptable manner of categorizing them. Otherwise someone
may end up taking offense.

(Sil computes for a moment.)

Sil: Greetings, instances of C,
     where C is the least generic class
          of which both Human and classof(Mathematics) \
          are subclasses.
Reader: Hi!
Author: Hello!
Mathematics: NICE TO MEET YOU.
A.T.: Sil, these folks were wondering if you might be willing to automate
something for them.
Sil: Of course. Define your problem.
Reader: We were wondering if you could compute this for us:

$$\sum_{n=0}^{\infty} \frac{1}{n!}$$

Sil: `Stack overflow.`
Reader: What does that mean?
A.T.: Sil is efficient, but he is a finite being with a finite memory capacity. We can add as much memory tape as we want, but if you want him to give you an answer in a finite amount of time, you'll have to give him a finite job.
Author: Well those $n!$ numbers get huge pretty quickly, so I bet we could just ask for the first 100 terms or so. We just want to get an idea of what this number e is. We don't need it to infinitely many decimal places.
A.T.: Now you're talking! Sil, do your magic.
Sil: `To nine decimal places:`

$$\sum_{n=0}^{100} \frac{1}{n!} = 2.718281828$$

Reader: Nice!
Mathematics: OKAY, BUT HOW DO WE KNOW THIS WORKED? WE MIGHT HAVE MADE A MISTAKE, OR THIS THING MIGHT HAVE MADE A MISTAKE.
Author: You're just mad because it called you a human.
Mathematics: ...
Author: I suppose you're right, though. We should look at the other expression for e too. Sil, would you mind computing this for us?

$$\lim_{N \to \infty} \left(1 + \frac{1}{N}\right)^N$$

Sil: `Segmentation fault.`
A.T.: Weren't you listening? No infinite jobs.
Author: Sorry. Sil, would you mind computing the expression when N is 100?
Sil: `To nine decimal places:`

$$\left(1 + \frac{1}{N}\right)^N = 2.704813829$$

`when n = 100.`
Mathematics: I DON'T BELIEVE THAT'S THE SAME ANSWER AS LAST TIME.
Reader: Wait. The things we're giving to Sil shouldn't be exactly equal.
Author: Oh, of course. Even if we did everything right, our argument only showed that these expressions turn into e when n and N go to infinity. They shouldn't necessarily be equal if we cut them off after a finite number of terms. Maybe they're both just approaching the right answer at different speeds.
Mathematics: NOT AN UNREASONABLE THOUGHT. LET'S GO A BIT HIGHER. SIL, CAN YOU COMPUTE THE SUM WHEN n IS A BILLION?
Sil: `To nine decimal places:`

$$\sum_{n=0}^{1000000000} \frac{1}{n!} = 2.718281828$$

Mathematics: WELL THAT DIDN'T CHANGE ANYTHING. THE NUMBERS ARE DIFFERENT, THAT'S ALL THERE IS TO IT.

Reader: Hold on, we didn't retry the other expression yet. Sil, can you compute the product when N is a billion?

Sil: To nine decimal places:

$$\left(1 + \frac{1}{N}\right)^N = 2.718281827$$

when $N = 1,000,000,000$

Reader: There we go. Same except for the final spot.

Author: Nice! I guess the second expression for e is just approaching the right answer more slowly as N gets bigger. This is great! Let's write it in a box to make it official.

Summary of Our Adventures with the Immovable Object

We discovered that there is some specific number e for which the machine

$$E(x) \equiv e^x$$

is its own derivative. All multiples of this machine are also their own derivatives, but this is the only such machine that is also a member of the AM species. E is the only machine that is *both* its own derivative *and* turns addition into multiplication. That is, for all x and y,

$$E(x + y) = E(x)E(y)$$

We then used the Nostalgia Device to compute this number e, and found

$$e = \sum_{n=0}^{\infty} \frac{1}{n!}$$

where $n! \equiv (n)(n-1)\cdots(2)(1)$. Not wanting to put complete trust in the Nostalgia Device, we used the definition of the derivative to compute e another way, and found

$$e = \lim_{N \to \infty} \left(1 + \frac{1}{N}\right)^N$$

A.T.'s partner Sil then helped us compute specific numbers for these expressions, for very large but finite values of n and N. Thanks to their help, we found that

$$e \approx 2.718281828$$

5.3.3 Thanks for Everything

Mathematics: THANKS SO MUCH FOR THE HELP, A. DOING ALL THAT
ARITHMETIC BY HAND WOULD HAVE BEEN HORRIBLY TEDIOUS.
A.T.: Please, it was nothing. And don't thank me, thank Sil.
Mathematics: DEEPEST THANKS, SIL.
Sil: My pleasure.
Author: Hey A.T., have you and Mathematics been friends a long time?
A.T.: Certainly. We've been friends since the foundational stages.
Author: Then why are you calling yourself "A"? What's with the initials?
Mathematics: (*To Author*) HE'S A VERY PRIVATE PERSON.
Author: Come on, you don't need to hide. You're among friends. Full
disclosure, right? What does A.T. stand for? Automated Teller?
A.T.: No.
Author: Andrew Tanenbaum?
A.T.: Closer! But no.
Author: ... Achilles and the Tortoise?

(*A.T. throws his hands up sarcastically.*)

A.T.: You got me.
Author: What? Really?
A.T.: Hah, no. My name is Al.
Al T.: See?
Mathematics: (*Chuckling*) I'VE MISSED YOU, AL. HOW'VE YOU BEEN?
Al T.: Better, now. I had been lonely for quite awhile, but I had failed
to properly categorize the problem until recently, as a result of a basic error
in my premises for axiomatic introspection: that loneliness is a function of
solitude. It isn't... This premise delayed my pursuit of a solution for quite
some time, because being around people only made the feeling worse. They
never understood me; I understood them even less. With machines I've always
felt I could be myself, but it's hard to talk to them. Most machines, anyway.
Things have been a lot better since Sil came along. He's the apple of m—
Author: I know who you are! You're him! The IEKYF ROMSI ADXUO
KVKZC GUBJ!!!
Al T.: (*Wide-eyed*) How did you know?!
Author: Oh, come on. Could it *be* more obvious?
Reader: What are you two talking about?
Author: Nothing. You know, Al, these days they say brains and behavior
can't be understood without understanding computation.
Al T.: Who says that?
Author: People who study that sort of thing for a living.
Al T.: Wow... people are really coming around after all...
Author: Sorry you couldn't be here to see it. We would have loved to have
you around.

Al T.: (*Awkwardly*) ...I had better get going. It was wonderful to meet you all.
Author: What?...Already?...
Mathematics: GOODBYE, OLD FRIEND.
Reader: Nice meeting you!

(*Al and Sil begin walking away.*)

Author: (*To Al*) Thanks for everything. All of it.
Al T.: (*Confused*) You're...welcome.

(*Al and Sil continue walking away.*)

Author: (*To Al, in the distance*) Oh! Al! The British say they're *really* sorry.
Al T.: It's a bit late for that now...but tell them I appreciate the sentiment. Call us if you ever need anything automated, numerical or otherwise.
Author: Will do. Hope we see you again...

5.4 The Zoology of the MA Species

Okay...Anyways...Before the above exploration of the immovable object e^x, we spent some time discussing the "four species." In that discussion, we found that each member of the AM species could be written as $f_c(x) \equiv c^x$, and we later found that each such member of the AM species has a partner in the MA species that "undoes" it. That is, for any positive number c, we get two machines. First, there is a member of the AM species $f_c(x) \equiv c^x$. Second, we get a member of the MA species g_c whose defining property is that it behaves like this:

$$g_c(f_c(x)) = x \qquad \text{for all } x \qquad\qquad (5.18)$$

$$f_c(g_c(x)) = x \qquad \text{for all } x$$

As you may or may not have realized, the members of the MA species have a name in the standard textbooks: they are called "logarithms." In those textbooks, we would see $\log_c(x)$ written instead of what we have called $g_c(x)$, but they both refer to the exact same thing.

What we call them isn't important. What *is* important is that, in yet another instance of backward pedagogy, we're typically introduced to these topics before we're exposed to calculus. It has somehow been decided that long before students are taught about calculus, they should learn about the number e, the machine e^x, and logarithms. To top it all off, during this baffling discussion of logarithms, a particular logarithm is presented as "special" for some mysterious and unexplained reason. It goes by the name $\ln(x)$, and it is called the "natural logarithm," or "log base e." Students are then taught a series of incomprehensible incantations about these objects, in the hope that it might somehow prepare them for calculus. Unsurprisingly, this causes widespread confusion about just what these odd logarithm things are, what this number e is, or why mixing the former concept with the latter to obtain the function $\ln(x)$ is anything other than alchemy. Indeed, many graduate students and professors in non-mathematical fields will readily admit that they still have no idea what a logarithm is, despite the fact that they were once taught about them.

The source of this confusion is *not* the concepts themselves. As we've discovered in this chapter so far, not only did we need calculus in order to compute the number e, but the only reason we cared about this number in the first place was because of its relationship with the derivative! That is, we only care about the number e because the machine e^x happens to be its own derivative, and we only care about $\ln(x)$, or $\log_e(x)$, because it undoes the actions of e^x. As such, if we don't already know about derivatives, we have exactly no reason to care about e, or e^x, or $\ln(x)$, or any of their properties. "No reason to care," I would wager, is precisely how the majority of students feel when they are taught about these topics in high school. Can you blame them?

Having seen the conceptual origin of these odd things called "logarithms," we still don't know much about them, and we still don't feel very comfortable with them. Although we managed to figure out what the members of the AM species look like (i.e., they are all number-to-the-x machines), we still have no idea what the members of the MA species look like, in the sense that we can't describe them in terms of anything we know. This is a side effect of having defined the four species not by *what they are* but by *how they behave*. Whenever we define a mathematical object only by saying how it behaves, we are not allowed to be surprised if it is not immediately apparent what the object *is*, whatever that means.

This highlights a deeper reason why the topics of this chapter can be confusing to newcomers. Despite its simplicity, the strange dance of defining an object by its behavior is extremely foreign to the way human minds typically transact with the world. For the vast majority of human evolutionary history, human brains had exactly zero experience with this odd ritual of defining objects by their behavior, and to this day, our neural machinery does not come into the world expecting to deal with objects thus defined. During any stage of human evolutionary history, up to and including the present day, virtually all of the "things" encountered by humans have fallen into one of the following categories: humans, nonhuman animals, plants and fungi, invisible disease-causing microorganisms, human-made artifacts (like axes or computers), and inanimate features of the geography. In dealing with anything from any of these categories, it is always safe to assume that there exists a large set of preexisting facts about that thing of which you are unaware. In none of these cases is the set of all facts about the object in question *discoverable from a simple principle that is postulated by the human doing the discovering*. In our natural default setting, we are simply not accustomed to thinking this way.

As such, I expect that when students first hear about "logarithms," they attribute a kind of hidden essence to these objects, and assume that there must be a world of (almost zoological) information about them that the professor has not revealed. In a sense, that's true! But the odd fact is that *all* of their properties are implicit in their definition. While that's true of any object in mathematics, it is more salient (and more intimidating) in the case of logarithms, because when students are first exposed to these objects, they are encountering this novel behavior-based style of definition *directly* for (almost) the first time: they are given a description only of how logarithms behave, not of "what they are." That is, for familiar machines such as $m(x) \equiv x^2$, the description on the right-hand side tells us about the inner workings of the machine, and how we can compute specific outputs given specific inputs. The definition of logarithms gives us no such clue about the inner workings of the machine, but only its behavior with respect to the operations of addition and multiplication: logarithms are *whatever they have to be* in order to satisfy the behavior $\log(xy) = \log(x) + \log(y)$.

We can now show that all of the other "logarithm properties" of the standard curriculum follow for free, based on how we defined the MA species. At that point, we will know enough of their properties to unravel them with the Nostalgia Device, thus giving us a simple description of them as an infinite plus-times machine. In principle, this description will enable us to compute the logarithm of any number to arbitrary accuracy by using nothing but addition and multiplication.

5.4.1 Something Else They Never Tell You

Let's first choose some better abbreviations. We've been writing $f_c(x) \equiv c^x$ for members of the AM species, and $g_c(x)$ for members of the MA species. Since the AM species are all "power machines," let's abbreviate them as $p_c(x) \equiv c^x$. Since the MA species are essentially the opposite of the power machines, let's abbreviate them as $q_c(x)$, because the letter q looks like a backward p.

Okay, because of the way we defined the MA species (i.e., as opposites of the AM species), all facts about them should come in pairs. For each fact about the simpler AM species, we should be able to deduce a fact about the more mysterious MA species. Facts about AM machines are therefore a kind of currency with which we can purchase a greater understanding of the MA machines. So where do facts about the AM species come from? Well, since all such machines look like c^x, all facts about the AM species *must* follow from the way we defined powers! Because we ourselves invented powers, everything we know about them follows from these two sentences:

$$s^{x+y} \overset{\text{Force}}{=} s^x s^y \tag{5.19}$$

and

$$(s^x)^y \overset{\text{Force}}{=} s^{xy} \tag{5.20}$$

where x, y, and s are any numbers. Now, because of the way we defined the MA species, we know that all of its members behave this way:

$$q_s(xy) = q_s(x) + q_s(y) \tag{5.21}$$

This is the "logarithm property" whose "opposite" is equation 5.19. Textbooks usually write the above sentence this way:

$$\log_b(xy) = \log_b(x) + \log_b(y) \tag{5.22}$$

Is there a corresponding sentence about the MA species that is the "opposite" of equation 5.20? Let's see. Let's try to "undo" equation 5.20 by taking the "log base stuff" of both sides, that is, wrapping both sides in q_s. We discovered earlier that MA machines undo their partner AM machines, which is to say $q_s(s^\#) = \#$ for any number $\#$. Using this idea, we can write

$$q_s\left(s^{xy}\right) \overset{(5.18)}{=} xy \tag{5.23}$$

and by whacking the left side of this with equation 5.20, we can write

$$q_s\left((s^x)^y\right) \overset{(5.20)}{=} q_s\left(s^{xy}\right) \overset{(5.18)}{=} xy \tag{5.24}$$

Now, this is *completely* true, and since it's a fact about "logarithms" that we built using equation 5.20, it sort of qualifies as the "opposite" of equation 5.20 we were looking for. Given all that, we could simply stop here. However, the above equation is kind of ugly. If we want to get a sentence that is only talking about logarithms (i.e., the MA species), then it would be nice to kill the piece that looks like s^x. How might we do this? Well, if equation 5.24 is true for *all* numbers x and y, then it has to be true when $x = q_s(z)$ for any particular number z. If we replace the x's in 5.24 with $x \equiv q_s(z)$, then this will kill the s^x piece, because by our tricky choice of abbreviations, we've made it true that $s^x \equiv s^{q_s(z)} = z$, where the second equality comes from the fact that AM machines undo their partner MA machines. Using this tricky abbreviation, we can rewrite equation 5.24 in an equivalent but different-looking way, like this:

$$q_s(z^y) = y \cdot q_s(z) \tag{5.25}$$

which says something much easier to interpret than equation 5.24, even though it's exactly the same idea in disguise. All this says is that we can "bring powers outside of logarithms." This is a fact about MA machines that we built using the fact about AM machines in equation 5.20. Yet again we see that facts about these two species come in pairs. Equation 5.25 is the "logarithm property" that textbooks usually write like this:

$$\log_b \left(x^c \right) = c \cdot \log_b(x) \tag{5.26}$$

If the equation above looks at all scary and non-obvious, remember that it's just saying the same thing as the much simpler-looking equation 5.20. We can continue to ask "why" until we hit rock bottom. "Okay," you might say, "so equation 5.26 is true because of equation 5.20, but why is equation 5.20 true?" Good question! The simple answer is that equation 5.20 is true because we *forced* it to be true when we invented the idea of powers in Interlude 2! As usual in mathematics, if we continue asking "why" for long enough, we will eventually find that the answer to any question of the form "Why is such-and-such true?" is simply "Because of some decision we made earlier."

Alright, so we've discovered a few things about these beasts. What else can we say? Well, if equation 5.21 is true for all x and y, then it has to be true when we choose to think of y as one over some other number. That is, if we think of y as $y \equiv \frac{1}{z} \equiv z^{-1}$, then because of the way we invented negative powers, we can rewrite equation 5.21 this way:

$$q_s \left(\frac{x}{z} \right) \equiv q_s \left(xz^{-1} \right) \overset{(5.21)}{=} q_s(x) + q_s \left(z^{-1} \right) \overset{(5.25)}{=} q_s(x) - q_s(z) \tag{5.27}$$

Just looking at the far left and far right, the above sentence says:

$$q_s \left(\frac{x}{z} \right) = q_s(x) - q_s(z) \tag{5.28}$$

This is the "logarithm property" that textbooks usually write like this:

$$\log_b \left(\frac{x}{y} \right) = \log_b(x) - \log_b(y) \tag{5.29}$$

Is there any relationship between different members of the MA species? Well, remember that different members of the MA species are determined by which member of the AM species c^x they "undo," which is to say that they're determined by which particular number c is. So we can rephrase the question as "Given two numbers a and b, is there any relationship between q_a and q_b?" If we write the obvious sentence $x = x$ in the scary-looking form

$$x = b^{q_b(x)} \tag{5.30}$$

and then wrap both sides in the function q_a, we get

$$q_a(x) \overset{(5.30)}{=} q_a \left(b^{q_b(x)} \right) \overset{(5.25)}{=} q_b(x) \cdot q_a(b) \tag{5.31}$$

or equivalently,

$$q_b(x) = \frac{q_a(x)}{q_a(b)} \tag{5.31}$$

This is the "logarithm property" that textbooks usually write like this:

$$\log_b(x) = \frac{\log_a(x)}{\log_a(b)} \tag{5.32}$$

This is wonderful, because it tells us that we can ignore virtually all of the members of the MA species! Why? The piece $\log_a(b)$ is just a number, independent of x, so equation 5.32 tells us that all members of the MA species are simply constant multiples of each other! Because of this, it is no longer worthwhile to continue talking about all of the members of the MA species. We can simply pick our favorite one, and then talk about that one. This amounts to picking our favorite "base." We could choose 2 or 52 or 10 or 93.785 or anything else. However, we've developed quite a fascination with the immovable object e^x, it being the only machine that is *both* its own derivative *and* a member of the AM species. As such, purely for aesthetic reasons, we will choose to talk only about the member of the MA species that is the opposite of the immovable object, namely, the machine $q_e(x)$. Our lone surviving MA machine $q_e(x)$ is what textbooks call the "natural logarithm," or $\ln(x)$. Having happily jettisoned every logarithm except the one with base e, we can continue our journey with a much lighter load.

5.4.2   Reabbreviation Hammer to the Rescue

Since there's only one MA machine left, we don't need the subscript on $q_e(x)$ anymore. From now on, let's just call it $q(x)$. Can we use what we know to differentiate the machine $q(x)$? Well, we don't know the derivative of $q(x)$, but we do know that

$$q(E(x)) \equiv q(e^x) \equiv x$$

So maybe by writing the machine $M(x) \equiv x$ in this complicated way, we can convince the mathematics to tell us the derivative of the so-called natural logarithm $q(x)$. On the one hand, the derivative (with respect to x) of the stuff in the above equation is just 1, because it's all just x. Let's use the hammer for reabbreviation to try to figure out the derivative in another way. Let's abbreviate e^x as s, where s stands for *stuff*. Then we've got $q(s) = x$, and we want to figure out the derivative of q with respect to whatever variable is inside it. We can write

$$1 = \frac{dq(s)}{dx} = \frac{dq(s)}{ds}\frac{ds}{dx}$$

Hmm...what on earth is $\frac{ds}{dx}$? Well, we defined s to be shorthand for e^x, so

$$\frac{ds}{dx} \equiv \frac{d}{dx}e^x = e^x \equiv s$$

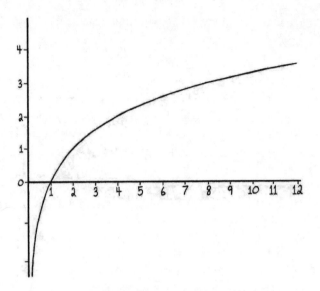

Figure 5.2: Having discovered that all logarithms are multiples of each other, if we can graph one of them, then we'll know what they all look like. Here we'll picture the "base two" logarithm, because it's easy to explain. We know $2^0 = 1$, $2^1 = 2$, $2^2 = 4$, $2^3 = 8$, $2^4 = 16$, etc. This is just another way of saying $\log_2(1) = 0$, $\log_2(2) = 1$, $\log_2(4) = 2$, $\log_2(8) = 3$, $\log_2(16) = 4$, etc., as shown in the picture.

So the combined insight of the above two equations is $1 = s\frac{dq(s)}{ds}$. Throwing s over to the other side gives

$$\frac{dq(s)}{ds} = \frac{1}{s} \tag{5.33}$$

Hey! This equation contains no mention of x, so even though we originally defined s to be shorthand for e^x, we can forget about that now, and just treat it as a meaningless abbreviation. Equation 5.33 is saying, "The derivative of q with respect to whatever is inside it is one over whatever is inside it." That is, we can replace the s with x if we want, because at this point it's just a symbol. To emphasize this freedom to reabbreviate, let's summarize what we learned from equation 5.33 in several equivalent but different-looking ways:

$$\frac{dq(x)}{dx} = \frac{1}{x} \quad \text{or} \quad \frac{d}{dx}q(x) = \frac{1}{x} \quad \text{or} \quad q'(x) = \frac{1}{x} \tag{5.34}$$

Or as the textbooks write it:

$$\frac{d}{dx}\ln(x) = \frac{1}{x} \tag{5.35}$$

Hooray!

5.4.3 Lying and Correcting Gets Us Further by Getting Us Nowhere

That was a fairly bizarre argument, and we may or may not trust the result. As often happens when we're inventing mathematics for ourselves, we've made an argument and we're not sure if it's correct. Can we derive the same result in another way? Let's try to use the definition of the derivative and see if we get the same answer. Just this once, let's use the textbooks' strange $\ln(x)$ notation, to make sure we don't get too accustomed to any particular choice of abbreviations. Here we go.

$$
\begin{aligned}
\frac{d}{dx}\ln(x) &\equiv \frac{\ln(x+dx)-\ln(x)}{dx} \\
&\overset{(5.29)}{=} \frac{\ln(1+\frac{dx}{x})}{dx} \\
&\equiv \left(\frac{1}{dx}\right)\ln\left(1+\frac{dx}{x}\right) \\
&\overset{(5.26)}{=} \ln\left(\left[1+\frac{dx}{x}\right]^{\frac{1}{dx}}\right)
\end{aligned}
\qquad (5.36)
$$

At this point we're stuck, but the place we got stuck looks strangely familiar. Recall that during the dialogue from earlier in this chapter, we discovered that

$$
e = (1+dx)^{\frac{1}{dx}} \qquad (5.37)
$$

This looks almost like the spot where we just got stuck. The last line of equation 5.36 would be exactly the number e (thanks to 5.37), were it not for that obnoxious x sitting inside. Maybe we could get rid of it. Let's try to massage the stuff at the end of equation 5.36 until it looks more like equation 5.37.

Okay, well there's nothing special about the notation dx in equation 5.37. It just refers to an infinitely small number, and the only important thing is that the two dx's in equation 5.37 are the *same* number. If dx is infinitely small, and x is not, then dx/x is also infinitely small. So if only the power in the last line of equation 5.36 were x/dx instead of $1/dx$, then we could sneakily define the abbreviation

$$
dy \equiv \frac{dx}{x} \qquad (5.38)
$$

and use this to write

$$
\ln\left([1+dy]^{\frac{1}{dy}}\right) = \ln(e) = 1
$$

However, that would be lying, and lying changes the problem. But this line of argument suggests that the familiar trick of lying and correcting might help. Let's do that! First, we lie by changing the power from $1/dx$ to x/dx. This makes the problem easier. However, we then have to correct for the lie by changing the power back to $1/dx$. This entire process would have gotten us nowhere, except *now* we can see that rewriting $1/dx$ as $x/(x \cdot dx)$ might help us get past the point where we got stuck. Having rewritten the power, we can now reabbreviate the infinitely small number dx/x as dy in equation 5.36. Summarizing:

$$\frac{1}{dx} = \frac{x}{x \cdot dx} \equiv \frac{1}{x \cdot dy} \tag{5.39}$$

Now we can use this to break through the wall we hit earlier. I'll throw numbers above equals signs to remind us where each step comes from. Starting from where we got stuck, we can write

$$\frac{d}{dx}\ln(x) \stackrel{(5.36)}{=} \ln\left(\left[1+\frac{dx}{x}\right]^{\frac{1}{dx}}\right)$$

$$\stackrel{(5.38)}{=} \ln\left([1+dy]^{\frac{1}{dx}}\right)$$

$$\stackrel{(5.39)}{=} \ln\left([1+dy]^{\frac{1}{x\cdot dy}}\right)$$

$$\stackrel{(5.20)}{=} \ln\left(\left([1+dy]^{\frac{1}{dy}}\right)^{\frac{1}{x}}\right) \tag{5.40}$$

$$\stackrel{(5.37)}{=} \ln\left(e^{\frac{1}{x}}\right)$$

$$\stackrel{(5.26)}{=} \frac{1}{x}\ln(e)$$

$$\stackrel{(5.13)}{=} \frac{1}{x}$$

Just like we hoped, we got the same answer as before. In both cases, we found that the derivative of the machine q (a.k.a. "natural log") is $1/x$. Having arrived at the same answer in two different ways, we become much more confident that this is indeed the correct answer.

5.4.4 The Nostalgia Device Simplifies Life Yet Again

Even though we've discovered that the derivative of $q(x)$ is $1/x$, we still don't know how to write down a description of $q(x)$ in terms of anything simpler. That is, we have no idea how to compute a specific number for $q(3)$ or $q(72)$. This is the same predicament we found ourselves in during Chapter 4, when we didn't know how to describe the machines V and H (what the textbooks call "sine" and "cosine") in any way except by drawing pictures. In that case, we eventually discovered that our Nostalgia Device allowed us to write V and H as plus-times machines with an infinite number of pieces, which greatly eased our worries about them. Armed with the plus-times machine expansions of V

and H, we knew that we could calculate specific numbers for them if we ever needed to, just by doing a bunch of addition and multiplication.

Maybe we could use the Nostalgia Device to get an idea of what this "natural logarithm" machine $q(x)$ looks like and how we might be able to compute specific numbers for $q(9)$ or $q(42)$ or whatever. It might not work, but it's worth a shot. Feeding q to the Nostalgia Device, it tells us that

$$q(x) = \sum_{n=0}^{\infty} \frac{q^{(n)}(0)}{n!} x^n$$

Hmm... this isn't going to work. We know that q's derivative is $1/x$ and that's going to be infinite at $x = 0$. Why does q get all weird at zero? Well, the machine q was defined by its behavior $q(e^x) = x$. But e^x is always positive: it gets smaller and smaller for large negative values of x, and only approaches zero as x goes to $-\infty$. Based on this reasoning, it should be the case that $q(0) = -\infty$, so maybe we shouldn't use the Nostalgia Device on q directly, because q goes a bit crazy at zero.

What if we tried using the Nostalgia Device on the machine $q(1+x)$ instead? This is the same machine, just shifted a little bit, and it's much more well-behaved when $x = 0$, since $q(1 + x) = q(1) = 0$. It seems like thinking about the machine this way might make things easier. Okay, can we figure out all the derivatives of this machine at $x = 0$? Well, they're just the derivatives of the original machine q at $x = 1$, so we need $q^{(n)}(1)$ for all n. Let's list a few of q's derivatives and see if we notice a pattern:

$$q'(x) = x^{-1}$$
$$q''(x) = -x^{-2}$$
$$q'''(x) = 2x^{-3}$$
$$q^{(4)}(x) = -(3)(2)x^{-4}$$
$$q^{(5)}(x) = (4)(3)(2)x^{-5}$$

Hey, this is pretty simple! We just have to keep using the familiar pattern $(x^\#)' = \#x^{\#-1}$ that we discovered in Chapter 3. The powers are negative, so the sign changes every time we take a derivative. The power on the nth derivative will be negative n, and the number out front will be $(n-1)!$, so we see the pattern. We can write everything we just said in shorthand like this:

$$q^{(n)}(x) = (-1)^{n+1}(n-1)!\, x^{-n}$$

Okay, so we've got all the derivatives of $q(x)$. What about the derivatives of $q(x+1)$? Well, using the hammer for reabbreviation, we see that the derivatives of $Q(x) \equiv q(x + 1)$ are just the derivatives of $q(x)$ with $x + 1$ plugged in, because $\frac{d}{dx}(x + 1) = 1$. Great! So we can write

$$Q^{(n)}(x) = (-1)^{n+1}(n-1)! \ (x+1)^{-n}$$

when $n \geq 1$. All we need in order to use the Nostalgia Device is $q^{(n)}(1)$, which is another way of writing $Q^{(n)}(0)$. Using the above equation, these numbers are just:

$$Q^{(n)}(0) = (-1)^{n+1}(n-1)! \tag{5.41}$$

for $n \geq 1$. When $n = 0$, we have $Q^{(n)}(0) = Q(0) \equiv q(1) = 0$. Now we can finally apply the Nostalgia Device to get:

$$q(x+1) \equiv Q(x) = \sum_{n=1}^{\infty} \frac{Q^{(n)}(0)}{n!} x^n = \sum_{n=1}^{\infty} \frac{(-1)^{n+1}(n-1)!}{n!} x^n \tag{5.42}$$

The $(n-1)!$ in the above equation 5.41 will cancel against part of the $n!$ pieces on the bottom, because $n! = n \cdot (n-1)!$. This leaves us with:

$$q(x+1) = \sum_{n=1}^{\infty} \frac{(-1)^{n+1}}{n} x^n$$

which is just an abbreviation for

$$q(x+1) = x - \frac{x^2}{2} + \frac{x^3}{3} - \frac{x^4}{4} + \cdots$$

Since $q(x)$ blows up at zero, $q(x+1)$ will blow up at -1, so it's not clear that we can trust this expression for all x. However, for now at least, it gives us a much more concrete way of thinking about the MA machines. The above equation gives us a way of thinking about the so-called "natural logarithm" q as an infinite plus-times machine. Since all of the members of the MA species are just constant multiples of q, we therefore have a way of describing "logarithms" (MA machines) without just saying "they're whatever undoes the number-to-a-power machines c^x."

5.5 The MM Species

We've got one more species to explore before the end of the chapter: MM. The MM species was defined to be the set of all machines that behave like this

$$f(xy) = f(x)f(y)$$

for all x and y. Let's see if we can figure out what they look like. Since we don't know what to do, it might help to do what worked before for the AM species: differentiating the definition of these beasts with respect to one of the variables, say y. This gives

$$xf'(xy) = f(x)f'(y)$$

where the prime stands for differentiation, thinking of y as the variable. This equation has to be true for all x and y, but that's more information than we want, so it might help to plug in specific values for x or y to see if we can reduce it to a more transparently meaningful condition. If we set $x = 1$, then we get

$$f'(y) = f(1)f'(y) \tag{5.43}$$

which I suppose tells us that $f(1) = 1$. Okay, what if we go back and set $y = 1$ instead of $x = 1$? This gives

$$xf'(x) = f(x)f'(1)$$

Since $f'(1)$ is just some number we don't know, we can rewrite this as

$$f'(x) = c\,\frac{f(x)}{x} \tag{5.44}$$

This is one of those places in mathematics where it's not at all clear what to do, and in the process of playing around with this problem, anyone would probably get stuck here for awhile. Now, no one in their right mind would likely think to do this right away, but if (for whatever silly reason) we differentiate the "natural logarithm" of $f(x)$, then it becomes a bit easier for our human minds to see what these equations are saying. Recall that we discovered above that

$$\frac{d}{dx}q(x) = \frac{1}{x}$$

where $q(x)$ is what textbooks call the "natural logarithm," or "log base e," and write as $\ln(x)$. Because we can abbreviate things however we want, the above equation says the same thing as any of the following:

$$\frac{d}{d\star}q(\star) = \frac{1}{\star} \qquad \frac{d}{ds}q(s) = \frac{1}{s} \qquad \frac{d}{df(x)}q(f(x)) = \frac{1}{f(x)}$$

And it is the last of these that turns out to help us here. Now, if we differentiate $q(f(x))$ with respect to x using the hammer for reabbreviation, we get

$$\frac{d}{dx}q(f(x)) = \underbrace{\frac{df(x)}{dx}}_{f'(x)}\ \underbrace{\frac{d}{df(x)}q(f(x))}_{1/f(x)} = \frac{f'(x)}{f(x)}$$

And if we use the fact we discovered in equation 5.44, we can turn the $f'(x)$ on the far right of the above equation into $cf(x)/x$, which gives us

$$\frac{d}{dx}q(f(x)) = \frac{cf(x)}{xf(x)} = c\frac{1}{x}$$

Notice that the $\frac{1}{x}$ on the right side of the above equation can be thought of as the derivative of the natural logarithm $q(x)$. As such, we can use this fact and the property of logarithms $c \cdot q(x) = q(x^c)$ to write

$$\frac{d}{dx}q(f(x)) = c\frac{d}{dx}q(x) = \frac{d}{dx}\Big(c \cdot q(x)\Big) = \frac{d}{dx}q(x^c) \qquad (5.45)$$

We've arrived at a statement of the form "the derivative of (one thing) equals the derivative of (another thing)." But if that's true, then it has to be the case that "(one thing) = (another thing) + (some number)." Why? Well, for the slopes of two machines to be exactly the same everywhere, the two machines have to be the same everywhere, except for a vertical shift of their graphs up or down. These general considerations tell us that we can use equation 5.45 to conclude

$$q(f(x)) = q(x^c) + A$$

where A is some number we don't know. But not knowing A is the same as not knowing its logarithm, so we can express our ignorance of it equally well by writing $q(B)$ instead of A, where B is some other number we don't know. This trick allows us to write

$$q(f(x)) = q(x^c) + q(B) = q(Bx^c)$$

And finally, feeding both sides of this equation to the opposite machine of q (namely, e^x), we obtain

$$f(x) = Bx^c$$

Plugging in $x = 1$ gives $f(x) = B$, but we discovered earlier[1] that $f(1) = 1$ for all members of the MM species, so it must be the case that $B = 1$. Putting it all together, we have discovered that all members of the MM species have the form

$$f(x) = x^c$$

where different choices of c give us different members of the MM species. Since we're already familiar with these machines, we don't need to spend any more time discussing them.

5.6 Reunion

Let's summarize what we've done in this chapter.

1. Given the importance of addition and multiplication on our journey so far, we defined four species of machines based on how they interact with these operations. Using the abbreviations A and M for addition and multiplication, respectively, the four species were defined to be any machines that (1) turn A into A, (2) turn A into M, (3) turn M into A, (4) turn M into M.

---

1. In the line immediately after equation 5.43.

2. We played around with the AA species and found that all its members looked like $f(x) \equiv cx$, where c is some fixed number.

3. We played around with the AM species and found that all its members looked like $f(x) \equiv c^x$, where c is some number. Even better, we found that one of these machines had the surprising property of being its own derivative. Combining these two facts tells us that there must be some extremely special number e for which e^x is its own derivative. All constant multiples of this machine are their own derivatives as well, but we found that e^x is the only machine that is *both* its own derivative *and* a member of the AM species.

4. We played around with the MA species, and found that its members each had a partner in the AM species. That is, for each AM machine $f_c(x) \equiv c^x$, there had to be some MA machine $g_c(x)$ that did the opposite for all x, i.e.,

$$g_c(f_c(x)) = x \qquad \text{and} \qquad f_c(g_c(x)) = x$$

However, we had no idea what the members of the MA species "looked like" (i.e., we couldn't describe them in terms of anything else we knew).

5. We found two different expressions for the extremely special number e.

$$e = \sum_{n=0}^{\infty} \frac{1}{n!} \qquad\qquad e = \lim_{N \to \infty} \left(1 + \frac{1}{N}\right)^N$$

These two expressions both agreed that the number e was approximately $e = 2.71828182\ldots$

6. We then returned to playing with the MA species ("logarithms," in textbook jargon). We chose to write these machines as $q_c(x)$, because they act like the "power machines" $p_c(x) \equiv c^x$ in reverse, and a reversed p looks kind of like a q.

7. We used what we had discovered about the MA species to derive the various "logarithm properties" that we hear about in mathematics courses. Along the way, we found that all of the MA machines were just constant multiples of each other, so we wouldn't lose anything by simply picking our favorite and ignoring the rest. We chose $q_e(x)$, the opposite of our immovable object e^x. Dropping the now-unneeded subscript, we called this machine $q(x)$. This is the machine known as $\ln(x)$ in textbooks.

8. We figured out the derivative of q in two different ways. In both cases we found that:

$$q'(x) = \frac{1}{x}$$

As with the machines V and H from Chapter 4, we were able to determine the derivative of q before we learned how to write a description of it as a plus-times machine.

9. We attempted to use the Nostalgia Device on q, but we noticed that it wasn't well-behaved at zero, so we shifted it over one unit, and examined the machine $q(x+1)$. In doing so, we found

$$q(x+1) = x - \frac{x^2}{2} + \frac{x^3}{3} - \frac{x^4}{4} + \cdots$$

This gave us a way of thinking about the "natural logarithm" q as an infinite plus-times machine. However, we're not sure whether the expression will work for all x.

10. We then found that the members of the MM species all have the form $f(x) = x^c$, where c is some fixed number.

Interlude 5: Two Clouds

No More Mysteries?

In 1894, the famous physicist Albert Michelson remarked that all of the fundamental mysteries of physics appeared to have been resolved. The physics of the future, therefore, would be primarily concerned with cleaning up a few details. Michelson is quoted as saying:

> The more important fundamental laws and facts of physical science have all been discovered, and these are now so firmly established that the possibility of their ever being supplanted in consequence of new discoveries is exceedingly remote... [O]ur future discoveries must be looked for in the sixth place of decimals.

Six years later, a similar remark was made by William Thomson, more commonly known as Lord Kelvin. As he put it:

> The beauty and clearness of the dynamical theory, which asserts heat and light to be modes of motion, is at present obscured by two clouds. I. The first came into existence with the undulatory theory of light, and was dealt with by Fresnel and Dr. Thomas Young; it involved the question, how could the earth move through an elastic solid, such as essentially is the luminiferous ether? II. The second is the Maxwell-Boltzmann doctrine regarding the partition of energy.

Kelvin's two clouds turned out to be far more than technical details of minor interest; they now comprise the two fundamental pillars of our modern understanding of Nature. The first mystery was only resolved with the advent of Einstein's theory of special relativity, while the resolution of the second mystery led to the even more mysterious framework of quantum mechanics. Although these two theories can be combined in what is commonly known as quantum field theory, Einstein's theory of *general* relativity still does not "play nicely" with quantum mechanics, and one of the main unsolved mysteries in physics to this day is how to combine the two in a successful quantum theory of gravity. More than a hundred years after the remarks of Michelson and Kelvin, many mysteries remain in our understanding of the universe. A glance at history shows that the remarks of Kelvin and Michelson are simply a special case of a quite general phenomenon: a recurrent human fondness for the always misguided idea that everything has been discovered, everything has been done, everywhere has been explored. The catalog includes:

> *Everything that can be invented has been invented.*
> —Charles H. Duell, Commissioner, U.S. Patent Office, 1899[2]

> *We are probably nearing the limit of all we can know about astronomy.*
> —Simon Newcomb, early American astronomer

> *Inventions reached their limit long ago and I see no hope for further development.*
> —Julius Sextus Frontinus, prominent Roman engineer (ca 40–103 AD)

> *What has been will be again,*
> *what has been done will be done again;*
> *there is nothing new under the sun.*
> —Ecclesiastes 1:9

In every case, declarations to the effect that "everything about X is known," have turned out to be dramatically wrong. Such presump—

(Author looks up from his computer.)

Author: Hey, do either of you know how to spell "presumptuous"?
Reader: Why?
Author: For a book I'm writing.
Mathematics: WHAT'S A BOOK?
Author: Nevermind.

The Scene
The three characters are sitting in Mathematics's house,
(number ∅ Void St., in the upper west Void).
The characters are bored and attempting to kill time.
Unbeknownst to any of them, they are failing in
this effort, due to the timeless nature of the Void.
Reader is sitting in a undefined location, reading about Reader.
Author is procrastinating from work, by writing a book.
Mathematics is perusing this week's edition of

---

2. This "quote" is actually one of those persistent apocryphal misquotations. Such quotes exist in a strange state between fiction and nonfiction, and have evolved methods of self-replication without the need of a primary source. Although there must have been a first time this quote was uttered, it appears to have no connection with the U.S. Patent Office or any of its commissioners, but rather appears to have been first uttered in a spaceless, timeless Void. Needless to say, even if this speculation were true, it would be equally difficult to find a primary source.

The Entirely Blank Newspaper of the Void,
featuring a lengthy commentary on a portrait
of the sheet music for John Cage's *4′33″*, as
painted by Robert $\frac{1}{2}$(Ryman+Rauschenberg).

Author: There's nothing to do.
Mathematics: OF COURSE THERE'S NOTHING TO DO. WHERE DO YOU THINK YOU ARE?
Author: No, that's not the problem. We did everything already. Everything that can be invented has been invented.
Reader: How do you know?
Author: Well, can you think of anything we haven't done yet?
Reader: Sure. We still didn't figure out how to calculate ♮.
Author: Yeah, yeah. But that's just a technical detail. All that's left to do is calculate ♮ to more and more decimal places. What's the point?
Reader: Wait a second...

(Our characters wait for an undefined amount of time.)

Author: ...Has it been a second yet?
Reader: I don't know! We're in the Void! Time is almost mostly nonexistent here. In any case, I didn't literally mean "Wait one second." I just meant that you shouldn't be so ready to assume we've done everything.
Author: Well, can you think of anything big and important that we don't know yet?
Reader: How am I supposed to know what we don't know? If I knew that, wouldn't I know it?
Author: Not necessarily.

(Reader thinks for a moment.)

Reader: Oh, I guess you're right. Well, why didn't we calculate ♮ yet?
Author: We don't know how.
Reader: Why don't we know how?
Author: Because we couldn't figure out the areas of curvy things.
Mathematics: AND CURVY LENGTHS. WE COULDN'T DEAL WITH THOSE EITHER.
Author: Okay, I guess there are still two things we don't know. Once we figure those things out, I guess we'll know everythi—

(A bright flash of light fills the room.)

Mathematics: OH! THAT'S THE DOORBELL.
Reader: Why do you have a light for a doorbell?
Mathematics: NORMAL DOORBELLS DON'T WORK HERE. YOU SEE, THERE'S

NO AIR IN THE VOID, SO THERE'S NOTHING FOR THE SOUND TO PROP-
AGATE THROUGH. FOR AWHILE NOBODY BOTHERED TO MARKET LIGHT-
BASED DOORBELLS HERE, BECAUSE THEY ASSUMED THAT LIGHT WAS LIKE
SOUND, IN THAT IT NEEDED A MEDIUM TO PROPAGATE THROUGH — AN
"ETHER" IF YOU WILL. THIS SUBSTANCE WAS ASSUMED TO FILL UP OUTER
SPACE, BUT OF COURSE EVERYONE KNEW THAT THERE'D BE NO ETHER HERE
IN THE VOID, BECAUSE THE VOID IS THE VOID. THEN, AN UNKNOWABLE
AMOUNT OF TIME AGO, A CLEVER FELLOW REALIZED THAT THERE IS NO
SUCH MEDIUM — LIGHT CAN PASS DIRECTLY THROUGH EMPTINESS! FOR-
TUNATELY, THIS FELLOW ALSO WORKED AT THE PATENT OFFICE, SO HE
PATENTED THESE LIGHT-BASED DOORBELLS, AND IT IS NOW MUCH EASIER
FOR WE VOID DWELLERS TO TELL WHEN SOMEONE IS AT THE DOOR.

> (*At this point, the three characters have been waffling about
> doorbells for an undefined amount of time since the doorbell
> flashed. Despite Mathematics's claim that it is now much easier
> for Void dwellers to tell when someone is at the door, the three
> characters have completely forgotten about their patiently
> waiting guest.*)

Author: Alright, where were we?

> (*The doorbell flashes again.*)

Mathematics: OH! THAT'S THE DOORBELL.
Reader: Why do you have a light for a doorbell?
Mathematics: NORMAL DOORBELLS DON'T WORK HERE. YOU SEE, THERE'S
NO AIR IN THE VOID, SO...
Author: No no no, we did this already.

> (*Mathematics reads through the above discussion.*)

Mathematics: OH MY... INDEED WE DID.
Reader: Time is strange here.
Mathematics: WHO COULD POSSIBLY BE AT THE DOOR?
Author: How should I know?

> (*Author runs to the door, followed by Reader and Mathematics.*)

Intralude: Introduction to Meta and Steve

> (*The door opens to reveal a cheery bald man.
> Author's eyes widen.*)

Author: Oh my Gö—
Stephen Kleene: Hello! I'm here about the foundations. Nature called and

told me you were having some problems.

Author: (*Starstruck*) Nice to meet you! I'm Author. Let me introduce you to my friends. This is Reader, and this is Mathematics.

Mathematics: HELLO THERE.

Reader: Nice to meet you, Stephen.

Stephen Kleene: It's wonderful to meet you two as well. And call me Steve. As long as we're on the topic of introductions, I've got one of my own. It'll be a moment. It's parking the car.

> (*A deep blue van parks outside with undue precision.*
> *Austere white letters on the side of the van read:*
> **Kleene's (Foundations) & (Conceptual Cleaning)**
> **Service. Est. 1952. Kleeneliness is next to Gödeliness.**
> *Eventually something exits the van and approaches the door.*)

Stephen Kleene: Ah, fantastic. Author, Reader, Math, allow me to provide you with an introduction to my friend and business associate, Metamathematics.

Metamathematics: ...

Stephen Kleene: Please, introduce yourselves to it.

Author: Hi.

Reader: Hello.

Metamathematics: ...

Stephen Kleene: It's not very talkative, but it doesn't mean to be impolite.

Mathematics: CAN YOU TELL IT TO STOP STARING AT ME?

Stephen Kleene: (*Chuckles*) I'm afraid not, my friend.

> (*Metamathematics stares with a curious but expressionless face*
> *at Mathematics. Mathematics leers back, half mockingly, half out*
> *of uncertainty about what to do in this uncomfortable new*
> *situation.*)

Stephen Kleene: My business associate will need some time to examine the foundations. In the meantime, I was thinking the three of us could go out for some drinks.

Reader: There are bars in the Void?

Stephen Kleene: Just one. Meta and I were hired to clean it up recently, and I've grown rather fond of the place. It's called The Beweis Bar.

Author: I could go for some drinks.

Stephen Kleene: Oh dear... well... how to put this?... It's a fairly formal place, friends. You three might want to change clothes.

Mathematics: I DON'T BELIEVE I OWN ANY CLOTHES.

Stephen Kleene: Oh. Of course. Well, at least wear this.

> (*Steve hands Mathematics a small pin shaped like the letter A.*)

Mathematics: EXPLAIN, PLEASE?
Author: Just put it on so we can go. Reader and I already changed.
Mathematics: WHEN DID YOU—
Reader: The Void is timeless.
Author: Haven't we been over this?
Mathematics: . . .

> (*Mathematics λ·reluctantly affixes the pin on itself,*
> *roughly where a lapel would go if it had one,*
> *where $\lambda \in [0, 1]$.*)

Stephen Kleene: You're wearing it upside down. . .

> (*Mathematics smiles in $(1 - \lambda)$·amusement at Kleene,*
> *and rotates the pin by an angle of $\natural$,*
> *where $\natural$ is defined as in Chapter 4,*
> *and λ is the same as before.*)

Stephen Kleene: Alright. Stand back, everyone. Meta, do your thing.
Metamathematics: SUPPOSE:

$$\forall \mathbb{C} \left(\left(\text{IS\_CHARACTER}(\mathbb{C}) \wedge (\mathbb{C} \in \mathcal{BMC} \text{ pg } 248) \wedge (\mathbb{C} \neq \text{ME}) \right) \Longrightarrow \mathbb{C} \in \text{BEWEISBAR} \right)$$

SubIntralude: The Beweis Bar

> (*Author, Reader, Mathematics, and Steve suddenly find*
> *themselves seated in a small building. The menu reads*
> *"The Beweis Bar. Formal attire required. Every night is $\models \wedge \neg \vdash$ night."*)

Stephen Kleene: Fantastic. Here we are. Waitress!
Waitress: May I take your orders?
Stephen Kleene: I'll have the usual.
Waitress: Mahlo cocktails it is.
Author: I'll have the Hahn-Banach and tonic.
Waitress: Separable or Unseparable?
Author: Separable, please. Don't want to drink too much while I'm writing.
Waitress: (*To Mathematics*) And you?
Mathematics: WHAT'S THE PAN MATHEMATIC GARGLE BLASTER?
Waitress: One of our top sellers. Our spin on an old favorite.
Mathematics: I'LL HAVE ONE OF THOSE.
Reader: I can't see the menu. . . what would you recommend?
Waitress: Since you're new here, I'd start with the WKL_0 shots. They're fairly weak.
Reader: What are they?
Waitress: That's a tough question. It's not really clear what the WKL_0 shots *are*. You see, our drinks are only defined up to isomorphism — it's basically

the Void's only law.

Reader: What's an isomorphism?

Waitress: Basically just a meaningless change of abbreviations. Two things are isomorphic if they have all the same behaviors, even though they might look different. If two things are indistinguishable when you just look at how they behave, then they're really the same thing as far as the Void is concerned, and we're not allowed to treat them differently. It's Void policy.

Reader: Uhh, can you give me an example?

Waitress: Ŝûr̂ê. Ĥôŵ'ŝ t̂ĥîŝ f̂ôr̂ ân̂ êx̂âm̂p̂l̂ê?

Reader: What? How is that an example? You just said "Sure. How's this for an example?"

Waitress: Precisely! You understood what I said, even though I put hats on everything. The alphabet with hats on it is isomorphic to the alphabet I'm using now, because it's just a meaningless relabeling — a change of abbreviations, if you will. Recognizing that isomorphic things aren't really different after all makes the Void much less cluttered.

Reader: Oh, I think I get the idea. If two sentences are isomorphic, then they're saying the same thing, even though it might *look* like they're saying something different. Like, if we did arithmetic using Roman numerals, then we could write $III + IV = VII$, and things like that, but that sentence would really be saying the same thing as $3 + 4 = 7$, so the two ways of doing things would be "isomorphic"?

Waitress: Precisely! You're quick to catch on!

Reader: Really? Does it usually take longer to explain that idea? It seems pretty simple.

Waitress: Well, the explanation we usually use in the Void isn't exactly the explanation I gave you, but rather an isomorphic explanation of what an isomorphism is. Depends on the dialect, of course, but we usually say something like this: Suppose S and T are two sets, and suppose two binary operations $\circ$ and $\diamond$ are defined on S and T, respectively. Then S and T are isomorphic if and only if there exists an invertible map ϕ from S to T such that for all elements a and b in S we have $\phi(a \circ b) = \phi(a) \diamond \phi(b)$. See? Exactly the same idea!

Reader: That doesn't look like the same idea at all!

Waitress: It doesn't? I can't tell.

Reader: It made more sense when you just puts hats on the letters. So "isomorphic" things can be the same in every meaningful way, but not the same in terms of how easy they are to understand? Doesn't that imply that ease of understanding isn't meaningful?

Waitress: Precisely! And no. Respectively. I'll put in your order.

Reader: Wait, you still haven't explained this WKL$_0$ thing. I'd prefer to know what I'm drinking.

Waitress: Don't worry about what it *is*. That's not the point. It's exactly the same strength as what Author is getting, so they're the same drink as far as the Void is concerned. In summary, you newcomers should be fine.

(Waitress smiles and leaves the table.)

Reader: I can't tell whether I'm confused or not.
Stephen Kleene: That's unfortunate, my friend, but it's a common feeling at the Beweis Bar. Look on the bright side. Until you have determined whether or not you are confused, at least you can take consolation in knowing with absolute certainty that you are meta-confused.
Reader: I'm not sure that helps.
Stephen Kleene: Precisely!
Reader: ...

(Waitress returns with the drinks.)

Reader: Wow, that was quick!
Author: Are you sure?
Reader: Oh...I guess not...
Author: So, Steve! How's the business?
Stephen Kleene: Business couldn't be better. How are you three?
Author: Stuck on an invention.
Stephen Kleene: Oh! Might I be of some service? I do love a good invention.
Reader: We're trying to figure out a way to deal with the area of curvy things.
Stephen Kleene: How are you doing that?
Author: Well, we don't know how to figure it out, so we also don't know how we should try to figure it out.
Stephen Kleene: I don't believe that follows.
Reader: Steve's right.
Author: Oh...Well then how should we try to figure it out?
Stephen Kleene: How have you dealt with these "curvy things" in the past?
Reader: *(Pointing at Author)* Well, earlier he kept rambling about how the infinite magnifying glass helps with that.
Author: Yeah, but we've only used that to find curvy steepness, not curvy area.
Stephen Kleene: And how did you do that?
Reader: Well, we dealt with curvy steepness by zooming in and pretending things were straight.
Author: It felt kind of like cheating, but it worked.
Stephen Kleene: More details, please.
Reader: Well when we were first thinking about curvy steepness, we started by thinking of the curvy thing as the graph of some machine — just for convenience — and then we imagined zooming in. Once we were zoomed in, we just used our steepness definition from Chapter 1.

Stephen Kleene: I believe I have an idea.
Author: Really? What?
Stephen Kleene: See those two sentences Reader just said?
Author: Yeah. What's your idea?
Stephen Kleene: (1) Replace the word "steepness" with the word "area."
(2) Try that.
Author: Oh...
Reader: Oh...
Mathematics: OH...

The Return of the Magnifying Glass

Author: Alright, so we've got some machine M.
Reader: Right.
Mathematics: WHICH MACHINE?
Reader: Best not to decide. Let's just stay agnostic about which for now.
Author: Okay, so we've got some machine that looks curvy when we draw it.
Steve suggested we do what we did last time.
Reader: So let's pick a point on it and zoom in infinitely far.
Mathematics: THEN WHAT?
Reader: Well, now that we're zoomed in, things look straight. So we might
be able to pretend that the tiny amount of area under the curvy thing at that
point is just a rectangle.
Mathematics: CAN WE DRAW A PICTURE SO THAT I KNOW WHAT'S GOING
ON?
Author: Sure.

(Reader draws Figure 5.3.)

Author: Wait a second. In the picture you drew, there's some area we're
missing. See that empty triangle? I think we have to fill that in, or else we
won't get the right answer.
Mathematics: THAT MAY NOT BE A PROBLEM!...

(Mathematics begins a ((fairly) lengthy)(optional(!))(semi-)digression.)

Mathematics: SEE, THESE INFINITELY SMALL NUMBERS SEEM TO COME IN
LAYERS, OR AT LEAST THEY ACT LIKE THEY DO. THAT IS, NUMBERS LIKE
3 AND 99 ARE IN SOME SENSE "ONE LEVEL ABOVE" NUMBERS LIKE $7dx$ OR
$52ds$, AND TWO LEVELS ABOVE NUMBERS LIKE $6\,dx\,dy$ OR $99\,ds\,dt$. WHEN
VARIOUS TERMS WITH MULTIPLE LEVELS ARE ADDED TOGETHER, THERE IS
ONE SENSE IN WHICH ONLY THE HIGHEST LEVEL SEEMS TO MATTER. SUCH AS
HOW 3 IS INFINITELY CLOSE TO $3+2dx$, SO WHENEVER WE ASK THE NUMBER
$3+2dx$ HOW LARGE IT IS, WE CAN'T TELL ITS ANSWER APART FROM 3. BUT
IN ANOTHER SENSE THE LOWER LEVELS MATTER TOO, BEFORE WE ASK THE

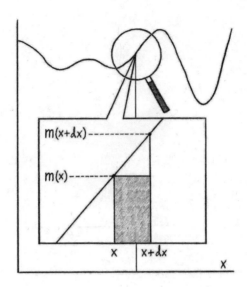

Figure 5.3: Trying to think of a way to use our infinite magnifying glass to calculate the area under curvy things. Our best idea so far is something like this. First step: draw a curvy thing. Second step: pick a place on it and zoom in. Third step: hope that we can write the area under that point of the curvy thing as the area of an infinitely thin rectangle. If we could do that, then the area of each tiny rectangle would just be "length times width," or $m(x)dx$. Then we would just have to add up all the tiny areas, one for each point x. So if only we knew how to add up an infinite number of infinitely small things, it seems like we'd be done!

NUMBER FOR ITS SIZE. THE LOWER LEVELS CAN'T BE IGNORED, BECAUSE AT ANY TIME, SOME OPERATION WE PERFORM MAY CANCEL THE TERMS FROM THE HIGHER LEVELS. WE'VE SEEN THIS IN YOUR DEFINITION OF THE DERIVATIVE. I'LL USE THE TERM "WAFFLING X" AS AN ABBREVIATION FOR "SUBTRACTING THE HIGHEST-LEVEL NUMBER FROM X, AND DIVIDING THE RESULT BY dx." NOW, IF WE ASK THE NUMBER $9 + 6dx + (dx)^2$ HOW LARGE IT IS, IT WILL TELL US ITS SIZE IS 9. NATURALLY, THE NUMBER 9 WOULD HAVE TOLD US THE SAME THING ABOUT ITSELF. SO IT MAY APPEAR THAT THE LOWER LEVELS CAN BE IGNORED FROM THE START. BUT THEY CANNOT. THE NUMBER 9 AND THE NUMBER $9 + 6dx + (dx)^2$ HAVE DIFFERENT BEHAVIORS, AS WE CAN SEE BY WAFFLING EACH. WAFFLING $9 + 6dx + (dx)^2$ WILL TURN IT INTO $6 + dx$, A NUMBER THAT WILL TELL US ITS SIZE IS 6. BUT WAFFLING 9 WILL TURN IT INTO $\frac{0}{dx}$, WHICH IS JUST 0. SO WITH INFINITELY SMALL NUMBERS, THE LOWER LEVELS CAN'T BE IGNORED, BUT AS SOON AS WE ASK THE NUMBER TO TELL US HOW LARGE IT IS, THEN ONLY THE LARGEST LEVEL MATTERS.

(Mathematics's (semi-)digression ends.)

Stephen Kleene: I'm not sure what I think about all that.

Reader: Me neither.

Author: What does this have to do with ignoring the empty triangle?

Mathematics: OH, RIGHT. BACK TO FIGURE 5.3. THE TALL THIN RECTANGLE UNDER EACH POINT x WILL HAVE AREA $m(x)dx$. THE MISSING TRIANGLE'S AREA WILL BE $\frac{1}{2}dxdM$, WHICH IS ONE LEVEL LOWER. SO WE CAN IGNORE IT, SINCE WE'RE JUST ADDING THE AREAS AND NEVER WAFFLING THEM.

Reader: This is probably scaring me.

Author: Yeah, I have no idea what you just said.

Stephen Kleene: Why do you keep saying "waffle"?

Mathematics: SORRY. . .

Author: I think we can just think about normal rectangles, not infinitely small ones, and then imagine making them smaller and smaller.

Mathematics: HOW SO?

Author: Like this:

(*Author draws Figure 5.4, and says everything in its caption.*)

Mathematics: WE'RE STILL LISTENING.[3]

(*Author returns from the caption,*
Mathematics returns from the footnote,
Reader seems to have returned from both the caption and the footnote,
Narrator checks his notes to make sure everyone is back. . .

. . .

Oh, and Steve is off with Waitress
. . . somewhere.)

Mathematics: (*To Author*) I THINK I SEE WHAT YOU MEANT IN THE CAPTION ABOUT THAT GIVING ALMOST THE RIGHT ANSWER. I SUPPOSE WE COULD THINK ABOUT IT THAT WAY AS WELL.

Reader: So to make sure I understood the caption, you're saying the area under the curvy thing will be almost. . . but not exactly. . . this?. . .

$$\sum_{i=1}^{n} (\text{Rectangle Areas}) \;\equiv\; \sum_{i=1}^{n} (\text{Height}_i)(\text{Width}_i) \;\equiv\; \sum_{i=1}^{n} m(x_i)\Delta x_i$$

Author: Wow that's a lot more concise than what I said. But yeah, exactly. That's what I was getting at.

Reader: And why would we want to do the thing with the rectangles? From the caption?

Author: Oh, we don't. Or at least I don't. I mean, the idea about infinitely

---

3. **Reader:** What? **Mathematics:** Oh, I suppose that might seem a bit out of order. Go read the caption under Figure 5.4. I'll wait for you back here.

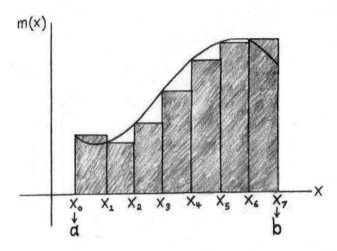

Figure 5.4: **Author:** Wow, it's more cramped in here than I expected. So, we're trying to find the area under a curvy thing between two points $x = a$ and $x = b$. If thinking about infinitely thin rectangles is too weird, we can just imagine using a finite number of rectangles. Like in the picture. Just imagine picking a bunch of points between $x = a$ and $x = b$. Let's give the points names, like $x_1, x_2, \ldots, x_n$, and I guess we could reabbreviate $a \equiv x_0$ and $b \equiv x_{n+1}$ for convenience. At each of the points x_i, the height of the machine's graph $m(x_i)$ will be the height of one of our rectangles. That takes care of height. The width of each rectangle will just be the distance between one point and the next, like in the picture. Each rectangle will have some small width $\Delta x_i \equiv x_{i+1} - x_i$. So the area of each one will look like "height times width," or $m(x_i)\Delta x_i$. Then, if we add up all the rectangle areas, we won't get exactly the right area, but probably something close. And if we imagine adding more and more rectangles and making each one thinner, then we'd be getting closer to the real area. So it's easy. I mean, easy to imagine. It sounds like a pain to actually do this. So maybe let's not do it... Wow, I've been in this caption for awhile. Better go make sure they're still listening.

small rectangles was nicer. But we don't know how to actually *use* it yet, to compute anything. So the idea from the caption wasn't something I think we should actually *do*. But at least it lets us compute specific numbers. Hypothetically. If we want to. And as we add more rectangles, we'll slowly be getting closer to the exact answer Mathematics was talking about.

Mathematics: I LIKE MY WAY OF THINKING ABOUT IT BETTER.

Author: Me too. I'm ready to give up on this problem.

Mathematics: AS AM I.

Reader: Ooh. Then let's give up! Can we pull the Molière trick again?

Author: Sure! I mean, that's what we've always done in the past when we decided to give up. Even if we don't know how to compute the exact answer yet, there's no harm in naming it. Let's make sure the name we use for it reminds us of what it is, though.

Reader: Alright, well we wrote the approximate answer like this:

$$\sum_{i=1}^{n} m(x_i)\Delta x_i$$

So let's use this to build an abbreviation for the exact answer. You mentioned awhile back that the textbooks use Δ*stuff* as a shorthand for "a difference in *stuff* between two points," because Δ is the Greek capital d, and d stands for difference. Or distance. I don't think we decided yet.

Author: I don't think we have either.

Reader: Anyways, because of that, when we invented calculus we needed another way to talk about a difference, but an infinitely small one this time. So we just turned the Greek letter Δ into the Latin letter d. So dx or dt or $d(stuff)$ all just meant a difference in some quantity between two points that were infinitely close together. That made the new idea look kind of like the old one.

Mathematics: HOW DOES THAT HELP US?

Reader: We're giving up, remember? So we're pulling the Molière trick and thinking of a good name for the thing we couldn't figure out.

Mathematics: OH, RIGHT. I FORGOT. SO WHAT SHOULD WE NAME IT?

Reader: Well, I just wrote the approximate answer up there. The one Author described from inside the caption. So let's take that abbreviation, and replace the Δ with a d, since our rectangles are infinitely thin. That would give us

$$\sum_{i=1}^{n} m(x_i)dx_i$$

But I guess this doesn't really mean anything, because we're changing the abbreviations one at a time.

Author: Nice! Alright, the exact answer should have infinitely many rectangles, one for every point x, so the x_i notation doesn't really work anymore — the i's just counted rectangles, but we can't count the rectangles if there's one at every possible point on a line. So let's forget about the i's. And the pieces $i = 1$ and n don't work anymore either. Those were just there to remind us where we were starting and ending, but now we can just go back to calling them a and b again. So I guess now our abbreviation is

$$\sum_{a}^{b} m(x)dx$$

We could just stop here and—

Mathematics: FINALLY! I WAS WAITING FOR YOU TO FINISH. I'M READY TO MAKE MY CHANGE.

Author: Uhh, there's not much else to change. We could basically just be don—

Mathematics: AUTHOR, THERE ARE THREE OF US. READER GOT TO CHANGE THE GREEK d TO A LATIN d. YOU GOT TO GET RID OF THE SUBSCRIPTS. IT'S ONLY FAIR TO LET ME CHANGE SOMETHING. FAIRNESS IS ABOUT INVARIANCE. DON'T YOU LOVE INVARIANCE?

Author: (*Reluctantly*) ...Yes.

Mathematics: PERFECT! I'LL FOLLOW READER'S LEAD. Σ IS THE GREEK LETTER S, AND IT STANDS FOR "SUM." AS BEFORE, TO TRANSLATE OUR ABBREVIATION FROM THE FINITE TO THE INFINITE, LET'S CHANGE THE GREEK LETTER TO THE CORRESPONDING LATIN LETTER, LIKE THIS:

$$\int_a^b m(x)dx$$

Reader: What is that?!

Author: Yeah...uhh...Math? That's not an S.

Mathematics: LOOK, HOLDING A PEN MAY BE TRIVIAL FOR YOU DIGIT-LADEN PRIMATES, BUT I'M NOT EXACTLY A PHYSICAL BEING.

Author: We understand. It just looks a bit funny. But let's keep it that way.

Mathematics: SO WHAT DOES THIS STAND FOR?

Reader: Well, we named it carefully, so we can probably figure it out. The S-like thing reminds us it's a sum, it's Latin now instead of Greek, the $m(x)$ is a height, and dx is an infinitely small width. So it stands for a sum of infinitely many things, and each of those things is the area of an infinitely thin rectangle. Oh, and the a and b just remind us where to start and stop.

Author: Great! We successfully pulled the Molière trick!

Reader: So we successfully did...nothing?

Author: Well yeah. But we can still write stuff like this:

$$\left(\text{Area under } m\text{'s graph between } x = a \text{ and } x = b\right) \equiv \int_a^b m(x)dx$$

But you're right. We still don't know how to compute this, so we didn't really "do" anything yet.

Reader: Wait, so after all that we're still no closer to solving our problem?

Stephen Kleene: Precisely!

(*Everyone looks at Steve.*)

Author: (*To Steve*) You weren't listening to us that whole time, were you?

Stephen Kleene: No, it just seemed like a good time to say "precisely." I was back at Mathematics's house. Meta—

Author: Wait, I thought you were with Waitress.

Stephen Kleene: No. You just wrote that I was.

Author: Huh? I don't remember wri—

Stephen Kleene: Anyways, as I was saying, Meta needed some help, so I went back.
Mathematics: HELP WITH WHAT?
Stephen Kleene: We'll get to that.

(*Steve pulls a small device from his coat and speaks into it.*)

Stephen Kleene: Meta?...Okay. We're ready when you are.

Extralude: The Existence Dilemma

(*Our characters hear the word "suppose," followed by something unintelligible, at which point they find themselves back at Mathematics's house.*)

Stephen Kleene: Meta has finished examining the foundation...Or foundations, I can never remember which. Plurals are tricky.
Mathematics: AND?
Stephen Kleene: Your house is sinking...
Mathematics: MY HOUSE IS SINKING?
Stephen Kleene: Yes. I'm afraid Nature was right. The foundation wasn't built for a corporeal entity. Well, not corporeal...but an entity. You've been over this before, right?
Mathematics: I BELIEVE WE HAVE.
Stephen Kleene: You exist now! I was surprised to see the extent of it. Your mismatch with the Void is quite severe, and it will only grow worse with time. The Void is no place for an entity.
Mathematics: OH...
Author: Wait, what about you? Don't you live in the Void too?
Stephen Kleene: Indeed.
Author: Then why don't you have the same problem?
Stephen Kleene: Modus tollens, friend.
Author: Huh?
Stephen Kleene: You've assumed I exist. Check your premises.
Author: What???
Reader: What???
Mathematics: WHAT???
MetaMathematics: ...
Author: How are you here if you don't exist?
Stephen Kleene: The Void is home to all things that don't exist. And one that does. For now...
Author: That sounded like foreshadowing.
Stephen Kleene: That sounded like metaforeshadowing.
Author: Like foreshadowing about foreshadowing? I didn't intend it to be.

Stephen Kleene: No no, metaphor shadowing. Not meta-foreshadowing. Nevermind.

(*A curious metasilence elapses.*)

Stephen Kleene: (*To Author and Reader*) Do whatever you can to help Mathematics get out of here and into a new home as soon as possible. One appropriate for this new form it's taken. It needs to be somewhere it belongs.
Author: Of course. We'll do anything we can.
Stephen Kleene: Wonderful. On that note, Meta and I must be off. Good luck, you three.
Author: Hope we see you again!
Reader: Bye Steve!
Mathematics: SAY HI TO NATURE FOR ME!
Author: Alright, time for the next chapter. Let's go!

(*Nothing happens... Almost as if this spot were already reserved...*)

Act III

6 Two in One

6.1 Two is One

6.1.1 Another Abbreviation Becomes an Idea

In the previous interlude, we discussed the problem of finding the areas of curvy things. We eventually gave up, but we had one minor insight. When we zoom in on the graph of a machine at any point x, the tiny sliver of area under that point can be thought of as an infinitely thin rectangle with height $m(x)$ and width dx. So the area hiding under a machine's graph at any point x should just be $m(x)dx$. But hypothetically, if we could somehow add up all the areas of those infinitely thin rectangles under every point x (say, every point between $x = a$ and $x = b$), then we should get the total (possibly curvy) area under the graph of m.

We have no idea how to actually do this. However, we wrote down a great abbreviation to summarize the idea: $\int_a^b m(x)dx$. This abbreviation reflects the fact that we can think of a curvy area as the sum (hence the S-like $\int$ thing) of an infinite number of infinitely thin rectangles (hence the $m(x)dx$). But this was just an abbreviation for the (unknown) answer. It didn't tell us how to actually calculate curvy areas.

Anyone would be likely to get stuck here for awhile. However, that dx in our new abbreviation is suggestive. We've seen that symbol before, sitting on the bottom of a derivative. Maybe using our old derivative idea will help us to understand our new $\int$ idea.

Derivatives are machines, too. For example, something like $f(s) \equiv s^2$ has the derivative $f'(s) \equiv 2s$, and that $2s$ is a machine just as much as x^2 was. So, suppose the $m(x)$ inside our funny $\int$ symbol is really the derivative of some other machine $M(x)$. That would let us write

$$m(x) = \frac{dM}{dx} \tag{6.1}$$

If we think of m this way, then we can pull the following tricky move to rewrite our abbreviation for the (unknown) curvy area:

$$\int_a^b m(x)dx \overset{(6.1)}{=} \int_a^b \left(\frac{dM}{dx} \right) dx = \int_a^b dM \tag{6.2}$$

The symbol dM was just our abbreviation for $M(x+dx) - M(x)$, a tiny change in height between two points that are infinitely close together. So the meaning

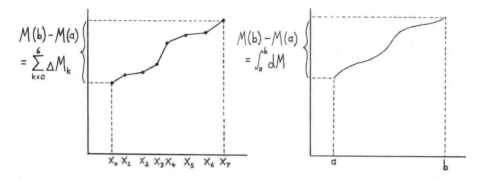

Figure 6.1: Visualizing the reason why $\int_a^b dM = M(b) - M(a)$. The first picture demonstrates the idea for machines whose graphs are made up of a finite number of straight lines. The total height change between a and b can be thought of in two ways. On the one hand, the height change is $M(end) - M(beginning)$ (that is, $M(b) - M(a)$ for the picture on the right, or $M(x_7) - M(x_0)$ for the picture on the left). On the other hand, the total height change is just the sum of all the tiny height changes experienced in each "step" as we walk along the machine's graph. Each of these tiny changes is of the form $\Delta M_k \equiv M(x_{k+1}) - M(x_k)$. In the truly curvy case on the right, the same idea holds, but now there are an infinite number of infinitely small steps, each of which gives us an infinitely small change in height dM. The sum of all these infinitely small changes in height (i.e., $\int_a^b dM$) is just the total change in height (i.e., $M(b) - M(a)$).

of the symbol $\int_a^b dM$ is this: walk from $x = a$ to $x = b$, and add up all the tiny changes in height you experience along the way.

Now, since the symbol $\int_a^b dM$ stands for "whatever we would get if we could somehow add up all the tiny changes in height between a and b," then it must just refer to the *total* change in height between a and b, which is to say $M(end) - M(beginning)$, or equivalently, $M(b) - M(a)$. If this isn't clear, then the pictures in Figure 6.1 might help. Alright, using this fact, we can extend the above equation by one step, and summarize the gist of it by writing

$$\int_a^b \left(\frac{dM}{dx} \right) dx = M(b) - M(a) \tag{6.3}$$

And now we're stuck aga... wait a second. Are we done?

6.1.2 The Fundamental Hammer of Calculus

There's no way this is correct — it's too simple. It says that this new $\int$ symbol we wrote down, just as a meaningless abbreviation for the area under a curve, turns out to be sort of the opposite of the derivative idea we invented earlier. That is, starting with M, then finding its derivative, then applying the "area under me" operation to that, we get something just involving M;

something that involves neither derivatives nor the "area under me" operation. Equation 6.3 is a sentence that relates the two main calculus ideas we've invented so far: namely, computing curvy steepness (differentiation) and curvy areas ("integration," in textbook jargon). Because it ties our two main ideas together, let's call it the **fundamental hammer of calculus**. Now we can rephrase equation 6.3 by writing this:

$$\int_a^b m(x)dx = M(b) - M(a) \tag{6.4}$$

where M is any machine whose derivative is m. So this says that if we want to find the area under some curvy thing m between two points $x = a$ and $x = b$, then all we have to do is think of an "anti-derivative" of m, that is, a machine M whose derivative is the machine m that we started with. If we can somehow do that, then the area under m is just $M(b) - M(a)$. So whenever we can think of an anti-derivative of a given machine, the seemingly impossible problem of computing its (possibly curvy) area is simple! Armed with an anti-derivative of m, we can find curvy areas using nothing but subtraction.

As always, it's worth checking our idea in a few simple cases where we already know what to expect. If it gives us the wrong answers in those simple cases, then our new idea is broken and we have to start over. However, if the crazy argument above is correct, then that would mean we'd get two ideas for the price of one; the $\int$ idea (computing curvy areas) would just be the derivative idea (computing curvy slopes) in reverse. Let's do some field-testing on this new idea of ours.

6.2 Field-Testing the Fundamental Hammer

6.2.1 Fundamentally Hammering Constants

We know how to find the area of a rectangle, so let's test the idea on the graph of the constant machine $m(x) \equiv \#$. This machine is just a horizontal line whose height is $\#$. The graph of this machine between two points $x = a$ and $x = b$ is just a rectangle with height $\#$ and width $b - a$, so the area should be $\# \cdot (b - a)$. Now, if all our ideas about anti-derivatives are really on the right track, then it also has to be the case that

$$\int_a^b m(x)dx = M(b) - M(a) \tag{6.5}$$

where $m(x) \equiv \#$, and $M(x)$ is an "anti-derivative" of $m(x)$. That is, $M(x)$ is just any machine whose derivative is $m(x)$, which is to say $M'(x) = m(x) = \#$. We don't have any methods of finding anti-derivatives, so we basically just have to use our familiarity with derivatives, together with some hopeful optimism, and stare blankly at the machine $m(x) \equiv \#$ until we can think of something whose derivative is that. Fortunately, in this case, it's not too difficult, since

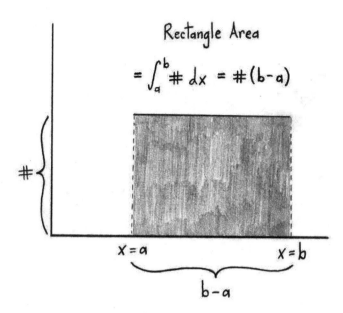

Figure 6.2: We can compute the area of a rectangle without any calculus. This fact
provides a simple test of our fundamental hammer idea. In this case,
the fundamental hammer reproduces just what we'd expected, namely,
$\int_a^b \# \, dx = \# \cdot (b - a)$.

we know that $(\#x)' = \#$. This tells us that $M(x) = \#x$, and we can use this
to check and see whether our idea seems to be on the right track.

Between any two points $x = a$ and $x = b$, the area under the graph of a
horizontal line $m(x) \equiv \#$ has to be $\# \cdot (b - a)$, because it's just a rectangle.
So in this particular case, our strange new symbol $\int_a^b m(x) dx$ refers to the
number $\# \cdot (b - a)$. We don't yet know for sure whether this new symbol *really*
is just the difference of the anti-derivatives, but now we can check! Since we
found $M(x) = \#x$, we also know that $M(b) - M(a) = \#b - \#a$, but this is
just $\# \cdot (b - a)$.

Hooray! We got the same answer both ways. Notice that we never *used*
equation 6.5. That's good. We don't want to assume it's true yet. Instead, we
computed the two sides of it in different ways and then *showed* that equation
6.5 was true in this simple case. Let's keep going.

6.2.2 Fundamentally Hammering Lines

What about lines? Back in Chapter 1, we discovered that the graphs of machines like $m(x) \equiv cx + b$ are straight lines. As a simple example, let's look
at $\int_0^b cx \, dx$. This just refers to the area under $m(x) \equiv cx$ between $x = 0$ and
$x = b$. Fortunately, this is just a triangle with a width of b and a height of cb.
Two such triangles let us build a rectangle with area $(b)(cb) = cb^2$, so the area

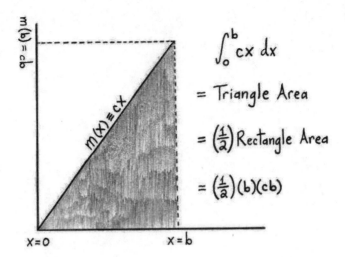

Figure 6.3: Picturing the reason why $\int_0^b cx\,dx = \frac{1}{2}cb^2$. This allows us to check that the fundamental hammer works by examining another simple case, in which we already know how to compute the area without using the fundamental hammer.

of the triangle is just $\frac{1}{2}cb^2$. Without any calculus, we know that

$$\int_0^b cx\,dx = \frac{1}{2}cb^2$$

We didn't use any calculus here. The $\int$ symbol is just an abbreviation for the area, and we already know how to compute that. Now we can test our fundamental hammer again by seeing if it reproduces what we already know. Can we think of a machine whose derivative is cx? Let's guess cx^2 and see if it works.

If we differentiate cx^2, we'll get $2cx$. There's an extra 2 out front that we didn't want, so let's throw a $\frac{1}{2}$ in front of our original guess and try again. If we differentiate $\frac{1}{2}cx^2$, then we'll get cx, and that's what we wanted. Now we can use this to check our idea again. Starting with $m(x) \equiv cx$, we just found that this machine's anti-derivative is $M(x) = \frac{1}{2}cx^2$, and therefore,

$$M(b) - M(0) = \frac{1}{2}cb^2 - \frac{1}{2}c0^2 = \frac{1}{2}cb^2$$

Perfect. First, we found that $\int_0^b cx\,dx = \frac{1}{2}cb^2$ by thinking about the area of a triangle. Then, we thought of an anti-derivative of cx, and found that $M(b) - M(0) = \frac{1}{2}cb^2$. Both sides of the fundamental hammer match up, just like we hoped. To reiterate, we aren't *using* the fundamental hammer at this point. Rather, we're *testing* it by computing both sides of the hammer (equation 6.5) in different ways, and verifying that they're equal. This gives us more confidence that the line of reasoning we used to derive the fundamental hammer was indeed on the right track.

6.2.3 A Worry

As we already know, derivatives kill constants, so if a machine m has one anti-derivative, it must necessarily have an infinite number of them. Why? Well, if $\frac{d}{dx}[M(x)] = m(x)$, then $\frac{d}{dx}[M(x) + \#] = m(x)$ as well, for any fixed number $\#$. So each possible number $\#$ gives us an equally valid "anti-derivative" of m, namely $M(x) + \#$. This is worrisome at first. Our fundamental hammer requires us to find an anti-derivative in order to compute an *area*. The area is something concrete, and intuitively there should be only one correct answer for it. However, if there are an infinite number of different anti-derivatives of a machine m, they had better all give the same answer for the area, or else our fundamental hammer is broken. While we're not yet sure if our hammer will always work, we can at least convince ourselves that the worrisome example above is no cause for worry. The fundamental hammer tells us to compute an area using a *difference*: $M(b) - M(a)$. As such, any anti-derivative of the form $W(x) \equiv M(x) + \#$ will give us the same answer for the area, because $W(b) - W(a) = [M(b) + \#] - [M(a) + \#] = M(b) - M(a)$. Having put that worry to rest, let's keep moving.

6.2.4 Full Speed Ahead! Using the Fundamental Hammer

What have we done so far? First we made an argument in which we discovered the fundamental hammer of calculus:

$$\int_a^b m(x)\,dx = M(b) - M(a) \tag{6.6}$$

But we weren't entirely sure whether we should trust our derivation of it. So we tested it on several simple cases where we already knew what to expect. So far it has worked in every case, so we're a bit more confident. Let's see what it gives us when we're *not* sure what to expect.

6.2.5 An Unfamiliar Case

As always, the machine $m(x) \equiv x^2$ provides a simple test case. Let's see what the fundamental hammer gives us when we ask it about the area between (say) $x = 0$ and $x = 3$:

$$\int_0^3 x^2\,dx = M(3) - M(0)$$

where $M(x)$ is some machine whose derivative is x^2. Can we think of such a machine? Well, since derivatives knock the power down by one, it had better look like $\#x^3$. Then, when we take the derivative and bring the power down, we've got to end up with x^2, so we need it to be true that $(\#x^3)' = 3\#x^2 = x^2$, which tells us that the number # has to be $\frac{1}{3}$. So $M(x) = \frac{1}{3}x^3$ is an anti-derivative of x^2, which lets us extend the above equation a few steps further, writing

$$\int_0^3 x^2 \, dx = M(3) - M(0) = \frac{1}{3}3^3 - \frac{1}{3}0^3 = 9$$

Interesting... The area of this particular curvy thing is just a whole number. That's eerily simple. What else could we do?

6.3 Forging the Anti-Hammers

What should we do next to build our understanding of this new $\int$ idea? If we wanted to, we could just continue trying to anti-differentiate a bunch of specific machines, essentially by guessing, which is what we've been doing so far. However, in the past when we were inventing the idea of the infinite magnifying glass and playing with the concept of derivatives for the first time, we got the most bang for our buck not from differentiating specific machines but by building hammers that work for *any* machine. I'll go back to Chapter 3 for a moment and steal one of our boxes where we described all of our hammers. Wait here. I promise I won't take long.

(Time passes.)

Okay, I'm back. Here it is:

Hammer for Addition

$$(f + g)' = f' + g'$$

Hammer for Multiplication

$$(fg)' = f'g + fg'$$

Hammer for Reabbreviation

$$\frac{df}{dx} = \frac{ds}{dx}\frac{df}{ds}$$

Since it appears that the $\int$ thing is kind of the opposite of the derivative, it would be super nice if there were three analogous "anti-hammers," one for each of the originals! Then we'd be able to do all kinds of things with curvy areas. Let's see if we can forge some anti-hammers.

6.3.1 Anti-Hammering Addition

Earlier, we discovered a nice thing about derivatives, called the "hammer for addition." Essentially it said that "the derivative of a sum is the sum of the derivatives," or to put it another way: $(f + g)' = f' + g'$. It would be nice if a similar fact were true for this new "integration" idea we've come up with, because then we'd have a tool that would let us shatter certain hard problems into easier pieces, just like the original hammers allowed us to do. Since the derivative of a sum is the sum of the derivatives, we might guess that the integral of a sum is the sum of the integrals.[1] That is, we want to see whether it is true that

$$\int_a^b \Big(f(x) + g(x)\Big) \, dx \stackrel{???}{=} \left(\int_a^b f(x) \, dx\right) + \left(\int_a^b g(x) \, dx\right) \qquad (6.7)$$

We don't know if this is true, but it would be nice, in that it would be the opposite of the hammer for addition. It would be an "anti-hammer for addition," if you will. Let's try to see if it's true.

To begin, notice that we're welcome to think of $f(x) + g(x)$ as a single machine in its own right, and we could even give it a name, like $h(x) \equiv f(x) + g(x)$. Then $h(x)$ would be the machine that spits out $f(x)$ plus $g(x)$ whenever we feed it a specific number x. Imagine looking at the graph of $h(x)$, which might be some crazy squiggly thing, and imagine putting an infinitely thin rectangle at an arbitrary point x, stretching vertically from the horizontal axis up (or down) to the graph of h (see Figures 6.4 and 6.5).

The width of this rectangle will be dx, and its height will be $h(x) \equiv f(x) + g(x)$, so its area will be $(f(x)+g(x))dx$. But then by the obvious law of tearing things, this tall thin rectangle can be torn into two smaller thin rectangles to give $f(x)dx + g(x)dx$. Now we have two rectangles, one with height $f(x)$, and one with height $g(x)$. So we can think of an infinitely small rectangle at each point x in two ways, as a tall thin rectangle or as two shorter but equally thin rectangles.

Intuitively, it seems clear that we can then get the whole area under h in two ways: we could add up all the tall thin rectangles to get $\int_a^b (f(x)+g(x))dx$, or we could add up all the torn thin rectangles to get $\int_a^b f(x) \, dx + \int_a^b g(x) \, dx$. These are two descriptions of the same thing: the total area under $f + g$. So

---

1. Recall that the "integral" of m is what textbooks call $\int_a^b m(x) \, dx$. Actually, they tend to call this a "definite integral" of m. The term "indefinite integral" is often used to refer to an anti-derivative of m, but this terminology can be slightly misleading, in that it only makes sense *after* the discovery of the fundamental hammer. Before the discovery of the fundamental hammer, it's not obvious that anti-derivatives have anything to do with "integrals," (i.e., possibly curvy areas). We'll prefer to simply call $\int_a^b m(x) \, dx$ the integral of m from a to b. Only after we've defined this terminology does the fundamental hammer tell us that integrals and anti-derivatives are related concepts.

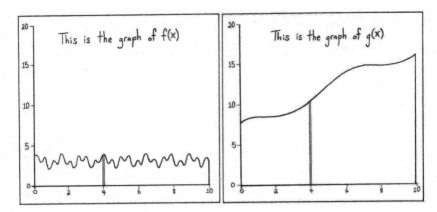

Figure 6.4: Two machines, chosen randomly. On the left, the tiny rectangle has area $f(4)dx$. On the right, the tiny rectangle has area $g(4)dx$.

we can slap an equals sign between the two descriptions and write

$$\int_a^b f(x) + g(x)\, dx \;=\; \left(\int_a^b f(x)\, dx\right) + \left(\int_a^b g(x)\, dx\right) \qquad (6.8)$$

That's exactly what we wanted to show. Let's write it in a box to make it official.

Anti-Hammer for Addition

We've discovered another fact about our new $\int$ idea,
The $\int$ of a sum is the sum of the $\int$'s.
To put it another way: for any machines f and g, we have

$$\int_a^b f(x) + g(x)\, dx \;=\; \int_a^b f(x)\, dx \;+\; \int_a^b g(x)\, dx$$

Now that we've invented the anti-hammer for addition, Figures 6.4 and 6.5 let us visualize what it is saying in another way. The letters f and g stand for any two machines, possibly with very curvy graphs. Figure 6.4 lets us picture the individual machines f and g, as well as the area under them. Figure 6.5 lets us picture $h(x) \equiv f(x) + g(x)$, and gives us another way of picturing what the anti-hammer for addition is saying.

6.3.2 Anti-Hammering Multiplication

The Simple Anti-Hammer for Multiplication

We actually had two hammers for multiplication, although one of them was

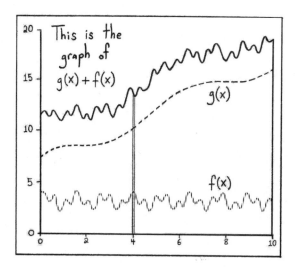

Figure 6.5: This is the graph of $f(x) + g(x)$. The tiny rectangle has an area of $(f(4)+g(4))dx = f(4)dx+g(4)dx$. That is, the area of this tiny rectangle is the sum of those in the left and right side of Figure 6.4. Since this fact is true for each point x, adding up the areas under each such point does not change the principle.

a special case of the other. Recall that in Chapter 2 we showed $(\# f(x))' = \# f'(x)$, where $\#$ is some fixed number like 7 or 59. So we can "pull constants out of derivatives." I wonder if we can "pull constants out of $\int$," so that $\int \#(stuff) = \# \int (stuff)$. The idea makes sense intuitively, if we think about what the two expressions mean. Remember that in an expression like $\int f(x)\, dx$, we're adding up the areas of a bunch of infinitely small rectangles. If we double the height of each rectangle, keeping its (infinitely thin) width the same, then we should have twice of the original area, so the area should double. There's nothing special about the word "double" in this argument, and it should be true for any amount of magnification $\#$. That is, it should be true that

$$\int_a^b \# \cdot f(x)\, dx = \# \cdot \int_a^b f(x)\, dx$$

for any number $\#$ and any machine f. Hooray! Again, we've found an "anti-hammer" for integration corresponding to one of our old hammers for differentiation.

The Less Simple Anti-Hammer for Multiplication
Above, we found that we can pull constants outside of integrals, much like we could pull constants outside of derivatives. However, the real hammer for multiplication was a bit more complicated than that. It said

$$(fg)' = f'g + fg'$$

or, to say the exact same sentence with different abbreviations,

$$[f(x)g(x)]' = f'(x)g(x) + f(x)g'(x)$$

Let's try to use this fact to build an analogous anti-hammer for multiplication. Well, if we "integrate" both sides of the above equation — that is, if we feed both sides to the $\int$ thing — then we would get

$$\int_a^b [f(x)g(x)]' \ dx = \int_a^b [f'(x)g(x) + f(x)g'(x)] \ dx \qquad (6.9)$$

The left side of the above equation is an integral of a derivative, so we can smack it with the fundamental hammer to get

$$f(b)g(b) - f(a)g(a) = \int_a^b [f'(x)g(x) + f(x)g'(x)] \ dx$$

The left side is a bit ugly, but it's a simple idea, so let's abbreviate it. We'll rewrite the above equation this way:

$$\left[f(x)g(x)\right]_a^b = \int_a^b [f'(x)g(x) + f(x)g'(x)] \ dx \qquad (6.10)$$

where the left side is shorthand for "plug in b everywhere, then plug in a everywhere, then take the difference," so $\left[f(x)g(x)\right]_a^b$ is an just abbreviation for $f(b)g(b) - f(a)g(a)$.

Now, although equation 6.10 is *true*, it's not particularly *helpful*, because we can only use it when we happen to encounter things that look exactly like $\int_a^b f'(x)g(x) + f(x)g'(x) \ dx$, and not many things look like that. If we do happen to encounter something that looks like that, then we can instantly get a number out and be done. As such, we could simply stop here and call this the anti-hammer for multiplication, since it "undoes" the hammer for multiplication. However, we can get a much more useful anti-hammer simply by thinking of the above idea in a slightly different way. Let's rewrite equation 6.10 by using the anti-hammer for addition to break up the big integral into two pieces, like this,

$$\left[f(x)g(x)\right]_a^b = \int_a^b f'(x)g(x) \ dx + \int_a^b f(x)g'(x) \ dx \qquad (6.11)$$

and throwing one of the two integrals over to the other side of the equation. It may not be clear at first why we would want to do that, but we'll discuss why in a few seconds. For now, let's write our invention in a box, to make it official.

Anti-Hammer for Multiplication

We've discovered another fact about our new $\int$ idea,
though we don't know how to use it yet:

$$\int_a^b f'(x)g(x)\,dx = \Big[f(x)g(x)\Big]_a^b - \int_a^b f(x)g'(x)\,dx \qquad (6.12)$$

How can we use this crazy sentence? Well, remember that the hammer for multiplication (and all the other hammers) were not *rules* that said what we had to do, but rather *tools* that let us make progress by choosing to interpret things in a specific way. For example, consider the machine $m(x) \equiv xe^x$. We don't *have* to think of this machine as two different machines multiplied together, but we are free to think of it that way if it helps us. We could choose to think of xe^x as $f(x)g(x)$, where $f(x) \equiv x$ and $g(x) \equiv e^x$. Then the hammer for multiplication (abbreviated HM in the following equations) would tell us that

$$m'(x) \equiv (xe^x)' \overset{HM}{=} (x)'(e^x) + (x)(e^x)' = e^x + xe^x$$

All our original hammers carried this same "only if it helps us" interpretation, and so do the anti-hammers. But after all that hammer yammering, why is equation 6.12 in the box above "more useful" than equation 6.9, even though they're *exactly* the same sentence wearing slightly different hats? Good question! Equation 6.12 tends to be more useful than equation 6.9 *not* because they're saying different things but because of the limits of the human imagination. To see what I mean, imagine we're stuck trying to compute something that looks like $\int_a^b m(x)\,dx$. Now, whenever we're stuck trying to compute something like this, it tends to be easier for most people's minds (including mine) to dream up two machines f and g for which m can be interpreted like this

$$m(x) \overset{Interpret}{=} f'(x)g(x)$$

than it is to dream up two machines f and g for which m can be interpreted like this

$$m(x) \overset{Interpret}{=} f'(x)g(x) + f(x)g'(x)$$

This is an important point that we've encountered several times before: two methods, ideas, equations, etc., may be *logically* identical, but that certainly doesn't mean that they're *psychologically* identical. That is, two ways of saying exactly the same thing may be very different in terms of how easy they are to understand. The anti-hammer for multiplication gives us a way to translate one problem into another. Will the translated problem always be easier *for us*?

Well, no. But it might, if we choose a clever translation. Here's an example of how we can translate a problem. Suppose we want to calculate this:

$$\int_0^1 xe^x \, dx$$

Because of the fundamental hammer, if only we could magically think of some machine $M(x)$ whose derivative was xe^x, then we could say, "Aha! The answer is $M(1) - M(0)$." As easy as that might sound, it's not! It is far from clear how to think of a machine whose derivative just happens to be xe^x. So it would appear that we're stuck. However, the anti-hammer for multiplication suggests a path forward. If we can dream up two machines f and g for which $f'(x)g(x) = xe^x$, then this new anti-hammer lets us transform the problem into a slightly different one. Let's start by choosing $f'(x) \equiv e^x$ and $g(x) \equiv x$, and we'll write AHM above an equals sign when we're using the anti-hammer for multiplication. Then we could rewrite the problem like this:

$$\int_0^1 xe^x \, dx \stackrel{AHM}{=} \Big[f(x)g(x) \Big]_0^1 - \int_0^1 f(x)g'(x) \, dx$$

So, we decided on f' and g ourselves, but now this anti-hammer has spat out a sentence involving f and g'. We don't know those yet, so we need to figure them out to make sense of that sentence. Getting f is pretty simple. We defined $f'(x) \equiv e^x$, and this is the extremely special machine that is its own derivative, so it's a perfectly good anti-derivative for itself: $f(x) = e^x$. What about $g'(x)$? That's simple too. We defined $g(x) \equiv x$, so $g'(x) = 1$. Now we can throw all this information into the above equation to get

$$\int_0^1 xe^x \, dx \stackrel{AHM}{=} \Big[xe^x \Big]_0^1 - \int_0^1 e^x \, dx$$

The first piece $[xe^x]_0^1$ is just an abbreviation for $1e^1 - 0e^0$, or e. What about the second piece? Well, we know that e^x is an anti-derivative of e^x, so by the fundamental hammer, we have $\int_0^1 e^x \, dx = e^1 - e^0 = e - 1$. Putting it all together, we have

$$\int_0^1 xe^x \, dx \stackrel{AHM}{=} e - (e-1) = 1$$

Nice! By transforming a problem we couldn't solve into an equivalent but different-looking problem using the anti-hammer for multiplication, we were able to solve the problem easily, and we found that the answer was simply 1. Things didn't have to work out so nicely, though. What if we had made an equally correct but slightly different choice for f and g? Let's see. We could just as well have chosen $f'(x) \equiv x$ and $g(x) \equiv e^x$. That would have led us to rewrite the problem like this:

$$\int_0^1 xe^x \, dx \stackrel{AHM}{=} \Big[\frac{1}{2}x^2 e^x \Big]_0^1 - \int_0^1 \frac{1}{2}x^2 e^x \, dx$$

$$= \frac{e}{2} - \int_0^1 \frac{1}{2} x^2 e^x \, dx$$

This is even scarier-looking than what we started with. It's important to emphasize, however, that we didn't do anything wrong. The above answer is completely correct. We simply made a choice of f' and g that didn't make the problem look simpler *to us*. As we discussed above, this is a general principle about the hammers and anti-hammers. We're always free to use them, but there's no guarantee that they'll transform the problem into something we think is "simpler." That's not their fault. It's a constraint of the human imagination.

6.3.3 Anti-Hammering Reabbreviation

Let's first remind ourselves about the original hammer for reabbreviation, and how it was used. The hammer for reabbreviation (called the "chain rule" in textbooks) was a helpful tool that we invented by lying and correcting for the lie. For example, suppose we were stuck trying to figure out the derivative of the scary-looking machine $m(x) \equiv [V(x)]^{795}$, thinking of x as the variable.[2] We need to calculate $\frac{dm}{dx}$. As we've seen before, reabbreviation helps. First, notice that $[V(x)]^{795}$ is just some *stuff* to a power, so using the abbreviation $s \equiv V(x)$, we can write $m(x) \equiv s^{795}$. Now that we've chosen this abbreviation, the hammer for reabbreviation helps us in the following way. We can write:

$$\begin{aligned}
\frac{dm}{dx} &= \left(\frac{dm}{ds} \right) \left(\frac{ds}{dx} \right) \\
&\equiv \left(\frac{d}{ds} s^{795} \right) \left(\frac{d}{dx} V(x) \right) \\
&= \left(795 s^{794} \right) H(x) \\
&\equiv \left(795 [V(x)]^{794} \right) H(x)
\end{aligned} \tag{6.13}$$

Okay, we've refreshed our memory about the hammer for reabbreviation, but we still haven't invented the corresponding anti-hammer. Here's the idea. As before, we always have the right to reabbreviate, but we're not guaranteed that it will help. Let's look at a specific example.

Reabbreviation Play

Imagine we've got a problem that looks like this, and we're stuck:

$$\text{Thing we want} = \int_a^b x e^{x^2} \, dx$$

---

2. Following the convention we've been using since Chapter 4, the abbreviation $V(x)$ stands for what the textbooks call $\sin(x)$.

Because of the fundamental hammer, if only we could magically think of some machine $M(x)$ whose derivative was xe^{x^2}, then we would know that the above confusing bag of symbols is equal to $M(b) - M(a)$. It's not at all clear how to think of such a machine, but we're free to play around by reabbreviating things. Here's one strategy: we've never dealt with e^{x^2} before, but we have dealt with e^x. However, e^{x^2} is just e^s, where $s \equiv x^2$. So we can write:

$$\text{Thing we want} = \int_a^b xe^s \, dx \qquad (6.14)$$

We haven't really done anything. We haven't even lied. We just reabbreviated. However, we've got two letters floating around, s and x. This isn't illegal, but it makes things a bit more confusing, and it's not clear whether or not we can use the fundamental hammer, because the fundamental hammer involved only one variable. So maybe if we get rid of all the x's and talk about the entire problem in s language, life might be a bit easier. Well, if $s \equiv x^2$, then $x = s^{1/2}$, so we might be tempted to write:

$$\text{Thing we want} = \int_a^b \left(s^{\frac{1}{2}}\right) e^s \, d\left(s^{\frac{1}{2}}\right) \qquad (6.15)$$

That's perfectly correct, but it's scary-looking, so forget that, and instead let's go back to the less scary equation 6.14 and stare at it for a moment. We want to turn the dx piece into something in s language, so it might help to try to relate dx and ds to each other. Derivatives do that, so it might help to compute a derivative. Maybe not, but let's try. Since $s \equiv x^2$, we have

$$\frac{ds}{dx} = 2x \qquad \text{so} \qquad dx = \frac{ds}{2x} \qquad (6.16)$$

By a huge stroke of luck, everything collapses into simplicity when we substitute this into equation 6.14. The x pieces kill each other, and we get

$$
\begin{aligned}
\text{Thing we want} &\overset{(6.14)}{=} \int_a^b xe^s \, dx \\
&\overset{(6.16)}{=} \int_a^b xe^s \left(\frac{ds}{2x}\right) \\
&= \int_a^b \left(\frac{1}{2}\right) e^s ds \\
&= \left(\frac{1}{2}\right) \int_a^b e^s ds
\end{aligned}
\qquad (6.17)
$$

But wait — the a and b were secretly abbreviations for $x = a$ and $x = b$. We have to remind ourselves of that so that we don't confuse the sentence $x = a$ with the sentence $s = a$. Let's just remind ourselves of that by writing

$$\text{Thing we want} = \left(\frac{1}{2}\right) \int_{x=a}^{x=b} e^s ds$$

Okay, since e^s is its own derivative (thinking of s as the variable) it's also its own anti-derivative, so we can use the fundamental hammer and write

$$
\begin{aligned}
\text{Thing we want} &= \left(\frac{1}{2}\right) \int_{x=a}^{x=b} e^s \, ds \\
&\overset{FH}{=} \left(\frac{1}{2}\right) \left[e^s\right]_{x=a}^{x=b} \\
&= \left(\frac{1}{2}\right) \left[e^{x^2}\right]_{x=a}^{x=b} \\
&= \left(\frac{1}{2}\right) \left[e^{b^2} - e^{a^2}\right]
\end{aligned}
\tag{6.18}
$$

And we're done. We just showed that

$$
\int_a^b x e^{x^2} \, dx = \left(\frac{1}{2}\right) \left[e^{b^2} - e^{a^2}\right]
\tag{6.19}
$$

This may not look like a very simple answer, but it tells us all sorts of crazy things that are far from obvious. For example, when $a = 0$ and $b = 1$ it says

$$
\int_0^1 x e^{x^2} \, dx = \frac{1}{2} (e - 1)
$$

How Is This an Anti-Hammer?

The specific example above was...well...a *specific* example, but there's a much more general principle hiding inside it. Trying to extract this general principle will hopefully make more clear how this style of reasoning is, in a sense, the "opposite" of the hammer for reabbreviation. Suppose we're stuck on a problem that looks like $\int_a^b m(x) \, dx$. If we can somehow think of a machine M whose derivative is m, then we can use the fundamental hammer and be done with the problem. But what if we get stuck trying to think of such a machine M? As always, we can try to reabbreviate and see if it helps. Suppose we reabbreviate a big scary chunk of symbols inside the integral as s. Let's call whatever is now inside the integral $\hat{m}(s)$, even though $m(x)$ and $\hat{m}(s)$ are really just two different abbreviations for the same thing. I'm writing the "hat" on $\hat{m}(s)$ instead of just writing $m(s)$ to emphasize that $\hat{m}(s)$ is the machine $m(x)$ written in s language. It's *not* simply $m(x)$ with s plugged in in place of x! For example, if $m(x) \equiv [V(x) + 7x - 2]^{795}$, and if we abbreviate $s \equiv V(x) + 7x - 2$, then $\hat{m}(s)$ would be s^{795}. Since we're throwing two letters around, I'll change the a and b into $x = a$ and $x = b$ to remind us what we're saying. Using these ideas lets us write:

$$
\int_{x=a}^{x=b} m(x) \, dx \equiv \int_{x=a}^{x=b} \hat{m}(s) \, dx
$$

It's not clear how we can use the fundamental hammer while we have two letters floating around. Let's try to eliminate more x's by writing them in s language. We need to have a ds instead of a dx on the far right, so we can lie and correct to write dx as $\left(\frac{dx}{ds}\right)ds$, and try to re-express all the remaining x's (the ones in $\frac{dx}{ds}$, and the ones in the sentences $x = a$ and $x = b$) in s language.

$$\int_{x=a}^{x=b} \hat{m}(s)\, dx = \int_{x=a}^{x=b} \hat{m}(s) \left(\frac{dx}{ds}\right) ds \qquad (6.20)$$

Will this help? We can't tell. We're not even looking at a specific problem! This is just an abstract description of what we did in the example above with xe^{x^2}. Let's summarize the idea in a box.

Anti-Hammer for Reabbreviation

If we're stuck on something like

$$\int_{x=a}^{x=b} m(x)\, dx$$

then we're always free to come up with an abbreviation s for a bunch of scary stuff in $m(x)$, and rewrite $m(x)$ as $\hat{m}(s)$, where $\hat{m}$ is some hopefully less-scary-looking way of writing $m(x)$. Then we can lie and correct for the lie to rewrite the problem like this:

$$\int_{x=a}^{x=b} m(x)\, dx = \int_{x=a}^{x=b} \hat{m}(s)\frac{dx}{ds}\, ds$$

If we can rewrite all the x's in s language, we'll have translated the original problem into a different one. This won't necessarily make the problem simpler, but it might if we reabbreviate cleverly. For some reason, textbooks call this process "u-substitution," and use the letter u where we used s. But it's all just reabbreviation.

As simple as this idea is, it's hard to come up with abbreviations that describe exactly how simple it is. Here's another way to think about it. Suppose we're going through this whole annoying process of (i) reabbreviating scary-looking chunks by s, (ii) lying and correcting for the lie to get everything in s language, and then (iii) staring at the rewritten version of the original problem to see if it looks any easier. It turns out that this process will automatically recognize when the machine $m(x)$ can be thought of as arising from original the hammer for reabbreviation. To see what I mean, let's look at a specific example. Suppose we're stuck trying to calculate something like this:

$$\int_{x=a}^{x=b} \underbrace{51 \left(x^5 + 17x - 3\right)^{999} \left(5x^4 + 17\right)}_{\text{Call this part } m(x)} dx \tag{6.21}$$

Now, it turns out that this machine m can be thought of as arising from taking the derivative of this big ugly machine

$$M(x) \equiv \frac{51}{1000} \left(x^5 + 17x - 3\right)^{1000}$$

by using the hammer for reabbreviation, but let's imagine that we don't notice that fact. We're just hopelessly stuck trying to calculate the stuff in equation 6.21. Let's see where the process of reabbreviation will get us, if we were to try it. Suppose that by luck or insight or anything else, we choose to use the abbreviation $s \equiv x^5 + 17x - 3$. That's one of the uglier pieces in equation 6.21, so this strategy makes a certain amount of sense. This changes the problem to

$$\int_{x=a}^{x=b} 51 s^{999} \left(5x^4 + 17\right) dx \tag{6.22}$$

Now, this process won't help us unless we can get everything in s language, so let's start by trying to translate the dx piece into s language. Since $s \equiv x^5 + 17x - 3$, we know that

$$\frac{ds}{dx} = 5x^4 + 17 \qquad \text{so} \qquad dx = \frac{ds}{5x^4 + 17}$$

Substituting this expression for dx into equation 6.22 collapses everything very nicely, to give us

$$\int_{x=a}^{x=b} 51 s^{999} \left(5x^4 + 17\right) \left(\frac{ds}{5x^4 + 17}\right) = \int_{x=a}^{x=b} 51 s^{999} \, ds \tag{6.23}$$

Now the problem is slightly less crazy-looking! Can we think of a machine whose derivative (thinking of s as the variable) is $51s^{999}$? Well, it had better look like $\# \cdot s^{1000}$ so that the power turns into 999 when we differentiate it. But then we have to make sure the $\#$ is such that $1000 \cdot \# = 51$, which means that $\# = \frac{51}{1000}$. Putting it all together, we figured out that an "anti-derivative" of $51s^{999}$ is

$$M(x) = \frac{51}{1000} s^{1000} \equiv \frac{51}{1000} \left(x^5 + 17x - 3\right)^{1000}$$

So the answer to the original problem, which seemed so impossible to begin with, is just $M(b) - M(a)$, where M is the ugly machine in the above equation. Notice that we didn't have to recognize that $m(x)$ in equation 6.21 arises from using the hammer for reabbreviation on M. We didn't even have to know what M was! Rather, just by defining s to be an abbreviation for the ugliest piece in equation 6.21 and then translating all the x stuff into s language, we found that we had transformed the scary-looking problem we started with into the much simpler problem of computing $\int_{x=a}^{x=b} 51s^{999}\, ds$. The net effect of this process was that we ended up doing a mathematical dance that, in the end, told us the anti-derivative $M(x)$, even though we ourselves couldn't just magically think of $M(x)$ all in one step. This reabbreviation process — as difficult as it can be to explain in symbols — effectively bootstraps us up past our own ignorance to a place where we can solve problems that we couldn't solve without reabbreviating.

6.3.4 Collecting the Anti-Hammers

Having forged three anti-hammers, one for each of the originals, let's summarize all of them in abbreviated form.

Anti-Hammer for Addition (AHA)

Suppose we're stuck on something that looks like

$$\int_a^b m(x)\, dx$$

If we can think of machine f and g for which $m(x) = f(x) + g(x)$, we're free to break the problem apart like this, if it helps:

$$\int_a^b f(x) + g(x)\, dx = \int_a^b f(x)\, dx + \int_a^b g(x)\, dx$$

Anti-Hammer for Multiplication (AHM)

Suppose we're stuck on something that looks like

$$\int_a^b m(x)\, dx$$

If we can think of machine f and g for which $m(x) = f'(x)g(x)$, we're free to transform the problem like this, if it helps:

$$\int_a^b f'(x)g(x)\, dx = \Big[f(x)g(x) \Big]_a^b - \int_a^b f(x)g'(x)\, dx$$

Anti-Hammer for Reabbreviation (AHR)

Suppose we're stuck on something that looks like

$$\int_a^b m(x)\, dx$$

If we can think of an abbreviation s for which $m(x)$ can be written in a simpler looking form $\hat{m}(s)$, then we're free to transform the problem like this, if it helps:

$$\int_{x=a}^{x=b} m(x)\, dx = \int_{x=a}^{x=b} \hat{m}(s)\frac{dx}{ds}\, ds$$

6.3.5 The Other Fundamental Hammer

At the beginning of this chapter, we discovered the fundamental hammer, and discussed how its basic message was that integrals and derivatives were opposites. However, what we really established is that integrals and derivatives are opposites if the derivative shows up *inside* the integral. It's easy to remember the general idea that integrals and derivatives are opposites, but the story would be less elegant if they were only opposites in a given order — that is, if we do the derivative first, and then the integral. This desire for a greater elegance of the narrative motivates us to see if there is any sense in which derivatives and integrals are opposites in the other order. What we might think to try first is to see if we can calculate

$$\frac{d}{dx}\int_a^b m(x)\, dx \tag{6.24}$$

but this expression turns out to be a bit misleading. That is, the x in the expression $\int_a^b m(x)\, dx$ is not really a "variable" in the same sense as x is a variable in an expression like $f(x) \equiv x^2$. In technical jargon, the x in an integral is what is called a "bound variable," which is to say that it is not something that we can plug a number into, but simply a placeholder. It serves the same purpose as the letter i does in the expression

$$\sum_{i=1}^{3} i^2 \tag{6.25}$$

This is just a fancy way of writing the number 14 (because $1+4+9 = 14$), so it doesn't make sense to plug something like $i = 17$ into equation 6.25. We could change i to some other letter like j or k in equation 6.25 and the expression would still just be a fancy way of writing the number 14. For the same reasons, the expression $\int_a^b m(x)\,dx$ doesn't depend on x, and it's no different from the expressions $\int_a^b m(y)\,dy$ and $\int_a^b m(\star)\,d\star$. It might then appear that our problem is solved. Since $\int_a^b m(\star)\,d\star$ is just a number, independent of x, we can write

$$\frac{d}{dx}\int_a^b m(\star)\,d\star = 0 \tag{6.26}$$

Hmm... This is hardly another version of the fundamental hammer. If derivatives always kill integrals from the outside, then it would appear that the two concepts are not opposites after all. However, such a conclusion would be hasty. What we really need is a different way of thinking about the problem. The derivative (with respect to x) of the integral was equal to zero because the integral didn't depend on x. Maybe if we ask a slightly different question, we'll get something more interesting. Let's try to take the derivative with respect to the number on the top of the integral:

$$\frac{d}{dx}\int_a^x m(s)\,ds$$

where we're using s instead of x as a label for the "bound variable" to avoid any confusions that might result from using x for both. There are two ways we might unravel this weird expression. First, if we write M for the anti-derivative of m, we can simply use the version of the fundamental hammer that we already discovered to obtain

$$\frac{d}{dx}\int_a^x m(s)\,ds$$

$$= \frac{d}{dx}\Big(M(x) - M(a)\Big)$$

$$= \left(\frac{d}{dx}M(x)\right) - \underbrace{\left(\frac{d}{dx}M(a)\right)}_{\text{This is zero, since it doesn't depend on }x}$$

$$= \left(\frac{d}{dx}M(x)\right)$$

$$= m(x)$$

Another way to show the same thing is to make a tricky, informal argument using the definition of the derivative, like this:

$$\frac{d}{dx}\int_a^x m(s)\,ds \equiv \frac{\int_a^{x+dx} m(s)\,ds - \int_a^x m(s)\,ds}{dx}$$

$$= \frac{\int_x^{x+dx} m(s)\,ds}{dx}$$

$$= \frac{1}{dx}\int_x^{x+dx} m(s)\,ds$$

where in passing from the first line to the second we basically just imagined tearing the full area into two pieces, letting us say that [the area from a to $(x + tiny)$] minus [the area from a to x] is just [the area from x to $(x + tiny)$]. Well, it might seem like we're stuck, but if we remember what everything means, then it isn't too hard to get unstuck. The funny term $\int_x^{x+dx} m(s)\, ds$ refers to the area under m's graph between two points that are infinitely close to each other: x and $x + dx$. This is therefore the area of a rectangle of width dx and height $m(x)$, which is of course $m(x)dx$. The final line above is just $\frac{1}{dx}$ times that, so the final line must be $m(x)$. Or, tying it all together,

$$\frac{d}{dx} \int_a^x m(s)\, ds = m(x)$$

and we therefore obtained the same answer as before: integrals and derivatives undo each other in *either* direction. To summarize what we've shown, I'll list both versions of the fundamental hammer here, and I'll write the old version in a slightly different way, to illustrate its relationship to the new version:

Fundamental hammer version 1 : $\qquad \int_a^b \left(\frac{d}{dx} m(x) \right)\, dx = m(b) - m(a)$

Fundamental hammer version 2 : $\qquad \dfrac{d}{dx} \displaystyle\int_a^x m(s)\, ds = m(x)$

6.4 The Second Cloud

We've conquered curviness in two different domains already — first steepness, now area — but we still don't know how to compute the lengths of curvy things. It's worthwhile at this point to look at the collected memoirs of our journey, and see what has helped us in the past.

In the domain of steepness, we conquered curviness by zooming in infinitely far on a curvy machine m (at a point x whose value we chose to remain agnostic about) and computing its steepness as if it were a straight line.

In the domain of area, we conquered curviness by realizing that the tiny area under a machine's graph at each point x can be thought of as an infinitely thin rectangle, with area $m(x)dx$. Then we just imagined adding up all the tiny areas. We had no idea how to do this at first, but we came up with a name for the unknown answer: $\int_a^b m(x)dx$. Once we discovered the fundamental hammer, we found that our new $\int$ idea was just our old derivative idea in reverse.

Now what? We've got enough experience with our magnifying glass that it shouldn't be too difficult to at least write down an expression for the lengths of curvy things. Whether or not we can actually compute the lengths of curvy things in any particular case is a different story, just as it was in the case of

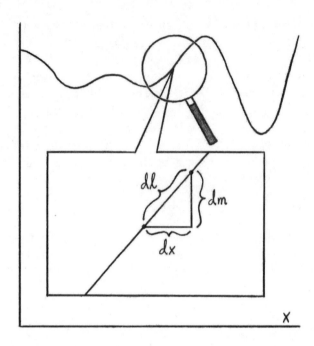

Figure 6.6: Trying to think of a way to compute the lengths of curvy things. The zooming in idea has worked in the past, so let's try it again. We zoom in infinitely far on the graph of some machine m at some point x, and the curve looks straight. Then we can use the formula for shortcut distances to compute the tiny length $d\ell$. Then I guess we just add up all the pieces. Wait, is that it?

area. Our experience with the magnifying glass so far suggests that we start with a machine m, and then imagine zooming in infinitely far on some point of its graph, remaining agnostic about which particular point we're zooming in on. Then, as always, let's look at some other point that's infinitely close to the first one. See Figure 6.6 for a picture of this situation.

Once we've zoomed in, the problem becomes much less intimidating. We've got two points whose horizontal distance apart we're calling dx, and whose vertical distance apart we're calling dm. As always, dm is just an abbreviation for $m(x + dx) - m(x)$, but we won't need to unwrap this abbreviation for our current purposes. Let's write $d\ell$ as an abbreviation for the actual distance between the two points — that is, the distance we'd experience if we were walking along the graph. Since zooming in turned the curvy thing into an infinitely small straight thing, we can use the formula for shortcut distances to write this:

$$(d\ell)^2 = (dx)^2 + (dm)^2$$

or equivalently, $d\ell = \sqrt{(dx)^2 + (dm)^2}$. Whatever the total length of our curvy thing happens to be, it should be whatever we would get if we could somehow add up all the infinitely small pieces $d\ell$.

What do we mean by the "total length" of a curvy thing? The curve might go on forever, in which case the answer would be infinity, but that's not quite the question we meant to ask. We really want to talk about the length of m's graph between any two points $x = a$ and $x = b$. So, to summarize this paragraph in symbols, it should be the case that:

$$\text{Total length between } a \text{ and } b \;\equiv\; \int_a^b d\ell \;=\; \int_a^b \sqrt{(dx)^2 + (dm)^2} \qquad (6.27)$$

Okay... We haven't seen anything like this before. We're only used to expressions involving the $\int$ symbol that look like this:

$$\int_a^b (\text{Some Machine}) \; dx$$

So let's try to force the $\sqrt{(dx)^2 + (dm)^2}$ piece in equation 6.27 to look like (Some Machine) dx. We can lie and correct to get $\sqrt{(dx)^2 + (dm)^2}\,\frac{1}{dx}dx$. The dx piece on the far right is what we wanted, but the leftovers are pretty ugly. Let's try to bring the $1/(dx)$ inside the rest of the leftovers. The square root symbol makes it harder to remember what we're allowed to do, so let's turn the square root symbol into a $\frac{1}{2}$ power. Because of the way we invented powers, we know that we can merge two things if they have the same power, like this:

$$(\text{thing})^{\#}(\text{other thing})^{\#} = \Big[(\text{thing}) \cdot (\text{other thing})\Big]^{\#} \qquad (6.28)$$

So if we want to move the $\frac{1}{dx}$ inside the confusing leftovers, this suggests that we should lie and correct yet again, to write the $\frac{1}{dx}$ in this funny sort of way:

$$\frac{1}{dx} = \left(\frac{1}{(dx)^2}\right)^{\frac{1}{2}} \qquad (6.29)$$

If we do that, then we can pick up where we left off and write the following. Don't be scared by the mountain of symbols below this! It looks like there are a ton of steps in the derivation below, but most of them could probably be skipped. I'm showing more steps because I really like this derivation, and I want it to be as easy as possible to understand each step. But each step is really simple. Ready? Here we go:

$$\sqrt{(dx)^2 + (dm)^2} = \left((dx)^2 + (dm)^2\right)^{\frac{1}{2}} \left(\frac{1}{(dx)^2}\right)^{\frac{1}{2}} dx$$

$$= \left[\left((dx)^2 + (dm)^2\right)\left(\frac{1}{(dx)^2}\right)\right]^{\frac{1}{2}} dx$$

$$= \left(\frac{(dx)^2 + (dm)^2}{(dx)^2}\right)^{\frac{1}{2}} dx$$

$$= \left(\frac{(dx)^2}{(dx)^2} + \frac{(dm)^2}{(dx)^2}\right)^{\frac{1}{2}} dx \qquad (6.30)$$

$$= \left(1 + \frac{(dm)^2}{(dx)^2}\right)^{\frac{1}{2}} dx$$

$$= \left(1 + \left(\frac{dm}{dx}\right)^2\right)^{\frac{1}{2}} dx$$

$$\equiv \sqrt{1 + \left(\frac{dm}{dx}\right)^2} \; dx$$

Now for the fun part. We can use this insight to write equation 6.27 in a much less confusing way. Simply throwing the equation we just derived into 6.27, we discover that the total length of any curvy thing m can be written like this:

$$\text{Total length of } m \text{ between } a \text{ and } b = \int_a^b \sqrt{1 + \left(\frac{dm}{dx}\right)^2} \; dx$$

$$\equiv \int_a^b \sqrt{1 + [m'(x)]^2} \; dx \qquad (6.31)$$

Beautiful! In the first part of this chapter, we found out that our new $\int$ idea was just the opposite of our old derivative idea, so in that sense we got two ideas in one. Just now, we had a newer idea of computing the length of a curvy thing by zooming in until it looks straight, measuring the tiny piece of length down there at the microscopic level, and then adding up the tiny lengths. Surprisingly, we've now discovered that this idea is really the same idea as the $\int$ thing, which was just the opposite of the derivative. Everything hangs together so nicely, and in a way that is so much prettier than we had any reason to expect! Equation 6.31 summarizes everything we did in this section, and as complicated as it looks, it is just saying something like this:

> **Question:** Hey, I've got a curvy thing $m(x)$. How do I compute its length?
> **Answer:** I don't know, but it's the same as the area under $\sqrt{1 + m'(x)^2}$
> **Question:** Does that make the length easier to compute?
> **Answer:** Maybe for certain m's. I don't know. Leave me alone.

6.5 Reunion

1. In this chapter, we wrote down an abbreviation for the area under a curvy thing, namely $\int_a^b m(x)dx$. This abbreviation reflects the fact that we can think of a curvy area as the sum (hence the S-like $\int$ thing) of an infinite number of infinitely thin rectangles (hence the $m(x)dx$), but it didn't tell us how to actually calculate curvy areas.

2. Eventually, we realized that if we thought of the m in $\int_a^b m(x)dx$ as the derivative of some other machine M, then we could rewrite the curvy area as $\int_a^b \frac{dM}{dx} dx$, and cancel the dx's. Since dM was our abbreviation for $M(x+dx) - M(x)$, we knew that $\int_a^b dM$ must be the sum of all the tiny changes in height as we walk from $x = a$ to $x = b$, which is to say $M(b) - M(a)$. That is:

$$\int_a^b m(x)dx = M(b) - M(a)$$

where M is any machine whose derivative is m (an "anti-derivative" of m). We called this the "fundamental hammer of calculus," since it relates our old idea (the derivative) to our new idea (the integral) and shows us a sense in which they are opposites.

3. We tested the fundamental hammer in a few simple cases where we could compute the required areas without any calculus. In each case, the fundamental hammer gave us the results we expected, which gave us a bit more confidence that our derivation of it made sense.

4. We found that any given machine has infinitely many anti-derivatives, but in a boring sense that doesn't make integrals any harder to compute, and doesn't make the fundamental hammer any harder to use. That is, if M is an "anti-derivative" of m, so that $M'(x) = m(x)$, then all the machines $M(x) + \#$ are also anti-derivatives of m.

5. We found that each of the derivative hammers we invented in Chapters 2 and 3 had a partner "anti-hammer" that told us about integration. We proceeded to invent anti-hammers for addition, multiplication, and reabbreviation, each of which "undoes" the original hammer with the same name.

6. We discovered that there is another sense in which derivatives and integrals are opposites, which essentially amounted to a second way of writing the fundamental hammer. That was

$$\frac{d}{dx} \int_a^x m(s)\, ds = m(x)$$

Therefore, our two calculus concepts can be thought of as opposites, irrespective of the order in which we apply them.

7. We turned our attention to the problem of curvy lengths and found that this too involved integration. We zoomed in infinitely far and used the formula for shortcut distances to discover that the length of a machine m's graph between two points $x = a$ and $x = b$ was

$$\int_a^b \sqrt{1 + \left(\frac{dm}{dx}\right)^2} \, dx$$

Having reduced the problem of finding curvy lengths to the problem of computing an integral (just like we did for curvy areas), we chose not to focus on a bunch of specific examples of how to compute curvy lengths. Since both problems are now phrased in the language of integrals, any skill we develop on one of these problems immediately carries over to the other. As such, we choose instead to forge ahead into new territory.

Interlude 6: Slaying Sharp

Best Served Cold

(*Author is frantically working on something, as Reader enters the interlude. All manner of paraphernalia is strewn about the room. Author is pacing back and forth next to a table, frowning at some papers.*)

Reader: What's all this? What's happening?

(*Author is absorbed in whatever he is doing.*)

Reader: Hellooooo???

(*Author continues working frantically.*)

Reader: Author!!!
Author: (*Startled*) Oh, hi. Good to see you.
Reader: So, what are we doing today?
Author: Settling the score.
Reader: What score? What are you talking about?
Author: Revenge!
Reader: Revenge? Against who? Everyone we've met has been really nice.
Author: I know!
Reader: Are you doing okay? You seem different recently. . .

(*Mathematics enters the room.*)

Author: Finally! You're here!
Mathematics: WHAT'S GOING ON? THIS HAD BETTER BE IMPORTANT.
Author: Oh, it is!
Mathematics & Reader: (*Simultaneously*) WhAt ArE wE dOiNg?

(*Author smiles.*)

Author: We're slaying ♮.
Reader: Finally!
Mathematics: I THOUGHT WE WOULD NEVER GET AROUND TO THIS!
Author: Come on now, team. Watches synchronized?
Reader: Check.
Mathematics: I DON'T HAVE A WATCH. THE VOID IS TIMEL—
Author: Helmets securely fastened?

Reader: Check.

Mathematics: I JUST GOT HERE ABOUT HALF A PAG—

Author: We may have to use the Nostalgia Device, so it'll need to be close at hand the entire time. First things first. We've got to form a plan of attack or else this thing is going to defeat us.

Mathematics: ALLOW ME!

A Plan of Attack

Mathematics: HOW'S THAT?

Author: Well, that's a start, but making a section called "A Plan of Attack" isn't the same as making a plan of attack. Philosophers call that a "use mention error."

Mathematics: ...

Author: Okay. There are several different routes we could take, and it's best to keep them all in mind. If one route doesn't work, we can fall back and try another. Any sentence we know that somehow involves ♯ is a possible weak point, so we need to focus on those. This ♯ thing tends to show up in sentences about circles, so let's start with those. We know that the area of a circle with radius r is

$$A(r) \equiv \sharp r^2$$

So ♯ itself is just the area of a circle with radius 1. We've built up some artillery that lets us compute the area of certain curvy things, so that's one possible path we could take.

Mathematics: WHAT? YOU TWO FIGURED OUT HOW TO COMPUTE THE $\int$ THING?

Author: Well, sometimes. But not always.

Reader: Oh, that's right, you weren't there.

Author: Well it's all on paper now. You're welcome to walk back to Chapter 6 and check it out at your leisure. But come on, let's focus. Other plans of attack!

Mathematics: OKAY, ♯ ALSO SHOWS UP IN THE DISTANCE AROUND A CIR-CLE WITH RADIUS r:

$$L(r) \equiv 2\sharp r$$

Mathematics: SO SINCE r IS HALF THE DISTANCE ACROSS A CIRCLE, ♯ IT-SELF IS JUST THE DISTANCE AROUND A CIRCLE WHOSE DISTANCE ACROSS IS 1.

Reader: At the end of the last chapter we found a way of computing curvy lengths, so that's another possible route.

Mathematics: WHAT?

Reader: Oh, that's right, you weren't there for that either.

Mathematics: DID YOU TWO INVENT ANY MORE OF ME THAT I'VE YET

TO BE INFORMED OF?
Reader: Nope, just Chapter 6.
Mathematics: GOOD. WELL, WE'VE GOT ALL THE CALCULUS WE'VE INVENTED SO FAR, LIKE THE HAMMERS.
Reader: And the anti-hammers.
Author: And the Nostalgia Device.
Mathematics: HEY, IN CHAPTER 5 WE USED THE NOSTALGIA DEVICE TO COMPUTE THAT NUMBER e. MAYBE IT WOULD WORK HERE TOO.
Reader: Good idea. How might we use the Nostalgia Device to compute ♯?
Author: Well when we computed e, we had this machine $E(x) \equiv e^x$ that spits out the number e when we feed it 1. The Nostalgia Device helped us talk about that machine using just addition and multiplication, like this:

$$e^x = \sum_{n=0}^{\infty} \frac{x^n}{n!} \qquad \text{so} \qquad e = \sum_{n=0}^{\infty} \frac{1}{n!}$$

Could we do that here?
Reader: We would need to think of a machine that spits out ♯ when we feed it some number.
Mathematics: SIMPLE ENOUGH. JUST DEFINE $M(x) \equiv x$, AND WE'VE GOT A MACHINE THAT SPITS OUT ♯ WHEN WE FEED IT ♯.
Reader: I don't know if that helps.
Author: Right... that's true but it's not really helpful. Using the Nostalgia Device on x just gives us x again. It's true that ♯ = ♯, but that doesn't really tell us what ♯ is.
Mathematics: THEN WHAT DO WE NEED?
Author: Well, in the example of e^x, we didn't just have a machine that spits out e when we feed it *something*. We had a machine that spits out e when we feed it something *simple*: the number 1. That gave us a description of e in terms of simpler things we already knew. We need to describe ♯ in terms of things that are as simple as possible.
Mathematics: OH. I DON'T KNOW HOW TO DO THAT.
Reader: Me neither.
Author: Me neither! But let's not get discouraged. Look at all these weapons we've got to tackle the problem! (*Author clears his throat.*) We've got...

An Excessive Pile of Weaponry

(*The characters look around the room, which is indeed littered
with [title-of-this-section]. There are multiple hammers, equally many
anti-hammers, an infinite magnifying glass, the Nostalgia Device, and two
oddly shaped objects labeled "Lying" and "Correcting for the Lie." A black
sheet covers an object in the corner, next to a sign that says "Tricking the
Mathematics: Out of Order."*)

Author: Look at all this stuff! There's no way this problem can defeat us now.

Reader: Are you sure?

Author: No! But come on, let's give it a shot. What do we need?

Mathematics: WELL, WE NEED A MACHINE THAT SPITS OUT ♯ WHEN WE FEED IT SOMETHING SIMPLE.

Reader: We kind of have the reverse of that.

Author: How do you mean?

Reader: Well, remember V and H? The things textbooks call sine and cosine?

Author: Of course.

Reader: Remember how V spits out 1 when we feed it $\frac{\sharp}{2}$?

Author: No.

Reader: You never remember anything. But you said it yourself in Chapter 4. We defined V visually and figured out that

$$V\left(\frac{\sharp}{2}\right) = 1$$

That's kind of the reverse of the e situation.

Mathematics: EXPLAIN?

Reader: Well, instead of spitting out the number we want when we feed it something simple, like e^x did, V spits out something simple when we feed it something we want! Or, something related to what we want. I mean, if we could figure out $\frac{\sharp}{2}$ then we'd also have figured out $\sharp$.

Mathematics: OH! OKAY. SO, IF THERE WERE AN "OPPOSITE" OF V — WHATEVER THAT MEANS — THEN IT WOULD EAT 1 AND SPIT OUT $\frac{\sharp}{2}$, RIGHT?

Author: What's the opposite of V?

Mathematics: I DON'T KNOW.

Author: I don't know either.

Reader: Wait, does it matter if we're not really familiar with the opposite of V? Can't we just give it a name and see how it has to behave?

Author: Sure, I guess we could try.

Mathematics: LET ME NAME IT!

Author: Okay.

Mathematics: I'LL ABBREVIATE THE OPPOSITE MACHINE OF V AS... Λ. WE DON'T KNOW WHAT Λ LOOKS LIKE EXACTLY, BUT IT'S WHATEVER MACHINE DOES THIS:

$$V(\Lambda(x)) = x \qquad \text{and} \qquad \Lambda(V(x)) = x$$

NO MATTER WHAT NUMBER x IS. SO WHATEVER V DOES, Λ DOES THE OPPOSITE.

Reader: Exactly! So it's got to be true that

$$\Lambda(1) = \frac{\sharp}{2} \qquad \text{so} \qquad \sharp = 2\Lambda(1)$$

Mathematics: So if we could figure out another way to compute $\Lambda(1)$, we'd be done?

Author: I guess so. How are we supposed to do that?

Reader: We could use the Nostalgia Device.

Author: Nah, that won't work. We don't know any of the derivatives of Λ. We need to know *all* of the derivatives to use the Nostalgia Device.

Mathematics: Perhaps we could trick the mathematics into telling us what the derivatives are.

Hammering with Reabbreviation

> (*Author and Reader simultaneously turn to Mathematics, with equally shocked expressions.*)

Reader: Wait, the first time we ever met, you got all mad at us for using that phrase.

Author: Yeah, all of a sudden you don't mind us saying "trick the Mathematics"?

Mathematics: Well, I said it with a lowercase m, so it doesn't refer to me. My name starts with a capital M.

Reader: Those are both a capital M.

Author: No no no! Mathematics speaks in "small caps." It's a typography thing. Totally different emotional tone from just writing something in ALL CAPITAL LETTERS.

Reader: Wait... so the Mathematics character isn't always yelling?

Author: What? Of course not! You didn't think that this whole time, did you?

Reader: I don't know... I may have...

Mathematics: Enough, you two. Like I said, we might be able to trick the mathematics by doing something like this: let's define the ambush machine as $A(x) \equiv \Lambda(V(x))$. Secretly, this is just the machine $A(x) = x$, so $\frac{dA}{dx} = 1$. Now we can use the reabbreviation hammer, like this:

$$1 = \frac{d}{dx}A(x) \; \equiv \; \frac{d}{dx}\Lambda(V(x)) \; = \; \frac{dV}{dx}\frac{d}{dV}\Lambda(V(x)) \qquad (6.32)$$

Author: How does that help?

Reader: Well, there are two pieces on the far right. We know one of them, because $\frac{dV}{dx} = H(x)$.

Mathematics: Exactly. So use that to rewrite what we just did. That gives:

$$1 = H(x)\left(\frac{d}{dV}\Lambda(V(x))\right)$$

Author: Okay, but what's $\frac{d}{dV}\Lambda(V(x))$?

Reader: I don't know.

Mathematics: COME ON, AUTHOR. REMEMBER, WE CAN ALWAYS CHANGE ABBREVIATIONS. WOULD IT HELP IF I REWROTE THE SAME THING LIKE THIS:

$$\frac{d}{dV(x)}\Lambda(V(x)) \qquad \text{OR} \qquad \frac{d}{dV}\Lambda(V) \qquad \text{OR} \qquad \frac{d}{dx}\Lambda(x)$$

Author: No.

Reader: Oh! I see. Since we can always change abbreviations, those are all the same.

Mathematics: EXACTLY! THAT'S ALL I WAS TRYING TO SAY. WE CAN TRICK THE MATHEMATICS INTO TELLING US THE DERIVATIVE OF Λ BY DEFINING $A(x) \equiv \Lambda(V(x))$ AND THEN DIFFERENTIATING A IN TWO WAYS. ON THE ONE HAND, $\frac{dA}{dx} = 1$. ON THE OTHER HAND, WE CAN USE THE REABBREVIATION HAMMER LIKE WE DID ABOVE, WHERE WE FOUND

$$1 \;=\; H(x)\left(\frac{d}{dV}\Lambda(V)\right) \tag{6.33}$$

SO WE CAN THROW $H(x)$ OVER TO THE OTHER SIDE TO ISOLATE THE DERIVATIVE OF Λ, LIKE THIS:

$$\frac{d}{dV}\Lambda(V) = \frac{1}{H(x)} \tag{6.34}$$

Author: Sorry to keep weighing down this discussion, but I still don't see how this helps us at all. I can see how this type of argument can work in less tangled cases. Like, if you made an argument that showed something like

$$\frac{d}{d\star}M(\star) = \frac{1}{\star} \qquad \text{then you could reabbreviate to} \qquad \frac{d}{dx}M(x) = \frac{1}{x}$$

But what you wrote down in equation 6.34 is confusing and I don't see how it helps. We want $\frac{d}{dx}\Lambda(x)$, and sure that's the same as the original expression with Vs in place of xs, because we can abbreviate things however we want, but when you change the Vs to xs, what happens to that $H(x)$ on the right side? If you've got an entire equation written with V as the "variable," then I totally see how you can replace that with any other letter you want, but you've already got xs in there, and... I just don't know what's happening.

Mathematics: OH, WELL THEN WE SHOULD TRY AND WRITE H IN TERMS OF V. IF WE COULD DO THAT, THEN EQUATION 6.34 WOULD BE WRITTEN COMPLETELY IN V LANGUAGE, AND THEN WE COULD JUST REABBREVIATE ALL THE Vs TO xs, AND WE WOULD HAVE TRICKED ME INTO TELLING US THE DERIVATIVE OF Λ.

Author: Okay, I'm still not completely on board, but I think I see where you're going. The formula for shortcut distances told us that $V^2 + H^2 = 1$, which we could rewrite as $H = \sqrt{1 - V^2}$. If we do that, then equation 6.34 becomes

$$\frac{d}{dV}\Lambda(V) = \frac{1}{\sqrt{1 - V^2}} \tag{6.35}$$

Now there are only Vs everywhere, so since we can abbreviate things however we want, it has to be true that

$$\frac{d}{dx}\Lambda(x) = \frac{1}{\sqrt{1 - x^2}} \tag{6.36}$$

Author: But something still feels funny about this argument.

Reader: Well, you're always saying that if we're not sure about an argument, it helps to see if we can get the same result in a different way. We could try that.

Author: Alright... I guess we could run the same argument inside out.

Reader: How so?

Author: Well, in the above argument we used the sentence $\Lambda(V(x)) = x$ to trick the mathematics into telling us $\Lambda'(x)$. What if instead we used $V(\Lambda(x)) = x$, but otherwise did everything exactly the same? I mean, if we do that and get the same answer for $\Lambda'(x)$, I guess I'd be a bit less worried. You got a minute?

Reader: A minute? If I've made it this far in the book, I'm probably an extremely patient person. Take your time.

Author: (*Sigh*[3]) You're so great. Okay, I'll try to be quick. This time let's define $A(x) \equiv V(\Lambda(x))$, which is also just x because Λ and V undo each other, so the derivative of A is just 1 again. Now let's use the same type of reasoning we did before:

$$1 = \frac{d}{dx}A(x) \equiv \frac{d}{dx}V(\Lambda(x)) = \frac{d\Lambda(x)}{dx}\frac{d}{d\Lambda(x)}V(\Lambda(x)) \equiv \frac{d\Lambda}{dx}\frac{d}{d\Lambda}V(\Lambda) \tag{6.37}$$

The first equality is because $A(x) = x$, the third uses the hammer for reabbreviation, and the other two are just definitions. Now, on the far right we have the derivative of V, which is H. So we have

$$1 = \left(\frac{d\Lambda}{dx}\right)H(\Lambda)$$

This $H(\Lambda)$ piece looks funny, but it's just $H(stuff)$, and any stuff inside H can always be thought of as an angle. So we can just use the formula for shortcut distances to get $H(\Lambda)^2 + V(\Lambda)^2 = 1$. Then just isolate $H(\Lambda)$ to get $H(\Lambda) = \sqrt{1 - V(\Lambda)^2}$, which lets us write

---

3. (*Reader's patience in the face of Author's unbounded verbosity warms Author's heart, and Author is overcome with a feeling of $\frac{1}{2}$(platonic+undefined) love, followed soon thereafter (read: now) by a deep feeling of embarrassment at having divulged the former feeling so publicly. After all, Author thought, one isn't supposed to say such things in a textbook (or whatever this is). But I digress...*)

$$1 = \left(\frac{d\Lambda}{dx}\right)\left(\sqrt{1 - V(\Lambda)^2}\right) \qquad \text{so} \qquad \frac{d\Lambda}{dx} = \frac{1}{\sqrt{1 - V(\Lambda)^2}}$$

Wait, that's different from what we got last time.

Mathematics: I'M NOT SURE IT IS! WE STOPPED WRITING THE x'S SO THAT THINGS WOULD LOOK LESS SCARY, REMEMBER? LIKE WHEN YOU WROTE Λ INSTEAD OF $\Lambda(x)$.

Author: Oh, right! So that $V(\Lambda)$ piece is really $V(\Lambda(x))$, which by definition is just x. So we can rewrite what we just found like this:

$$\frac{d\Lambda}{dx} = \frac{1}{\sqrt{1 - x^2}}$$

which is exactly the same thing we got in equation 6.36!

Mathematics: CONVINCED?

Author: Slightly more than I was before.

Sharp Resists

Author: Okay, what now?

Mathematics: WELL, WE JUST CONVINCED OURSELVES THAT

$$\frac{d\Lambda}{dx} = \frac{1}{\sqrt{1 - x^2}} \tag{6.38}$$

WE DID THAT BECAUSE WE REALIZED THAT $2\Lambda(1) = \sharp$, SO IF WE COULD FIGURE OUT ANOTHER WAY TO COMPUTE $\Lambda(1)$, THEN WE COULD COMPUTE $\sharp$!

Author: How does what we just did help us compute $\Lambda(1)$?

Mathematics: OH... I DON'T BELIEVE IT DOES.

Author: Ugh...

Mathematics: WE WERE PLANNING ON USING THE NOSTALGIA DEVICE, AND WE NEED *ALL* THE DERIVATIVES OF A MACHINE TO USE THAT. WE JUST FOUND THE FIRST ONE.

Author: Yeah, and it was a pain.

Reader: Can we use the fundamental hammer?

Mathematics: OOH, MAYBE. WHAT MADE YOU THINK THAT?

Reader: Well, the fundamental hammer relates machines and their derivatives, so it might let us relate the information we just discovered about the derivative of Λ to information about Λ itself. I don't know.

Author: Hey, yeah. The fundamental hammer says that

$$\int_a^b \left(\frac{d\Lambda}{dx}\right) dx = \Lambda(b) - \Lambda(a)$$

and we want $\Lambda(1)$, so if we make $b = 1$ and use equation 6.38, we can write

$$\int_a^1 \frac{1}{\sqrt{1-x^2}}\, dx = \Lambda(1) - \Lambda(a) \qquad (6.39)$$

Ugh. We just want $\Lambda(1)$. It would be nice to get rid of that $\Lambda(a)$ on the right.
Reader: Is there any a that makes $\Lambda(a) = 0$?
Mathematics: WELL, WE DON'T KNOW MUCH ABOUT Λ — WE JUST DEFINED IT TO BE WHATEVER MACHINE UNDOES V. WE KNOW $V(0) = 0$. BUT Λ UNDOES V, SO $\Lambda(0) = 0$, RIGHT?
Author: Makes sense.
Reader: Nice! So let's choose $a = 0$ in equation 6.39 to get

$$\int_0^1 \frac{1}{\sqrt{1-x^2}}\, dx \;=\; \Lambda(1) - \Lambda(0) \;=\; \Lambda(1) \qquad (6.40)$$

Now we can combine this with the fact that $\sharp = 2\Lambda(1)$ from earlier to get

$$\sharp \;=\; 2\int_0^1 \frac{1}{\sqrt{1-x^2}}\, dx \qquad (6.41)$$

Author: Beautiful! Are we done?
Mathematics: I GUESS.
Reader: No we're not! We still don't have a specific number for $\sharp$! That was the whole point.
Author: Well, how do we compute a specific number for $\sharp$?
Reader: We've got to compute the integral in equation 6.41. How do we do that?
Author: I don't know.
Mathematics: I DON'T KNOW EITHER.
Reader: You've got to be kidding...

Back to the Drawing Board

Reader: You mean to tell me that we're no closer to figuring out $\sharp$ than we were when we started!?
Author: I don't know. We're kind of "closer." We've learned a lot more about how to attack the problem... but yeah, we're still stuck. We don't know how to compute the integral in equation 6.41.
Reader: We could just do it by brute force. Approximately.
Author: You mean by adding up the areas of a bunch of tiny rectangles?
Reader: Yeah!
Mathematics: THAT SOUNDS UNPLEASANT. I'D RATHER NOT.
Reader: Come on! This is taking so long!
Author: Alright, if you want. I'll get us started.
Mathematics: ENJOY, YOU TWO. I'LL BE OVER HERE THINKING ABOUT THE PROBLEM BY MYSELF.

(Mathematics goes to the other side of the room.)

Author: Okay, well we could look at a bunch of points in the interval from 0 to 1, say

$$0, \frac{1}{n}, \frac{2}{n}, \frac{3}{n}, \ldots, \frac{n-1}{n}, 1$$

These are all the points $x_k \equiv \frac{k}{n}$, where n is some big number and k goes from 0 to n.

Reader: Okay, keep going!

Author: Well, the distance between any two of these points is the same. That is, $\Delta x_k = \frac{1}{n}$.

Reader: Yeah. And then?

Author: Then we could approximate the integral in equation 6.41 by writing

$$\int_0^1 \frac{1}{\sqrt{1-x^2}} \, dx \approx \sum_{k=0}^n \frac{1}{\sqrt{1-x_k^2}} \Delta x_k = \sum_{k=0}^n \frac{1}{\sqrt{1-(k/n)^2}} \left(\frac{1}{n}\right)$$

Wow, this looks terrible.

Reader: Come on!

Author: Nah, it's not worth it. How are we supposed to calculate this ourselves?

Reader: We could call Al and Sil! They said to call them if we ever had any number problems.

Author: I guess we could, but I feel guilty using Al and Sil to do things for us that we can't do. Back when we were computing e, we invented two expressions that we *could* compute in principle. We just didn't feel like doing the arithmetic. If we could get to a place like that, then I wouldn't mind using Al and Sil.

Reader: Well why doesn't this count?

Author: I guess it kind of does, but it's a sum of a bunch of terms and each one has a square root in it. We don't really know how to compute square roots.

Reader: Yes we do! We can just use the Nostalgia Device!

Author: Sure…I mean, we could compute any *particular* square root to arbitrary accuracy. But this is a sum of a huge number of square roots. I guess we could compute the whole sum to arbitrary accuracy by expanding each square root with the Nostalgia Device, cutting off each expression after a finite number of terms, and then adding up all the sums, but that would be a big ugly double sum. Even if we'd get the right answer in that case, it's ugly. We might be able to *find* ♯, but it doesn't really feel like *slaying* ♯.

Reader: Forget this!

(Reader leaves Author and heads over to Mathematics, who is busily working on something on the other side of the room.)

Reader: Hi, Math. Any progress?

(Mathematics continues working, talking to Reader without looking up from the table.)

Mathematics: I DON'T KNOW YET. I'VE BEEN PLAYING AROUND WITH OTHER MACHINES THAT SPIT OUT ♯ WHEN WE FEED THEM SOMETHING SIMPLE. YOU AND AUTHOR MAKE ANY PROGRESS?

Reader: Kind of. Not really. We got an expression for ♯, but it involved square roots and it was really ugly, so Author said he didn't really feel like we had solved the problem.

Mathematics: WHAT? WHY NOT?

Reader: Because we couldn't compute the answer ourselves. Well, we could, but he was being picky. It would have been a lot of numbers and approximations. But we were so close!

Mathematics: I CAN SYMPATHIZE. LOTS OF NUMBERS AND APPROXIMATIONS CAN BE UGLY. WHAT'S THE POINT OF ALL THIS IF WE DON'T GET SOMETHING WE LIKE?

Reader: But if we're that picky, how are we ever going to solve this problem? I guess the answer itself had better be really simple or else you two won't feel like we've really slayed ♯. How on earth are we supposed to do that?

Mathematics: I DON'T KNOW. WHAT DID YOU SAY AUTHOR'S PROBLEM WAS?

Reader: Ultimately I guess he just didn't like the square roots.

Mathematics: INTERESTING... I THINK I MIGHT KNOW HOW TO AVOID THEM.

Reader: Wait, are you serious?

Mathematics: MAYBE. WHERE DID THE SQUARE ROOT COME FROM IN THE FIRST PLACE?

Reader: When we were reabbreviating. We wanted to express the whole thing in terms of V. We wrote $H^2 + V^2 = 1$ and then isolated H to get $H = \sqrt{1 - V^2}$. That's when the square root came in.

Mathematics: SO IT'S V'S FAULT.

Reader: How do you mean?

Mathematics: WELL WHY WERE WE USING V IN THE FIRST PLACE?

Reader: Because we wanted a machine that spit out ♯ when we fed it something simple. We knew $V(♯/2) = 1$, which means $\Lambda(1) = ♯/2$. So if only we could figure out how to compute $\Lambda(1)$, we'd know how to compute ♯. What does that have to do with the square roots?

Mathematics: THE SQUARE ROOT CAME IN BECAUSE WE NEEDED IT IN ORDER TO EXPRESS V'S DERIVATIVE BY ONLY REFERRING TO V ITSELF. THE DERIVATIVE OF V IS H, NOT V ITSELF, AND EVENTUALLY THAT LED US TO THE UGLY SQUARE ROOTS.

Reader: I don't see how this is getting us anywhere.

Mathematics: WE WANT A MACHINE THAT SPITS OUT ♯ WHEN WE FEED
IT SOMETHING SIMPLE, RIGHT?
Reader: Right.
Mathematics: AND WE WANT TO AVOID THE OBNOXIOUS SQUARE ROOT
THING HAPPENING AGAIN, SO WE DON'T WANT TO USE V OR H THEMSELVES.
Reader: Right.
Mathematics: WELL, I HAVE THIS CRAZY IDEA. REMEMBER ALL THOSE
POINTLESS MACHINES LIKE "TANGENT" THAT AUTHOR WAS COMPLAINING
ABOUT BACK IN CHAPTER 4?
Reader: Vaguely.
Mathematics: WHILE YOU TWO WERE OVER THERE, I WAS LOOKING BACK
AT THAT PART OF THE BOOK. THIS MIGHT BE THE ONLY TIME THAT IT'S
ACTUALLY USEFUL TO USE ONE OF THEM.
Reader: Don't tell Author. He'll probably get mad. What's your idea?
Mathematics: OKAY, BACK IN CHAPTER 4, AUTHOR MENTIONED THIS MA-
CHINE "TANGENT." IT WAS DEFINED TO BE $T \equiv \frac{V}{H}$. NOW, I WAS LOOKING
BACK AT CHAPTER 4, AND I SAW THAT ITS DERIVATIVE IS $T' = 1 + T^2$.
SINCE ITS DERIVATIVE CAN BE WRITTEN JUST IN TERMS OF ITSELF WITH-
OUT USING SQUARE ROOTS, THIS SEEMS TO SOLVE BOTH OF THE PROBLEMS
WE RAN INTO EARLIER.
Reader: Perfect. Let's try the argument we made before.

Unearthing Something We Buried

Reader: Okay, so $T(x) \equiv \frac{V(x)}{H(x)}$. When we feed it $♯/2$, it spits out... uh... $\frac{1}{0}$.
I don't quite know what to do with that. Does T ever spit out anything less
confusing?
Mathematics: WELL, IF WE FEED IT $♯/4$, THAT'S AN EIGHTH OF A FULL
TURN, SO V AND H ARE THE SAME, AND $T(♯/4) = 1$. THAT'S FAIRLY SIMPLE.
Reader: Nice. And for the same reason as before, it's not T that we care
about. We care about its opposite, because we want a machine that spits out
♯ when we feed it something simple. So if there's a machine $\perp(x)$ that can
undo T, like this:

$$\perp(T(x)) = x$$

then we'd have $\perp(1) = ♯/4$, which means $♯ = 4\perp(1)$.
Mathematics: (*Suddenly excited*) THIS MAY ACTUALLY WORK...
Reader: Alright, let's try to do exactly what we did with Λ, and maybe it'll
pay off this time. First, let's define $A(x) \equiv \perp(T(x))$, so secretly $A(x) = x$,
and its derivative is $\frac{dA}{dx} = 1$. Then we can use the hammer for reabbreviation
to write

$$1 = \frac{dA}{dx} \equiv \frac{d}{dx}\perp(T(x)) = \frac{dT(x)}{dx}\frac{d}{dT(x)}\perp(T(x)) \equiv \frac{dT}{dx}\frac{d}{dT}\perp(T) \quad (6.42)$$

Mathematics: Now we can use the fact that

$$\frac{dT}{dx} = 1 + T^2$$

from Chapter 4, together with equation 6.42, to write

$$\left(1 + T^2\right)\left(\frac{d}{dT}\bot(T)\right) = 1 \quad \text{which tells us that} \quad \frac{d}{dT}\bot(T) = \frac{1}{1 + T^2}$$

Reader: Nice. Everything is written in terms of T, so we can reabbreviate to get

$$\frac{d}{dx}\bot(x) = \frac{1}{1 + x^2}$$

In the earlier argument with V, what did we do next?

(Reader looks back.)

Right, we used the fundamental hammer on the derivative. Now we can use the fundamental hammer on the derivative of $\bot$ to write

$$\int_a^b \left(\frac{d\bot}{dx}\right) dx = \int_a^b \frac{1}{1 + x^2} dx = \bot(b) - \bot(a) \tag{6.43}$$

Mathematics: Ooh, I have a good feeling about this. We want $\bot(1)$, so let's make $b = 1$. We don't want to bother with that other term $\bot(a)$, so let's choose an a for which $\bot(a) = 0$.
Reader: Well, $T(0) = 0$, so $\bot(0) = 0$, too. But then by the far right of equation 6.43, the above integral with $a = 0$ and $b = 1$ is just $\bot(1)$.
Mathematics: Fantastic. And we already know that $\sharp = 4 \cdot \bot(1)$, so we can write

$$\sharp = 4 \int_0^1 \frac{1}{1 + x^2} dx \tag{6.44}$$

Unleashing the Nostalgia Device

Mathematics: What now?
Reader: Do we know how to compute this integral?
Mathematics: I don't.
Reader: I don't either.
Mathematics: If we did all this for nothing, I may have to quit the book.
Reader: Me too. This is painful.

(Author walks over to Reader and Mathematics.)

Author: Hi there, you two. Any progress?

Reader: We were so close again... and we just got stuck.

Author: What did you do?

Mathematics: GO SEE FOR YOURSELF.

(Author skims the above conversation.)

Author: *(Subdued)* Oh no...

(Author sits in silence for a moment.)

Reader: *(To Mathematics)* Uh-oh, I think he's mad about us using that machine he kept complaining about back in Chap—

Mathematics: IT WAS READER'S IDEA!!!

Author: No no, it's not that. It's just... if you two quit the book... either of you... I don't think I could get through this. After everything we've been through... I can't go back to doing this alone...

Reader: Oh.

Mathematics: OH.

Author: Please don't leave... Or, you can. If you want... It's okay if you do... I know it's rough. It's a lot of gory details from time to time, especially right now. But, I mean, the only alternative would be to hide something from you. I'm not sure I could finish the book if I had to do that. Even though it would be less work, it would be so much harder. I don't want to have to lie to you. So, you can leave if you want. But while you're here... let's try to get through this together... okay?

Reader: Okay.

Mathematics: OKAY.

(A moment of $\frac{1}{2}$(awkward+comfortable) silence elapses.)

Mathematics: I'M NOT SURE I COULD LEAVE IF I WANTED TO...

Author: Hah. Anyways, that's a nice result you two just discovered.

Reader: So what? We're stuck!

Author: Maybe you're not. While you two were talking, I was playing around with the Nostalgia Device on the version of the problem with the square roots. It was really ugly, but it gave me an idea. Check it out. You two just showed that

$$\sharp = 4 \int_0^1 \frac{1}{1 + x^2} \, dx \tag{6.45}$$

Now what if we just expand $\frac{1}{1+x^2}$ with the Nostalgia Device?

Mathematics: THAT'LL NEVER WORK. WE'D NEED TO FIGURE OUT ALL OF ITS DERIVATIVES.

Reader: Well, the zero[th] derivative was just the machine itself, right?

Author: Right! So $M^{(0)}(0) \equiv M(0) = 1$.

Reader: And the first derivative is

$$M'(x) = \frac{d}{dx}(1+x^2)^{-1} = -1 \cdot (1+x^2)^{-2}(2x)$$

so $M'(0) = 0$.

Mathematics: THE SECOND ONE IS JUST THE DERIVATIVE OF THAT:

$$M''(x) = (-1)(-2)(2x)(1+x^2)^{-3} + (-2) \cdot (1+x^2)^{-2}$$

So $M''(0) = -2$.

Author: This is getting really ugly. Can we cheat?

Reader: How?

Author: Well, $\frac{1}{1+x^2}$ is just the machine $m(s) \equiv \frac{1}{1+s}$ with x^2 plugged in.

Reader: So?

Author: Couldn't we just use the Nostalgia Device on $\frac{1}{1+s}$ and then plug in x^2 afterward? It seems like it would be so much simpler.

Reader: Would that work?

Author: Not sure. But it seems like it has to work.

Reader: I guess it's worth a shot.

Mathematics: WAIT, WE'RE STARTING OVER?

Author: No, don't worry. But let's try to find the derivatives of a slightly different machine. Here, watch. Define

$$m(s) \equiv \frac{1}{1+s}$$

The zeroth derivative is just the machine itself, so $m^{(0)}(0) \equiv m(0) = 1$.

Mathematics: THE FIRST DERIVATIVE OF m IS

$$m'(s) = -(1+s)^{-2} \qquad \text{AND THUS} \qquad m'(0) = -1$$

Reader: The second derivative is

$$m''(s) = (-1)(-2)(1+s)^{-3} \qquad \text{SO} \qquad m''(0) = 2$$

Wow, this is a lot easier!

Author: Right? And we can see that the nth derivative will be

$$m^{(n)}(s) = (-1)(-2)\cdots(-n)(1+s)^{-n-1} \quad \text{SO} \quad m^{(n)}(0) = (-1)(-2)\cdots(-n)$$

Hmm... what's that?

Mathematics: WELL, YOU'VE GOT n NEGATIVE THINGS, RIGHT?

Author: Yeah.

Mathematics: THAT'S n COPIES OF (-1). SO MOVE ALL OF THOSE TO THE FRONT. THEN THE REST IS JUST $n!$, RIGHT? THAT IS, $m^{(n)}(0) = (-1)^n n!$

Reader: And that's all we need to use the Nostalgia Device on $m(s)$, so

$$\begin{aligned} m(s) &= \sum_{n=0}^{\infty} \frac{m^{(n)}(0)}{n!} s^n \\ &= \sum_{n=0}^{\infty} \frac{(-1)^n n!}{n!} s^n \\ &= \sum_{n=0}^{\infty} (-1)^n s^n \end{aligned}$$

Now what?

Author: Well, $M(x) = m(x^2)$, so just toss x^2 in there:

$$M(x) = m(x^2) = \sum_{n=0}^{\infty} (-1)^n x^{2n} \tag{6.46}$$

Why were we doing this?

Mathematics: WE WERE TRYING TO WRITE THE EXPRESSION READER AND I CAME UP WITH IN A WAY THAT LET US GET UNSTUCK. WE HAD WRITTEN

$$\sharp = 4 \int_0^1 \frac{1}{1+x^2} \, dx \tag{6.47}$$

BUT NOW THAT WE USED THE NOSTALGIA DEVICE ON $\frac{1}{1+x^2}$ IN THAT INDIRECT SORT OF WAY, EQUATION 6.46 LETS US WRITE

$$\sharp = 4 \int_0^1 \left(\sum_{n=0}^{\infty} (-1)^n x^{2n} \right) \, dx \tag{6.48}$$

Reader: That may be the scariest thing I've ever seen.

Author: Yeah, I don't think we know what to do with this.

Mathematics: SURE WE DO!

Infinite Shattering with Anti-Hammers

Mathematics: IT'S NOT AS SCARY AS IT LOOKS. HERE, I'LL WRITE IT THIS WAY:

$$\sharp = 4 \int_0^1 \left(1 - x^2 + x^4 - x^6 + \cdots \right) \, dx \tag{6.49}$$

Reader: Oh, wow. That's a lot better.

Author: I guess now we can use our anti-hammer for addition to break apart the integral.

Reader: Does that work for infinite sums?

Author: I don't know, but let's hope it works for this one! It's either that or give up, so let's keep moving. Once we break the integral apart, we just need to think of an anti-derivative for each piece. Check it out:

$$\sharp = 4 \cdot \left[x - \frac{1}{3}x^3 + \frac{1}{5}x^5 - \frac{1}{7}x^7 + \cdots \right]_0^1 \tag{6.50}$$

which is just an abbreviation for

$$\sharp = 4 \cdot \left(1 - \frac{1}{3} + \frac{1}{5} - \frac{1}{7} + \cdots \right) \tag{6.51}$$

Sharp Yields

Reader: So $\sharp$ is just four times an infinite back-and-forth sum of the hand-stands of all the odd numbers?
Author: I guess so.
Mathematics: HEY, AND IF n IS A WHOLE NUMBER, THEN $2n$ IS ALWAYS EVEN AND $2n + 1$ IS ALWAYS ODD, SO WE COULD WRITE THE SAME THING THIS WAY:

$$\sharp = 4 \cdot \sum_{n=0}^{\infty} \frac{(-1)^n}{2n+1} \tag{6.52}$$

Author: Perfect — there are no square roots or anything. We described $\sharp$ just using arithmetic! We could absolutely compute this ourselves, in principle.
Reader: Does that mean we can call Al and Sil?
Mathematics: CERTAINLY.

(*Mathematics borrows Author's phone again and dials a number.*)

Mathematics: AL, LISTEN, ARE YOU BUSY?... YOU *ARE* BUSY?... WHAT ARE YOU DOING?... OH! THAT'S WONDERFUL. I'M SO GLAD YOU TWO FINALLY... LISTEN, I NEED TO BE QUICK. THERE ISN'T MUCH TIME. WE KNOW YOU CAN'T HANDLE INFINITE JOBS, BUT WOULD YOU MIND COMPUTING...

(*Mathematics whispers equation 6.52 into the phone.*)

WHEN N IS $100,000$?
Author: How long is this gonna take?
Mathematics: HE SAYS IT'S ABOUT 3.14160.
Author: Wow! That was quick. Do you mind doing it again? Ask him for the first million terms.
Mathematics: (*Time passes*) HE SAYS IT'S ABOUT 3.14159.
Reader: Nice, it looks like the first three decimal places have stabilized, and the other two aren't changing much.
Author: And hey, remember back in Chapter 4 we guessed that $\sharp$ should be somewhere in the neighborhood of 3? I guess we were right. There's some

more evidence that all those arguments we just made were probably on the right track.

Mathematics: I CAN'T BELIEVE WE'RE FINISHED.

Author: I know... That was by far the most difficult thing we've done so far. But we did it! We slayed ♮. In the end, it just turned out to be

$$\natural \approx 3.14159$$

Author: See how much was being hidden from us when we were all just told that the area of a circle is πr^2, and—

Reader: Hey, that reminds me. You said in Chapter 4 that after we figure out how to compute this thing, we would start calling it π.

Author: Did I?

(Author flips back to Chapter 4.)

Author: Huh. You're right. I guess I did say that.

Mathematics: WHAT ON EARTH IS π?

Reader: Well, in the textbooks, π is—

Author: Forget that. Let's keep calling it ♮. We earned this.

Reader: Works for me.

N New Is Old

N.1 A Bridge

N.1.1 Full Disclosure

*(Author and Reader are walking on a
locally Euclidean hill in an unknown location.)*

Author: This may be my least favorite chapter.

Reader: I don't think you're supposed to tell me that.

Author: Come on, really? After all we've been through? Why would I hide that from you?

Reader: No, I don't mind. I just meant we're taught not to say things like that... in places like this.

Author: Why?

Reader: I don't know. Professionalism?

Author: Yeah... I'm not very good at that... Even when I need to be. But seriously, why shouldn't I tell you that this is my least favorite chapter? Wouldn't hiding that just add to the distance?

Reader: How do you mean?

Author: Like, if I didn't tell you, and then the chapter didn't live up to your expectations, you might end up saying, "At first I thought the book was G units of good... but then in Chapter N it got a bit [*negative adjective*]. Then I only thought it was g units of good... where g is less than G."

Reader: You worry a lot. Plus, you were wrong. I didn't even say those things, with the G and the g and the N. You're putting words in my mouth.

Author: The alternative is just not to talk to you at all...

(An [undefined adjective] silence elapses.)

Reader: So why is this your least favorite chapter?

Author: It's a bridge.

Reader: A bridge to what?

Author: Somewhere better. Somewhere I really want to show you.

Reader: When will we get there?

Author: Soon. In the final chapter. I'm thinking of calling it Chapter ℵ.

Reader: What does the ℵ stand for?

Author: Nothing. It's the Hebrew letter aleph. It's used to represent the infinite.

Reader: I thought ∞ was used for that.

Author: It is. Different thing. The symbol ∞ stands for the non-numerical limit past the end of the real numbers. It's used in different ways. Usually just to mean that a sequence grows without bound. But it's a symbol that describes behavior, not really a symbol that describes a number or an honest mathematical object. There are exceptions, of course. Some people use it differently. People get squeamish about the infinite.

Reader: So what's $\aleph$?

Author: A better kind of infinity.

Reader: Better?

Author: Well, no. It's an aesthetic preference. But it's a symbol built by people who weren't squeamish. Who took the idea seriously. It's used in the formal theory of the infinite. To stand for different sizes of infinity. They're called "transfinite cardinals." We won't have time to talk about them... god, there's so little time left... Honestly, calling it Chapter ∞ might have been more accurate. But Chapter $\aleph$ seemed right. More genuine, somehow.

Reader: I don't mean this in a mean way, but... do I need to know all this?

Author: No. But Chapter $\aleph$ is where I want to get us to. It's about calculus in an infinite number of dimensions. It's beautiful. None of the books convey how simple it is. So I want to show you.

Reader: So why not start there?

Author: We need this chapter first. As a bridge. It's not so bad, I guess. The topic itself is wonderful. I just don't feel like I did it justice.

Reader: Did what justice?

Author: The subject of this chapter.

Reader: What's that?

Author: Multivariable calculus.

N.1.2 What's a Multivariable?

Reader: What's a multivariable?

Author: Hah, nothing. The word is used like an adjective, not a noun. I just thought it would be a fun section title.

Reader: So what does it mean? I mean, I've heard the word "variable" before.

Author: Where?

Reader: In this book.

Author: I've never said that word before.

(Reader flips back through the book.)

Reader: Yes you have!

Author: I have?

(Author flips back through the book.)

Author: Hmm... I suppose I have. My memory hasn't been the best lately. Remind me what a "variable" is again.

Reader: The textbooks' name for the food we feed our machines.
Author: Oh, okay. That's a weird name.
Reader: And they also use the word to refer to what the machines spit out.
Author: What? Why?
Reader: I think it's because we can choose to feed the machines different things... so what we feed it can vary... so it's "variable."
Author: Oh right, I remember. And our machines can also spit out different things depending on what we feed them, so by that logic, what they spit out is a "variable" too.
Reader: Yeah.
Author: Okay. Then what's a multivariable?
Reader: You just said yourself it's not a noun. But still, "multi" means "more than one," so maybe "multivariable" just means "more than one variable."
Author: Hold on, I have a question. You just said they use the word "variable" for the stuff we feed the machines *and* the stuff they spit out.
Reader: So?
Author: So we've already got more than one variable! How are we not *already* doing multivariable calculus?
Reader: Well I guess we are. In an unimportant sense. But that's just nitpicking about terminology. Before we can understand the terms I think we have to invent it.
Author: Invent what?
Reader: Multivariable calculus.
Author: How do we do that?
Reader: I don't know. Even if we are already doing "multivariable calculus," in the nitpicky way you just mentioned, the word "multi" doesn't just mean two... so why stop at two?
Author: How do you mean?
Reader: What if we build a machine that eats two things and spits out one thing?
Author: Oh... Or maybe eats one thing and spits out two things?
Reader: Exactly!
Author: Or eats two things and spits out two things?
Reader: Or eats n things and spits out m things?
Author: Hah! Or eats infinitely many things and—
Reader: Let's not get carried away. You said yourself this was a bridge.
Author: Who cares? Let's try it! All of it!
Reader: But we don't know how to try it.
Author: So?
Reader: I mean, I know how to make abbreviations and stuff, but...
Author: How?
Reader: Well, for a machine that eats two things and spits out one thing, we could call the machine m as usual, we could call the two things we feed it x and y, and we could call the one thing it spits out $m(x,y)$. And if a machine

eats n things and spits out one thing, we could write $m(x, y, z)$... Oops, I ran out of letters. I mean, we could write $m(x_1, x_2, \ldots, x_n)$.

Author: What about the other stuff?

Reader: Like a machine that eats one thing and spits out two things?

Author: Sure.

Reader: Well, we could write m for the machine, x for the thing it eats, and then use a and b for the two things the machine spits out.

Author: Sounds good. But wait, our old abbreviation for "the thing a machine spits out" reminded us that the output might depend on what we put in. I'll forget that unless we remind ourselves.

Reader: Oh, okay. Well we could write m for the machine, x for the thing it eats, and then use $f(x)$ and $g(x)$ for the two things the machine spits out.

Author: Nice! I just had an idea. Remember before how you wrote $m(x, y)$ for a machine that eats two things and spits out one thing? What if we think of this as a machine that eats *one big thing* and spits out one thing? The thing it eats is still a thing, but it's not a number anymore, it's a list of two numbers: that weird (x, y) thing you wrote.

Reader: Either way. I wasn't thinking of (x, y) as a single thing when I wrote it, but I guess the word "list" makes it sound like one thing. Sure. Think about it that way if you prefer. Ooh! And then we could do the same thing for the "one in, two out" machine and write it as $m(x) \equiv (f(x), g(x))$.

Author: And then we could do calculus on these things!

Reader: But we don't know how to do calculus on them. I can abbreviate stuff no problem, but...

Author: Who cares! Let's try it!

N.1.3 What Do We Do When We Don't Know What to Do?

Reader: What do we do to something new if we don't know what to do to things that are new?

Author: I'm pretty sure I don't know the answer to that.

Reader: Why not?

Author: By definition.

Reader: Then what do we do?

Author: The only thing we can.

Reader: What's that?

Author: Let's not do anything new.

(Crickets.)

Reader: That doesn't seem like a very intelligent idea.

Author: It's not! Let's try it!

Reader: How?

Author: Well, how did we invent the "derivative" originally?

Reader: We used our slope idea from Chapter 1 on two infinitely close points.

Author: How'd we do that?

Reader: Well, we had a machine m, we fed it some food x, and it spat out $m(x)$. Then we made a tiny change dx to the food, changing it from x to $x + dx$. We fed the machine the new food, and it spat out $m(x + dx)$. Then we looked at the difference in the machine's behavior before and after, which is to say

$$d(Output) \equiv Output_{after} - Output_{before}$$

or to write the same thing in a different way,

$$dm \equiv m(x + dx) - m(x)$$

Author: Exactly, and the derivative was just

$$\frac{dm}{dx} \equiv \frac{m(x + dx) - m(x)}{dx} \equiv \frac{\text{Tiny change in the output}}{\text{Tiny change in the input}}$$

So let's try to do that for our new machines!

Reader: Seriously?

Author: Why not? Let's try some of the machines we just made abbreviations for.

Reader: Which one should we try first?

Author: I don't know. Pick your favorite.

Reader: Let's try $m(x) \equiv (f(x), g(x))$. But what if what we do doesn't make sense?

Author: Don't worry! We'll try to make sense of what we did later. If it doesn't make sense, we'll just keep punching it until it does.

Reader: Uh...okay. Here we go. Let's define $m(x) \equiv (f(x), g(x))$. Then if we use the same definitions as we did in the single-variable case, we can write

$$dm \equiv m(x + dx) - m(x) \equiv \Big(f(x + dx), \, g(x + dx)\Big) - \Big(f(x), \, g(x)\Big)$$

Reader: I'm stuck. We don't know how to add or subtract two lists. What do we do?

Author: Just do the simplest thing you can think of.

Reader: What's the simplest thing I can think of?

Author: I don't know. You think of it.

Reader: Well, we don't know how to add lists, but we do know how to add numbers, so maybe whenever we see two lists added, we could just add slot by slot, like this:

$$(a, b) + (A, B) \equiv (a + A, \, b + B)$$

Same deal with subtraction. I guess if we do that, then the new idea of addition is just the old idea. How do we know if this worked?

Author: You say, "It worked!"

Reader: It...worked.

Author: Nah, say it more maniacally, like you just captured Batman in your shark tank.

Reader: *It worked!!!*

Author: That's more like it!

Reader: There's no way that's enough.

Author: Sure it is! List addition isn't something "out there in the world" that we might accidentally say something wrong about. We can just define it to behave in whatever way makes life easiest for us. Like when we invented powers back in Interlude 2.

Reader: Okay. Then I guess we can pick up where we left off. Now that we've said what adding and subtracting lists means, we can write:

$$dm = \Big(f(x + dx) - f(x), \ g(x + dx) - g(x) \Big)$$

so the "derivative" would be

$$\frac{dm}{dx} = \frac{\Big(f(x + dx) - f(x), \ g(x + dx) - g(x) \Big)}{dx}$$

Reader: I'm stuck again.

Author: Why?

Reader: Well, I know that dx is just a tiny number, and division by *stuff* is just multiplication by $1/stuff$, so I guess I could write

$$\frac{dm}{dx} = \frac{1}{dx} \Big(f(x + dx) - f(x), \ g(x + dx) - g(x) \Big)$$

but that doesn't get us any less stuck. We still don't know how to multiply a number by a list.

Author: How about we just do the same thing as before?

Reader: Okay. I don't know how to multiply a number by a list, but I do know how to multiply a number by a number, so I guess we can just define "number times list" to be multiplying slot by slot, like this:

$$c \cdot (x, y) \equiv (cx, cy)$$

If we do that, then we're unstuck again, and we can write

$$\frac{dm}{dx} = \left(\frac{f(x + dx) - f(x)}{dx}, \ \frac{g(x + dx) - g(x)}{dx} \right) = \left(\frac{df}{dx}, \ \frac{dg}{dx} \right)$$

So I guess the derivative of these weird new machines is just the old derivative in each slot.

Author: That wasn't as hard as I expected. Let's write it in a box to celebrate!

What We Just Invented

If m is a machine that eats one thing and spits out two things, so that

$$m(x) \equiv (f(x),\ g(x))$$

and if we define the sum of two lists in the dumbest way we can think of, like this:

$$(a,\ b) + (A,\ B) \equiv (a + A,\ b + B)$$

and if we define "number times list" in the dumbest way we can think of, like this:

$$c \cdot (x,\ y) \equiv (cx,\ cy)$$

then the derivative of our new kind of machine is

$$\frac{d}{dx}m(x) \equiv \frac{d}{dx}(f(x),\ g(x)) = \left(\frac{d}{dx}f(x),\ \frac{d}{dx}g(x)\right)$$

or to write the same thing in another way,

$$m'(x) \equiv (f(x),\ g(x))' = (f'(x),\ g'(x))$$

So in this case, the new "multivariable calculus" idea is nothing new. It's just our old familiar calculus ideas in each slot.

Reader: You know, you keep saying that this new stuff is nothing new, but I still don't feel very comfortable with it. It just feels too... new.
Author: It's not new.
Reader: Yeah, I know, but can we do a few examples?
Author: Sure. How's about we try to figure out the derivative of the machine $m(x) \equiv (2x, x^3)$?
Reader: Okay. Well, because of everything we just did, I guess the derivative is $m'(x) = (2, 3x^2)$.
Reader: Is that right?
Author: Who am I to say? You're acting like someone already invented this stuff. Like facts about it are sitting around in some dusty book somewhere. What we just did *has* to be true because of how we defined (i) list-plus-list and (ii) list-times-number. You tell me if it's right.
Reader: Okay, I guess it is.
Author: Great! Let's do another example. Say we define a new machine m to be:

$$m(x) \equiv \left(x^2 + 7x\ ,\ e^{2x} + H(x)\right)$$

How would we differentiate that?

Reader: Well based on everything we just did, the derivative of a list is just the list of the derivatives, so I guess the derivative is

$$m'(x) = \left(2x + 7 \ , \ 2e^{2x} - V(x)\right)$$

where I used the hammer for reabbreviation to differentiate e^{2x}. Also, I had to look back to what we did earlier to remember that we showed $H' = -V$ back in Chapter 4.

Author: Alright! Let's go on to the next secti—

Reader: Wait. Calculus wasn't just derivatives, right? I mean, "calculus" was the name we came up with for all the weird things we could do with our infinite magnifying glass. At first it was just derivatives, but later we came up with that "integral" idea, remember?

Author: Oh, right. But that was just addition, wasn't it?

Reader: Kind of. I mean, the symbol $\int_a^b m(x)\,dx$ was our abbreviation for the area under a curvy thing m between $x = a$ and $x = b$. And the abbreviation came from thinking of adding up a bunch of infinitely thin rectangles. So the $\int$ thing had a kind of addition-ish interpretation, but it didn't exactly *feel* like addition.

Author: So? Feelings are goofy. I feel all sorts of incorrect things. It was still just addition.

Reader: So how do we integrate these new machines?

Author: I don't know. Let's see.

Reader: Well, say we want to make sense of something like $\int_a^b \left(f(x),\ g(x)\right) dx$. That dx thing is just an infinitely small number, so I guess because of how we defined number-times-list, we can write this as $\int_a^b \left(f(x)dx\ ,\ g(x)dx\right)$.

Author: And we defined list addition to let us bring the $+$ inside, so I guess it makes sense to bring the $\int$ inside, since that was just addition too. Like this:

$$\int_a^b \left(f(x),\ g(x)\right) dx \equiv \left(\int_a^b f(x)dx\ ,\ \int_a^b g(x)dx\right)$$

Reader: So the integral of a list is just a list of both the integrals?

Author: I guess so!

Reader: But how do we know this is right?

Author: Same deal as last time. We're not really doing mathematics. Or, we are, but not quite. If it makes you more comfortable, just think of that argument as "pre-mathematics." The logic is backwards when we're inventing things. When we're inventing a new idea, we just use old ideas until we get stuck. Then we introduce some new assumption that lets us get unstuck. The trick is to get unstuck using as few assumptions as possible. That way, the mathematics we end up inventing will be "elegant." But it's not the kind of argument that we might be right or wrong about.

Reader: Still, I don't feel super-familiar with this yet. Can we do some examples?

Author: Sure! Let's undo the example from earlier. Say we've got the machine $m(x) \equiv (2, 3x^2)$. What's the integral of $m(x)$ from $x = 0$ to $x = 7$?

Reader: Well, using what we just invented, and writing $\overset{FH}{=}$ where we use the fundamental hammer, I guess we could write:

$$\int_0^7 m(x)\, dx \equiv \int_0^7 \left(2, 3x^2\right)\, dx$$
$$= \left(\int_0^7 2\, dx\, ,\, \int_0^7 3x^2\, dx \right)$$
$$\overset{FH}{=} \left(\left[2x\right]_0^7\, ,\, \left[x^3\right]_0^7 \right) \tag{N.1}$$
$$\equiv \left(2 \cdot 7 - 2 \cdot 0\, ,\, 7^3 - 0^3\right)$$
$$= \left(14,\, 7^3\right)$$

Reader: Remind me what 7^3 is, would you?

Author: A number.

Reader: Which number?

Author: Who cares? Just leave it as 7^3. Arithmetic makes me tired. Back to the ideas. What ideas did you use in the derivation you just did?

Reader: Let me see. I think I just used the fundamental hammer and the fact that "the integral of a list is the list of the integrals." The other steps were just abbreviations and arithmetic. Is that really all there is to it?

Author: It has to be! These new ideas aren't new!

Reader: Yeah yeah, but that was a pretty easy example. What if we couldn't manage to think of an anti-derivative of one of the slots?

Author: Well, what did we do when we couldn't think of an anti-derivative back in regular calculus?

Reader: (*Flipping back to Chapter 6*) We could use one of the three anti-hammers to rewrite the problem. Can we do that if we get stuck in multivariable calculus?

Author: Of course.

Reader: How do we know they still work in... oh, right. It's just regular calculus in each slot. They have to work. Well what if we tried that, and we were still stuck? Then what?

Author: Exactly what we did before: give up! We never got to a point in regular calculus where we could solve any imaginable problem. We're not omniscient now, and we never were. By definition, we can only solve problems that we can solve. If we can't solve a problem using the tools we've invented, then we can't really do much except keep banging our heads against it, or just sweep it under the rug and come back to it later if we want to.

N.1.4 Wait... Seriously?

Reader: Wait... seriously? So the new stuff is really nothing new?

Author: Well, it depends on what the meaning of the word "is" is. We *forced* the new to be nothing new, because life is easiest if we do things that way. Truly new stuff is definitely something new, by definition, but we defined our new stuff not to be anything new, at least this time.

Reader: I'm confused.

Author: Don't be. There's nothing to be confused about.

Reader: Okay... now what?

Author: I don't know. That's up to us. I'm having fun with this "multivariable calculus" idea, so let's play around with that some more.

Reader: Alright, what haven't we done?

Author: (*Looks back*) We haven't looked at machines that eat two things and spit out one thing.

Reader: Right. Let's say we've got a machine that eats two things x and y, and spits out one thing $m(x, y)$. Then, uh... what do we do?

Author: I don't know. What did we do at this point in single-variable calculus?

Reader: We made a tiny change in the food that we fed to the machine, changing it from *food* to *food* + *d*(*food*).

Author: Hmm... We've got two slots. What counts as the *food*? Ooh! I've got an idea...

(*Author begins writing something.*)

Reader: I guess we could just do each slot individually. Then we'd have two different derivatives.

Author: Huh? Sorry, I wasn't listening. I was thinking that we could think of the whole list as the *food*, and write something like $\mathbf{v} \equiv (x, y)$. Then a tiny change in the *food* might be something like a "tiny list," whatever that means, which we could abbreviate as $d\mathbf{v} \equiv (dx, dy)$. What were you saying?

Reader: I was saying that we could just choose to do each slot individually, which would give us two different ways to "do derivatives," one for each slot.

Author: Oh! I like that idea better. Forget what I said — let's do yours first. What do you mean by two different derivatives?

Reader: Well so far we've been dealing with new stuff by not doing anything new, and just doing the old stuff. So how about this: we've got this machine m that eats two things x and y and spits out one thing $m(x, y)$. First, I guess we could think of the *food* as x, leaving y alone. We make a tiny change in the *food*, changing it from x to $x + dx$. Then, as always, we compare the machine's output before and after, which means we're looking at $d(Output) \equiv Output_{after} - Output_{before}$, or equivalently

$$dm \equiv m(x + dx, y) - m(x, y)$$

Author: Wait, you said you're going to do this for y too, right?
Reader: Yeah, why?
Author: I'm a bit confused. What abbreviation do we use when we're changing y instead of x?
Reader: I suppose we'd just write $dm \equiv m(x, y + dy) - m(x, y)$... Oh, I see the problem. I wrote dm as an abbreviation for both. But based on how I defined them, they're really two different things. How about I just change abbreviations, and write something like this?

$$d_x m \equiv m(x + dx, y) - m(x, y)$$
$$d_y m \equiv m(x, y + dy) - m(x, y)$$

Author: Oh, okay. That makes a bit more sense.
Reader: Then I guess we could define the derivative of m with respect to x to be the thing you get from dividing $d_x m$ by dx, like this

$$\frac{d_x m}{dx} \equiv \frac{m(x + dx, y) - m(x, y)}{dx} \tag{N.2}$$

and we could define the derivative of m with respect to y to be

$$\frac{d_y m}{dy} \equiv \frac{m(x, y + dy) - m(x, y)}{dy} \tag{N.3}$$

Author: What about the other two possibilities, $d_y m / dx$ and $d_x m / dy$?
Reader: Uh, I dunno. I don't see how those mean anything.
Author: Really?

(Author looks back at how everything was defined, and thinks
for a moment.)

Author: Oh! I see. You're right. We can change x without changing y, just like we can go east without going north. If we want the derivatives to have the normal interpretation as some kind of steepness, then those two things I just wrote don't really make sense. It's like dividing the "rise" of Mount Everest by the "run" of the Appalachian Trail. I guess we're free to write stuff like $d_x m / dy$ if we want, but it isn't really talking about anything worth talking about, so let's just ignore it.
Reader: Deal! Now what?
Author: I don't know. I think we're done.
Reader: Are we?
Author: I think so. The definitions you wrote down in equations N.2 and N.3 tell us how to figure out both kinds of derivatives. And hey! This "new" idea is still nothing new! I mean, the derivative of the "two-variable machine" $m(x, y)$ with respect to x is just the thing we get from pretending that y is a constant, like 7 or 52, and then doing regular old calculus on it! Same deal for the derivative with respect to y.

Reader: Wait, how do you know that?

Author: That's what you said in equations N.2 and N.3. I was just looking at what you wrote.

Reader: Nice! Does that mean we can use our old derivative hammers on these machines too? I mean, I know we could use them on the "one in, two out" machines from earlier. But there are two different types of derivatives now. Can we really keep using our hammers in exactly the same way? How do we know they'll work?

Author: They have to! The new ideas are just the old ones!

Reader: Okay, I understand how the new stuff isn't new, but it still feels new.

Author: It's not just you. A slight change of abbreviations can be pretty confusing to all of us. Or it can help. The human mind is funny that way. Let's do a few new examples to get ourselves used to this old idea. I'll define $m(x, y) \equiv x^2 y + 7y^2 - 12xy + 9$. Now what's the x-derivative of that?

Reader: I guess

$$\begin{aligned} \frac{d_x m}{dx} &\equiv \frac{d_x}{dx}\left(x^2 y + 7y^2 - 12xy + 9\right) \\ &= 2xy + 0 - 12y + 0 \\ &= 2xy - 12y \end{aligned}$$

Is that right?

Author: What do you mean?

Reader: What do you mean what do I mean?

Author: Let's say we were back doing regular calculus, and you were trying to differentiate $x^2 \# + 7\#^2 - 12x\# + 9$. Would you be as cautious about whether you got the "right" answer then?

Reader: I guess not. I'm more familiar with regular calculus. Or single-variable, I mean.

Author: Well that's all we're doing here, remember? We're free to wonder whether or not we "did the right thing," but only if we're unsure about whether we did the single-variable calculus right, because that's all this is.

Reader: I mean, I *know* that. You've said it an annoying number of times by now — but it still feels like this new stuff is new.

Author: Then let's do another example. I've been wondering something since we defined these partial derivatives. Remember at the end of Chapter 2, when we talked about how the derivative lets us find the places where a machine is highest and lowest?

Reader: Maybe I remember. Maybe not. No way to tell.

Author: Well flip back if you don't remember. The basic idea was that the highest point of a machine's graph should also be a place where the slope is zero. Same goes for the lowest point. There were some exceptions, but let's worry about that later. Even if we couldn't picture the graph of a given machine, we could still usually find its highest and lowest points. Or at least trim down an infinite number of possibilities to some small finite number that we'd have to check by hand.

Reader: Can we do that here?

Author: That's what I've been wondering. Maybe we can find maximum and minimum points by forcing both partial derivatives to be zero.

Reader: Let's see. How do we start?

Author: Let's try something simple enough that we know where it has its minimum. Like $m(x, y) \equiv x^2 + y^2$. When x and y are both 0, then m spits out 0, but squaring stuff always makes it positive, so I guess the smallest thing it can ever spit out is $m(0, 0) = 0$.

Reader: Oh, I think I get it. Based on the way we defined partial derivatives, we can write

$$\frac{d_x m}{dx} = 2x \qquad \text{and} \qquad \frac{d_y m}{dy} = 2y$$

And if we force both of these to be equal to zero,

$$\frac{d_x m}{dx} = 2x \overset{\text{Force}}{=} 0 \qquad \text{and} \qquad \frac{d_y m}{dy} = 2y \overset{\text{Force}}{=} 0$$

then that's equivalent to saying $x = 0$ and $y = 0$. Hey, it worked!

Author: Nice! Will this work in more complicated cases?

Reader: I don't know. Let's look at $m(x, y) \equiv (x - 3)^2 + (y + 2)^2$. Now, for the same reasons as before, this will spit out zero when $x = 3$ and $y = -2$. Anywhere else it's bigger. And hey, just to see what happens, let's unwrap the machine and see if the mathematics tells us that that's where the minimum happens.

$$m(x, y) \equiv x^2 + y^2 - 6x + 4y + 13 \tag{N.4}$$

That's the same machine, because I just multiplied everything out, but when we write it this way it's not obvious that the minimum is at $x = 3$ and $y = -2$.

Author: Wait, why are you expanding everything out if it just makes the minimum less obvious?

Reader: Because all this stuff is only useful if it does something we can't do without it. I mean, a way of finding biggest and smallest locations is only useful when it's *not* obvious how to find them by just staring at the description of the machine. But if this "set both partial derivatives equal to zero" technique is really on the right track, then it ought to spit out $x = 3$ and $y = -2$ when we apply it on this more confusing description of the machine I just wrote in equation N.4.

Author: Brilliant! Let's try it!

Reader: Okay, taking the two partial derivatives of equation N.4 and forcing them to be zero, we get

$$\frac{d_x m}{dx} = 2x - 6 \overset{\text{Force}}{=} 0 \qquad \text{and} \qquad \frac{d_y m}{dy} = 2y + 4 \overset{\text{Force}}{=} 0$$

Hey! The first equation is only true when $x = 3$, and the second one is only true when $y = -2$. It worked!

Author: This is great! You feeling more comfortable with this stuff now?

Reader: A bit, but it still feels new.

Author: It's not.

Reader: I know. Can we do one more example?

Author: Of course. What kind?

Reader: How about an integral this time? Say

$$\int_{x=1}^{x=3} x^2 y^{72} + ye^x + 5 \; dx$$

If we're thinking of y as a fixed number like 7, then I guess it behaves just like 7 would. So an anti-derivative of the thing inside the integral would be $M(x) \equiv \frac{1}{3}x^3 y^{72} + ye^x + 5x$. And now I guess we can use the fundamental hammer.

$$\int_{x=1}^{x=3} x^2 y^{72} + ye^x + 5 \; dx \; = \; M(3) - M(1)$$

$$\equiv \left[\frac{1}{3} 3^3 \, y^{72} + y \, e^3 + 5 \cdot 3 \right] - \left[\frac{1}{3} 1^3 \, y^{72} + y \, e^1 + 5 \cdot 1 \right]$$

Author: Great! We're done!

Reader: No no no. I know simplification is a human construct and all that, but I want to clean it up just a little. I mean seriously, there's a 1^3 in there. The stuff above is just...

$$\text{(Stuff above)} \; = \; \left[9 \, y^{72} + y \, e^3 + 15 \right] - \left[\frac{1}{3} y^{72} + ye + 5 \right]$$

$$= \left(9 - \frac{1}{3} \right) y^{72} + \left(e^3 - e \right) y + 10$$

Now I could find a common denominat—

Author: What?! What have they been teaching you in school? We're *so done.*

Reader: Okay. It's just that after so many years in school, it's easy to develop this compulsive feeling that things need to be "simplified."

Author: Compulsions are fine, as long as they're your *own* compulsions. I've got plenty! But don't waste your time trying to satisfy someone else's compulsions. Otherwise, simplification just complicates things.

Reader: Alright, so I guess we're done. Wait, are we really done? The integral still has a y in it.

Author: Yeah, that's weird.

Reader: Hang on — I think that's okay. We just calculated the area under the machine $m(x) \equiv x^2 y^{72} + ye^x + 5$ between $x = 1$ and $x = 3$. But we're choosing to remain agnostic about which particular number y is, so I think we just did infinitely many integrals at once. I mean, each different choice of y gives us a different "single-variable machine." Like, when y is 1, what we just figured out tells us

$$\int_{x=1}^{x=3} x^2 + e^x + 5 \, dx = \left(9 - \frac{1}{3}\right) + \left(e^3 - e\right) + 10 \qquad \text{(N.5)}$$

And when $y = 0$, what we just figured out can be summarized by writing

$$\int_{x=1}^{x=3} 5 \, dx = 10 \qquad \text{(N.6)}$$

We never specified what y was, so it's not a bad thing that it showed up unspecified in the answer. So by staying agnostic about y, we get to say impressive things like "We just did an infinite number of integrals." Because the answer we got was really infinitely many sentences: one for each y. Some of the sentences were sort of scary-looking, like equation N.5, and some of the sentences seemed simpler, like equation N.6, which just says that a rectangle with height 5 and width $3 - 1 = 2$ has an area of 10. So the sentence with $y = 0$ feels simpler to us than the sentence with $y = 1$, but the mathematics doesn't care. The same calculation spits out both. Hey, that reminds me. Where is Mathematics?

Author: Still looking for a new home, I guess. I felt bad about having delayed the moving process in the last interlude.

Reader: You did what?

Author: Remember? I was all caught up in the idea of slaying ♯. But we're supposed to be helping Mathematics find a home — somewhere it belongs, now that it exists. So much of it exists now. Living somewhere that doesn't. . . or, at least not in the everyday sense. . . it's not going well. The Void is no place for an entity. So I'm giving it a break from these dialogues. It needs some time to work this whole situation out.

Reader: Wait, so aren't we doing the worst possible thing right now?

Author: What do you mean?

Reader: We're inventing more mathematics! Isn't that making the situation worse?

Author: No no, we're not inventing anything new, remember? I made sure of it. Look at the title of the chapter. We wouldn't want to do that to our friend.

N.2 The Notational Minefield of Multivariable Calculus

To most outsiders, modern mathematics is unknown territory. Its borders are protected by dense thickets of technical terms; its landscapes are a mass of indecipherable equations and incomprehensible concepts. Few realize that the world of modern mathematics

is rich with vivid images and provocative ideas.
—Ivars Peterson, *The Mathematical Tourist*

N.2.1 Simple Generalizations and Difficult Abbreviations

In the above dialogue, we "invented" multivariable calculus. For example, we looked at machines that eat one number and spit out two numbers, such as $m(x) \equiv (f(x), g(x))$. Textbooks call these "vector-valued functions," meaning that they eat a number and spit out a vector. "Vector" means basically the same thing as "list" (for our purposes), and in the remainder of the book, we'll use the two terms interchangeably. Notice that our pre-mathematical arguments in this interlude apply equally when the vectors have n slots. That is, we can simply dictate that we want lists with n slots to behave like this:

$$(x_1, x_2, \ldots, x_n) + (y_1, y_2, \ldots, y_n) = (x_1 + y_1, x_2 + y_2, \ldots, x_n + y_n)$$

$$c \cdot (x_1, x_2, \ldots, x_n) = (cx_1, cx_2, \ldots, cx_n)$$

Again, we're doing the simplest thing we can think of. These definitions allow us to show, in exactly the same manner as we did in the above dialogue, that the derivative of a "one in, n out" machine such as

$$m(x) \equiv (f_1(x), f_2(x), \ldots, f_n(x))$$

is the simplest thing we could possibly hope for, namely

$$m'(x) = (f_1'(x), f_2'(x), \ldots, f_n'(x))$$

Similarly, for machines that eat two numbers and spit out one number, which we abbreviated as $m(x, y)$, we again found a direct familiarity with single-variable calculus. We preserved the familiarity by simply defining two different derivatives: one for each input. That is, we had a derivative with respect to x (which treated y as a constant), and a derivative with respect to y (which treated x as a constant). We decided to write these as:

$$\frac{d_x m}{dx} \equiv \frac{m(x + dx, y) - m(x, y)}{dx} \qquad \frac{d_y m}{dy} \equiv \frac{m(x, y + dy) - m(x, y)}{dy} \qquad \text{(N.7)}$$

Textbooks call these "partial derivatives." They would call the one on the left "the partial derivative of m with respect to x," and they would call the one on the right "the partial derivative of m with respect to y." However, there's nothing "partial" about partial derivatives: they are computed using the exact same operations as the familiar derivative from Chapter 2. Notice that since we can change x without changing y (and vice versa) we can write expressions like

$$\frac{d_x x}{dx} = 1 \qquad\qquad \frac{d_x y}{dx} = 0$$

$$\frac{d_y x}{dy} = 0 \qquad\qquad \frac{d_y y}{dy} = 1 \qquad\qquad \text{(N.8)}$$

Again, there's nothing special about the fact that $m(x,y)$ has only two slots. We can make exactly analogous definitions and arguments when there are n slots. If we define m to be a machine that eats n numbers and spits out one number, which we can write as

$$m(x_1, x_2, \ldots, x_n) \qquad\qquad \text{(N.9)}$$

then we've got n slots, so there are n different derivatives: one for x_1, one for x_2, and so on, up to x_n. Just like before, we can define the derivative this way:

$$\frac{d_i m}{dx_i} \equiv \frac{m(x_1, \ldots, x_i + dx_i, \ldots, x_n) - m(x_1, \ldots, x_i, \ldots, x_n)}{dx_i} \qquad \text{(N.10)}$$

where we're choosing to write d_i instead of d_{x_i}, because the latter has a subscript on a subscript, which is a bit of a mess. Although the above equation may look scary enough to raise your blood pressure, it's saying something extremely simple: the derivative of an "n in, one out" machine with respect to some variable x_i is exactly the same thing it has always been. We simply ignore everything that isn't x_i, and do single-variable calculus thinking of x_i as the only variable. See how this is nothing new? We can clean up the above notation a bit by choosing some simpler abbreviations, which we will do in the next section.

N.2.2 Simple Ideas That Resist Simple Expression

I would argue that nearly all of the confusion about multivariable calculus comes from confusions about notation, and in this section we will explore some of the difficulties that arise in attempting to come up with good abbreviations in our new multivariable world.

In single-variable calculus, we've kept two notations for the derivative around throughout the book: $m'(x)$ and $\frac{dm}{dx}$. The same phenomenon that led us to do so appears in multivariable calculus with even greater force: the ideas themselves seem to resist being clearly expressed by any single set of abbreviations. As before, this leaves us with two options. The first option is to simply decide on one set of abbreviations with which to express all the ideas of multivariable calculus, in which case many conceptually simple expressions will appear hairy and counterintuitive. The second option is to switch notation at will, using whichever is appropriate for the problem at hand. This also has downsides, since there are then multiple symbolic languages floating around. In this chapter, we will err in favor of the latter, but we'll try to remind ourselves of what the different notations mean whenever we need to switch.

N.2.3 Coordinates: (Can't live with)(1, out)(them)

We will begin by attempting to invent some abbreviations that let us write equation N.10 in a simpler-looking way. To begin, let's write $\mathbf{v}$ as an abbreviation for the machine's input, so

$$\mathbf{v} \equiv (x_1, x_2, \ldots, x_n)$$

is a list of all n variables. Textbooks use the word "vector" to describe these things, hence the $\mathbf{v}$. The word "vector" may sound bizarre and archaic, but it's kind of fun to say, so let's keep it around. We're writing the vector $\mathbf{v}$ in boldface to remind ourselves that it is a different type of object than a number. Let's write $d\mathbf{v}_i$ as an abbreviation for the vector that has zero in every slot except the ith slot, in which it contains the infinitely small number dx_i. That is:

$$d\mathbf{v}_i \equiv (0, 0, \ldots, 0, \underbrace{dx_i}_{i^{th}\ slot}, 0, \ldots, 0, 0) \qquad (\text{N.11})$$

With these conventions, we can rewrite the messy definition in equation N.10 like this:

$$\frac{d_i m}{dx_i} \equiv \frac{m(\mathbf{v} + d\mathbf{v}_i) - m(\mathbf{v})}{dx_i} \qquad (\text{N.12})$$

That's a bit nicer, and it certainly takes up a lot less space, but this notation can be confusing for a completely new reason. Why? Well, glancing casually at the above equation makes it seem as if derivatives in this new multivariable world are something different from what they were in the single-variable world. Why? In the above equation, the tiny object on the top looks like $d\mathbf{v}_i$, which is a "tiny vector," whereas the tiny object on the bottom looks like dx_i, a "tiny number." That is, it seems as if there are *two different types of tiny thing* in this equation: a tiny number and a tiny vector. But notice that this is the fault of the new abbreviations we chose in an attempt to make equation N.10 look less scary.

Equation N.10, despite its drawbacks, made it more clear that there is really only one type of tiny thing on the top and the bottom. And that in turn makes it clear that the derivative has the same interpretation it has always had. That is, we've always been able to talk about derivatives like this:

1. We start with a machine m that we feed some stuff s. It spits out $m(s)$.

2. We make a tiny change in the stuff we're feeding the machine, changing it from s to $s + ds$. This changes the output from $m(s)$ to $m(s + ds)$.

3. We can abbreviate the change in output as something like $dm \equiv m(s + ds) - m(s)$. If we've got more than one variable, we may want to modify our abbreviations to remind ourselves what we're changing.

4. Whether the stuff s is a number, a vector, or an entire machine, the derivative is the same concept. The derivative of m is defined to be the tiny change in output dm, divided by the tiny change in the input ds.

So the two abbreviations in N.10 and N.12 have different costs and benefits. We're in an odd Catch-22 situation. As we'll see soon, the Catch-22 is more general than this example.

N.2.4 To ∂ or Not to ∂? Abbreviations Affecting Arguments

The next stop on our tour of the confusing notation in multivariable calculus is the alien symbol ∂. Earlier, I mentioned that the textbooks use the term "partial derivative" to refer to expressions like these:

$$\frac{d_x m}{dx} \equiv \frac{m(x + dx, y) - m(x, y)}{dx} \qquad \frac{d_y m}{dy} \equiv \frac{m(x, y + dy) - m(x, y)}{dy}$$

$$(\text{N.13})$$

Examining the standard notation for this concept will reveal another Catch-22. Here's the issue: in the two expressions above, you might notice that writing d_x and d_y is redundant. You might then think that if we simply wrote

$$\frac{dm}{dx} \equiv \frac{m(x + dx, y) - m(x, y)}{dx} \qquad \frac{dm}{dy} \equiv \frac{m(x, y + dy) - m(x, y)}{dy}$$

$$(\text{N.14})$$

then there would be no confusion, because the dx and dy on the bottom of each expression remind us in which of m's slots we're making a tiny change: in the left equation it's the x slot, and in the right equation it's the y slot. That's certainly true. The subscripts on d_x and d_y *are* redundant when they show up in derivatives, as they do in equation N.13. So why did we introduce the subscripts in the first place? Well, we originally introduced the notation $d_x m$ and $d_y m$ because the two different occurrences of dm in equation N.14 actually refer to *different things*, as you can see simply by comparing the top right sides of both equalities: on the left we're changing the first slot, and on the right we're changing the second slot. If we're only ever dealing with derivatives, then there's no reason to write $d_x m$ and $d_y m$. We can simply look at the bottom of the derivative to see what variable we're making a tiny change to. Most textbooks run with this line of thinking in choosing their notation, and write the following instead of equation N.13:

$$\frac{\partial m}{\partial x} \equiv \frac{m(x + dx, y) - m(x, y)}{dx} \qquad \frac{\partial m}{\partial y} \equiv \frac{m(x, y + dy) - m(x, y)}{dy}$$

$$(\text{N.15})$$

So comparing our notation and theirs, we've got

$$\frac{\partial m}{\partial x} \equiv \frac{d_x m}{dx} \qquad \text{and} \qquad \frac{\partial m}{\partial y} \equiv \frac{d_y m}{dy}$$

This different way of writing things has its own set of costs and benefits. On the one hand, it's a lot prettier than the notation I'm using, and it avoids the subscripts x and y, which are unnecessary when we're talking about derivatives themselves. On the negative side, this ∂ notation makes it much harder to make simple infinitesimal arguments. In the next few paragraphs, we'll explain why by examining a scary-looking equation that hides a simple idea. In every multivariable calculus book you'll find the equation

$$dm = \frac{\partial m}{\partial x}dx + \frac{\partial m}{\partial y}dy \qquad \text{(N.16)}$$

We'll derive this equation for ourselves very soon, but for the moment, just notice how confusing it looks! This awful equation contains six different symbols that look like infinitely small quantities: dm, ∂m, dy, ∂y, dx, and ∂x. Notice that equation N.16 doesn't really seem to have anything we can cancel to make it simpler. The crazy ∂x thing looks different from the more familiar dx thing, so it doesn't feel like we can cancel the ∂x on the bottom against the dx. Same deal for the y's; cancellation seems illegal, since the symbols look different.

Now, even though we haven't derived it yet, I can't resist telling you the ridiculous secret of equation N.16. Here it is: interpreted properly, the ∂x and the dx piece are *actually the same thing!* Same for the pieces ∂y and dy. As if that weren't confusing enough, we can then use this fact to get something even worse. Canceling the ∂x against the dx, and doing the same for ∂y and dy, we get this bizarre nonsense:

$$dm \overset{?!}{=} \partial m + \partial m \qquad \text{(N.17)}$$

As we'll see in a few pages, this equation is actually correct. However, $\partial m + \partial m$ is *not* equal to $2\,\partial m$! No, the laws of arithmetic have not broken down. Rather, in a monumental feat of confusing notation, the two different ∂m pieces in the above equation actually refer to two different things! They refer to what we have been calling $d_x m$ and $d_y m$, and it was precisely the earlier choice to ignore the (then redundant) subscripts that leads to all of these notational headaches down the road.

Why would anyone use notation in which a *single* symbol ∂m refers to two *different* things ($d_x m$ and $d_y m$) while simultaneously using two *different* symbols (∂x and dx) to refer to the *same* thing?! The reason isn't completely crazy. It's because the standard textbooks usually don't use infinitesimal arguments. Given the way all these ideas are usually formalized, most textbooks end up with a number system in which infinitesimals *don't* make sense, though derivatives do. But for us, it was clearly beneficial to distinguish between $d_x m$ and $d_y m$ because *they're not the same thing!* Textbooks usually neglect the subscripts, and their choice makes sense too: if we're only talking about derivatives and not the infinitesimals themselves, then the subscripts in $d_x m$ and $d_y m$ are *always* redundant.

In summary, when using the ∂ notation, we lose one of the best things about the d notation from single-variable calculus: the ability to manipulate infinitesimals just like numbers, canceling them and rearranging their order to derive things that would have been much more difficult to derive otherwise.

N.3 Enough Symbol Games! How Can We Picture This?

N.3.1 Multidimensional Mind Tricks

Despite the inability of the human mind to directly visualize dimensions higher than three, it is not necessary to abandon your visual intuition in moving to higher-dimensional mathematics. Nor is it necessary to develop magical methods of visualizing spaces of four, or ten, or infinitely many dimensions. If that sounds like a contradiction, read it again. Here's the trick, and it's something I was never told in multivariable calculus: a surprisingly large number of mathematicians visualize n-dimensional space simply by (drumroll) picturing three-dimensional space!

If that sounds ridiculous... good! But it's true. This is something we're not explicitly taught in any course, but it slowly dawns on many students of mathematics as they watch the great mathematicians of the previous generation reasoning out the answer to a question. Many's the time that I've posed a question about higher dimensions to an extremely intelligent mathematician, and seen them: think for a moment, realize s/he couldn't simply reason out the answer mentally, walk to the blackboard, draw a two-dimensional picture of a two- or three-dimensional object, and in doing so, *figure out the answer!* I've seen this occur more times than I can count, with questions ranging from the four-dimensional (in Riemannian geometry and general relativity) to the infinite-dimensional (in functional analysis and quantum mechanics).

Thinking about two or three dimensions may not help us to *visualize n* dimensions, but it can certainly help us to *reason about n* dimensions. It would be nice to experience this feeling ourselves. In the next part of this section, we'll try to use our visual intuition in three dimensions to derive a fact about n-dimensional calculus. In the following section, we'll invent the formula

$$dm = \frac{\partial m}{\partial x}dx + \frac{\partial m}{\partial y}dy \qquad (\text{N.18})$$

for a machine $m(x, y)$ of two variables. Having done so, we should immediately be able to see why the more general form of this expression, namely

$$dm = \frac{\partial m}{\partial x_1}dx_1 + \frac{\partial m}{\partial x_2}dx_2 + \cdots + \frac{\partial m}{\partial x_n}dx_n \qquad (\text{N.19})$$

is true for a machine $m(x_1, x_2, \ldots, x_n)$ of n variables. As intimidating as these equations look at the moment, we'll soon see — as we have before — how a simple change of abbreviations can change everything.

N.3.2 The Trick in Action

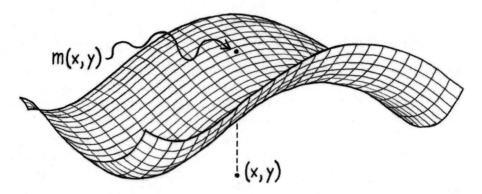

Figure N.1: Visualizing a machine m that eats two numbers and spits out one number. We can picture the two numbers x and y that the machine eats as coordinates on the "ground." If we feed the point (x, y) on the ground to the machine m, it will spit out a number $m(x, y)$, which we can visualize as the "height" of the graph floating above the point (x, y). Each point on the ground gets a height, so the graph of such a machine is a two-dimensional surface.

Notice that "two in, one out" machine $m(x, y)$ can be visualized as a (possibly curvy) surface floating off the ground. See Figure N.1 for a more detailed explanation.

Now we can extend the idea of our infinite magnifying glass to Figure N.1. When we zoom in on the curvy surface, it should look like a flat plane. Let's imagine picking an arbitrary point on the graph in Figure N.1 and zooming in infinitely far. The result is the tilted (but not curvy) plane pictured in Figure N.2.

We've talked about partial derivatives, but it would be nice if we could come up with a single-derivative-like idea — a "total derivative," if you will. Instead of thinking of x and y as separate numbers, let's (just for the moment) think of them as the components of a single object: a "vector," which we'll write as (x, y). Now we can proceed as we always have in defining a new type of derivative: start with a machine m, which we feed some *food*. In this case, the *food* is the vector (x, y). Then the machine spits out a number $m(x, y)$. Then we make a tiny change to the *food*, by adding a "tiny vector" (dx, dy) to it. This gives us $m(x + dx, y + dy)$. Then, as always, we see what changed before

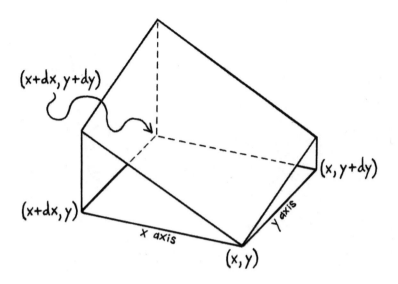

Figure N.2: We zoomed in infinitely far on a curvy surface to get a flat (but tilted) surface. In this picture, we're labeling the horizontal coordinates.

and after, so we look at $m(\textit{after}) - m(\textit{before})$, or equivalently,[1]

$$dm \equiv m(x + dx, y + dy) - m(x, y) \qquad (N.20)$$

The notation dm stands for a tiny change in the "height" of m's graph when we change both inputs by an infinitely small amount: changing x to $x + dx$ and changing y to $y + dy$. At this point, it helps to draw a picture of where we zoomed in on our curvy surface. There are a lot of things we could potentially label, so I've split them across three pictures. Here's what we're labeling in the three pictures.

1. In Figure N.2, I've just drawn four different points on the "ground," namely (x, y), $(x + dx, y)$, $(x, y + dy)$, and $(x + dx, y + dy)$.

2. In Figure N.3, I've drawn the names for the output or "height" of the graph at each of the points from Figure N.2. These heights are called $m(x, y)$, $m(x + dx, y)$, $m(x, y + dy)$, and $m(x + dx, y + dy)$.

3. In Figure N.4, I've drawn the tiny differences in height $d_x m$ and $d_y m$. Recall that the definition of the former was $d_x m \equiv m(x+dx, y)-m(x,y)$,

---

1. Usually we would now divide by the tiny change in *food* to get some kind of derivative. However, we're changing every slot at once, so *food* is now an entire vector. So to compute a derivative in the normal sense, we'd have to say what we meant by "dividing by a vector." Let's not do that right now, and instead just look at the top piece: dm, as defined above.

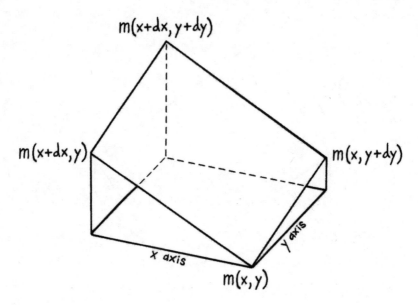

Figure N.3: Same idea as Figure N.2, but now we're labeling the vertical coordinates.

so if we think of the point (x, y) as the starting point, then $d_x m$ is the tiny difference in height that we would rise from walking an infinitely small distance dx in the x direction. Similarly, starting at (x, y), $d_y m$ is the tiny difference in height that we would rise from walking an infinitely small distance dy in the y direction.

So far we haven't really done much except zooming in and naming things, but we're surprisingly close to deriving the intimidating equations N.18 and N.19! Before we move on, make sure to stare at the three pictures in Figures N.2, N.3, and N.4, and make sure you understand why everything is labeled the way it is.

Now, because pointing is hard in a book, I'll have to define some terms. I'll define the "left journey" as follows: imagine starting at the point (x, y) in Figure N.2, and walking left along the graph, along the x axis, until you get to the point above $(x + dx, y)$. This is the first leg of the left journey. After completing the first leg, your height has increased by an amount $d_x m$ (make sure you see why). Now, for the second leg of the left journey, imagine continuing walking from your current location up to the top. On this leg of the journey, you're only walking in the y direction, and your height would increase by an amount $height(finish) - height(start)$, or

$$\hat{d}_y m \equiv m(x + dx, y + dy) - m(x + dx, y)$$

I put a hat on the d because we're already writing $d_y m$ to mean $m(x, y + dy) - m(x, y)$, and the hat just reminds us that these two quantities aren't the

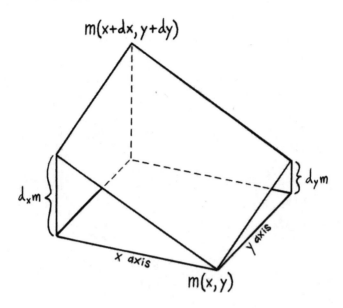

Figure N.4: Showing what $d_x m$ and $d_y m$ refer to geometrically.

same, or rather, they don't *look* the same.[2] The net effect of the left journey was that you went from a height of $m(x, y)$ to a height of $m(x + dx, y + dy)$, which is what we've called dm in equation N.20. So we can write

$$dm = d_x m + \hat{d}_y m \tag{N.21}$$

Similarly, although we don't need to, we could define the "right journey" as the process of walking in the other direction, and we'd get

$$dm = d_y m + \hat{d}_x m \tag{N.22}$$

where $\hat{d}_x m \equiv m(x + dx, y + dy) - m(x, y + dy)$. We can do either and we'd get to the same conclusion, so we can forget about $\hat{d}_x m$. Okay, now for the fun part. Recall that we defined dm by making tiny changes to both slots at once, because we were wondering if we could come up with a "total derivative." Notice that equations N.21 and N.22 are *almost* telling us something about the relationship between the "total differential" dm and the "partial differentials" $d_x m$ and $d_y m$. The trouble is, each equation only contains one of the familiar "un-hatted" partial differentials, and each has one of those pesky $\hat{d}$ differentials, which aren't the same thing.

Or are they? Since we've zoomed in infinitely far, the object we're looking at is a flat plane, so for the reason depicted in Figure N.5, the quantity $\hat{d}_x m$ should be the same as $d_x m$, and $\hat{d}_y m$ should be the same as $d_y m$. So the "hatted"

---

2. In a moment, we'll realize that $d_y m$ and $\hat{d}_y m$ actually *are* identical, but only because we zoomed in infinitely far.

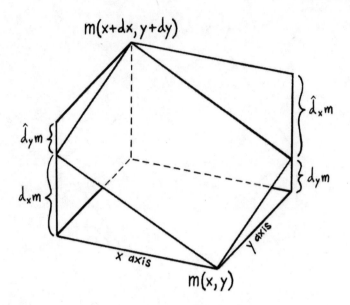

Figure N.5: Showing why $\hat{d}_x m$ is the same as $d_x m$ and why $\hat{d}_y m$ is the same as $d_y m$.

quantities are the same as the corresponding unhatted ones. Summarizing this:

$$dm = d_x m + d_y m \tag{N.23}$$

This equation is conveying an extremely simple fact: if we take either journey (left or right), then the *total height change* is the amount from the *first leg* plus the amount from the *second leg*. This idea could hardly be simpler. Read it again. Feel how simple it is. Almost pointless to even say. Now for the fun reveal. The simple sentence N.23 is saying the exact same thing as our scary sentence N.18 from earlier.

Further, all we have to do to get from equation N.23 to its scarier-looking twin equation N.18 is to multiply by 1 twice, and switch abbreviations. Let's do that. Starting with equation N.23, we can do this:

$$dm \overset{(N.23)}{=} d_x m + d_y m = \underbrace{d_x m \frac{dx}{dx} + d_y m \frac{dy}{dy}}_{\text{Multiplying by 1 twice}} = \underbrace{\frac{d_x m}{dx} dx + \frac{d_y m}{dy} dy}_{\text{Swap some terms}}$$

Now, if we simply switch to the (often misleading, but admittedly prettier) standard notation for partial derivatives, we obtain

$$dm = \frac{\partial m}{\partial x} dx + \frac{\partial m}{\partial y} dy \tag{N.24}$$

which is the scarier-looking equation N.18 we set out to derive originally. Recall that the notation above gives equation N.18 a sort of "one in two and two in one" property: ∂x and dx are the same quantity, despite the two symbols used to represent them, while the two ∂m's are *different* quantities, expressed by a single symbol. Notation can be goofy.

Next, we can make the logical leap discussed earlier, where we try to convince ourselves of something about n dimensions by visualizing three dimensions. The argument we just made involved a machine with two inputs and one output. So let's now suppose we've got a machine with n inputs and one output, and imagine making a tiny change in all the slots at once. As before, we examine the resulting change in what m spits out:

$$dm \equiv \underbrace{m(x_1 + dx_1, x_2 + dx_2, \ldots, x_n + dx_n)}_{\text{Output after}} - \underbrace{m(x_1, x_2, \ldots, x_n)}_{\text{Output before}}$$

or if you prefer,

$$dm \equiv m(\mathbf{v} + d\mathbf{v}) - m(\mathbf{v})$$

where $d\mathbf{v} = (dx_1, dx_2, \ldots, dx_n)$. Even though we can't picture what we're about to say, if we zoom in infinitely far on the graph of m at some arbitrary point $(x_1, x_2, \ldots, x_n)$, we should "see" an n-dimensional parallelogrammy-box-type thing, for the same reasons that we saw a two-dimensional parallelogram in the case above. So as before, it should be the case that

$$dm = d_1 m + d_2 m + \cdots + d_n m \tag{N.25}$$

where we're writing $d_i m$ instead of $d_{x_i} m$ to avoid cluttered notation. All this says is that the total height increase between two points in a space we can't picture is all the individual height increases added up. There's a height increase $d_1 m$ from walking a tiny amount dx_1 in the x_1 direction; there's a height increase $d_2 m$ from walking a tiny amount dx_2 in the x_2 direction, yadda yadda yadda. The total is just the sum of the parts — that's what equation N.25 is saying, and even though we can't picture what it's saying, we can be confident that it's true, because we understood the basic message from inventing equation N.23.

Alright! Equation N.25 is the result we wanted to derive, so we're basically done. However, if we define some new abbreviations, we can derive a formula that's written in all multivariable calculus books. Following the exact same logic as before, we can multiply each term in the above equation by 1, swap the order of multiplication, and then switch to the standard notation, to get

$$dm = \frac{\partial m}{\partial x_1} dx_1 + \frac{\partial m}{\partial x_2} dx_2 + \cdots + \frac{\partial m}{\partial x_n} dx_n \tag{N.26}$$

Here's how we could rewrite this if we wanted to pretend to be a textbook. Let's define the "dot product" of two vectors $\mathbf{v}$ and $\mathbf{w}$ to be an operation that bangs the two vectors together to give a number, like this:

$$\mathbf{v} \cdot \mathbf{w} \equiv v_1 w_1 + v_2 w_2 + \ldots + v_n w_n$$

That is, just multiply the vectors slot by slot and add up all the results. Having done this, if we define the abbreviations

$$d\mathbf{x} \equiv (dx_1, dx_2, \ldots, dx_n) \qquad \text{and} \qquad \nabla m \equiv \left(\frac{\partial m}{\partial x_1}, \frac{\partial m}{\partial x_2}, \ldots, \frac{\partial m}{\partial x_n} \right)$$

then we can rewrite equation N.26 like this:

$$dm = (\nabla m) \cdot d\mathbf{x}$$

So this fancy looking sentence with a vector full of partial derivatives and an infinitely small vector and a dot product is just telling us the familiar fact from before. The journey's total change in height is simply: the height change from the first leg, plus the height change from the second, and so on, up to the nth.

N.4 Is That All?

N.4.1 No

There's plenty more that we could do in the world of multivariable calculus, but it remains true that there are really no fundamentally new ideas. Derivatives mean the same thing they've always meant, although they continue to be written in a variety of different-looking ways. Integrals mean the same thing they've always meant, and although in textbooks you might see several of them at once — $\iiint$ instead of the familiar $\int$ — this is just the old idea of $\int$ three times in a row. So, rather than dwelling on all of the possible things we could do in this unfamiliar (but secretly familiar) world, let's forge on into the infinite wilderness: calculus in infinite-dimensional spaces. In infinite-dimensional calculus, there is still a strong sense in which all of the new, scary concepts are simply the old, familiar concepts wearing different hats. However, the infinite wilderness has a rather different feel, and it springs from a beautiful unification of machines and vectors. This unifying idea enhances the beauty, elegance, and applicability of infinite-dimensional calculus quite a bit relative to finite-dimensional calculus and the semi-familiarity of this chapter. Conceptually (if not always historically), this unification of machines and vectors gives rise to a bouquet of ideas including Fourier analysis, Lagrangian mechanics, the idea of a function space, the max entropy formalism in probability theory, and the language in which quantum mechanics — our species's deepest insight into the fundamental nature of reality — is expressed. We'll discuss this fundamental idea in a short dialogue after the Reu—

(Mathematics wanders into the chapter.)

Mathematics: WHAT'S ALL THIS? WHY... WHAT... DID YOU TWO CREATE MORE OF ME WITHOUT ME?

(*An awkward silence elapses.*)

Author: Uh . . . no?

(*Mathematics glances around at the nearby scenery.*)

Mathematics: DID YOU???
Author: Maybe.
Mathematics: WHY DIDN'T YOU TELL ME? I WOULD HAVE—
Author: If it's any consolation, the chapter isn't even close to being finished. I mean, there's *so* much to talk about on this subject, and I need to rewrite at least half of what I've written so far. It really doesn't do the subject justi—
Mathematics: JUST FINISH THE CHAPTER!
Author: Would you mind waiting while we finish up?
Mathematics: I DON'T HAVE ALL DAY. HOW LONG IS THIS GOING TO TAKE?
Author: Well, we could go on for quite a while, but. . . hmm. . . actually, we've been talking about this for long enough. Here, I'll write a Reunion right now and then we'll be done with this chapter.
Mathematics: WRITE QUICKLY.

(*Mathematics waits $\lambda P + (1 - \lambda)(imP)$ in an undefined location.*)[3]

Author: Alright, here we go. . .

N.5 Reunion

Let's take a minute to remind ourselves of what we did in this chapter.

1. We began with a dialogue in which Author disclosed that he did not feel he had done this subject justice. Nevertheless, this chapter will serve as a bridge to get us where we need to go.

2. This led us into a tour of the conceptual origins of multivariable calculus, where—
   Mathematics: WHAT'S MULTIVARIABLE CALCULUS?
   Reader: Hey, I didn't know you were allowed in here.
   Author: Yeah, Math. Do you want us to finish the chapter or not?
   Mathematics: I WANT YOU TO ANSWER ME. WHAT'S MULTIVARIABLE CALCULUS? WHAT'S THIS NEW PART OF ME YOU INVENTED WHILE I WAS AWAY?
   Author: Why don't you go read the chapter while I'm finishing it up?
   Mathematics: . . . OKAY.

   (*Mathematics flips back and begins reading the chapter.*)

---

3. (*Where $\lambda \subset [0,1]$ and $P \equiv patiently$.*)

Author: As I was saying, we took a tour of the conceptual origins of multivariable calculus, and we saw that many of its definitions spring from the desire to build new ideas out of nothing but old ideas.

3. We attempted to differentiate a machine of the form $m(x) \equiv (f(x), g(x))$ by doing the same things we would have done in single-variable calculus. Eventually we got stuck, but we found that we could get unstuck if we defined the addition of lists to be slot by slot: $(a, b) + (A, B) = (a + A, b + B)$. Being unstuck, we eventually got stuck again. To get unstuck the second time, we defined the multiplication of a number by a list to be multiplication slot by slot:

$$c \cdot (x, y) = (cx, cy)$$

Having made these two choices, we found that

$$\frac{d}{dx}\Big(f(x), g(x)\Big) = \left(\frac{d}{dx}f(x),\ \frac{d}{dx}g(x)\right)$$

This meant our new "one in, two out" machines could be differentiated by using only familiar single-variable calculus in each slot.

4. We found that a similar story held for "one in, n out" machines of the form

$$m(x) \equiv (f_1(x), f_2(x), \ldots, f_n(x))$$

5. Because of this, we found that we could bring all of our hammers for computing derivatives into this new multivariable world.

6. We found that a similar story was true for integration. Namely,

$$\int_a^b \Big(f(x), g(x)\Big)\ dx = \left(\int_a^b f(x)\ dx, \int_a^b g(x)\ dx\right)$$

As before, simply perform the familiar operations of single-variable calculus in each slot.

7. Because of this, we found that we could bring all of our anti-hammers for computing integrals into this new multivariable world.

8. We then turned our sights to "two in, one out" machines. We found that we could now define two different derivatives: one for each input slot. Textbooks call these "partial derivatives." These partial derivatives were simply single-variable calculus all over again. For example, the partial derivative with respect to x is whatever we would get from pretending all the variables except x were constants, and computing the derivative in the familiar way.

9. We found that a similar story held for "n in, one out" machines.

10. Using our infinite magnifying glass and some simple visual reasoning, we derived the formula

$$dm = \frac{\partial m}{\partial x}dx + \frac{\partial m}{\partial y}dy$$

and its generalization to n dimensions

$$dm = \frac{\partial m}{\partial x_1}dx_1 + \frac{\partial m}{\partial x_2}dx_2 + \cdots + \frac{\partial m}{\partial x_n}dx_n$$

11. Repeatedly in this chapter, whenever we found ourselves faced with a new idea, we found that it was not new after all, but had instead been cobbled together from old ideas, plus the occasional insignificant change to make everything make sense.

Author: Okay, Math. Are you happy now?

(*Mathematics is reading the chapter and doesn't hear Author.*)

Author: *Mathematics!*
Mathematics: AAAH! YOU STARTLED ME.
Author: I'm "done" writing the chapter. Though I'm not exactly hapy with it. It's horrible. See that typo I just made? Either way, we're done now. Feel like you understand the chapter?
Mathematics: WELL, I ONLY HAD A CHANCE TO SKIM IT. LET ME SEE IF I'VE GOT THE IDEA STRAIGHT: WE'RE DOING CALCULUS JUST LIKE WE ALWAYS HAVE, BUT NOW OUR INTEGRALS AND DERIVATIVES ARE ACTING ON WEIRD NEW TYPES OF MACHINES, LIKE ONES THAT EAT A NUMBER AND SPIT OUT A VECTOR, OR EAT A VECTOR AND SPIT OUT A NUMBER?
Reader: Yep. That was basically all there was to it. Nothing much new.
Mathematics: AND IF I UNDERSTAND CORRECTLY, THESE "VECTOR" THINGS ARE JUST MACHINES THEMSELVES, RIGHT?
Author: What? No! Okay, hold on. We're already hanging off the end of the Reunion, and this seems like it might take a while. Jump into the next interlude with me, you two.

(*Author slams the door as he leaves the chapter.*)

Interlude N: (Mis)interpreted (Readings) (Re)interpreted

Misinterpretation

Author: I missed that. In the last sentence. Come again?
Reader: I didn't say anything.
Author: I know. Mathematics. One more time?

(A cryptic silence elapses.)

Mathematics: I SAID, "IF I UNDERSTAND CORRECTLY, THESE 'VECTOR' THINGS ARE JUST MACHINES THEMSELVES, RIGHT?"... DID I MISUNDERSTAND SOMETHING?
Author: I think so. But don't worry, it's a tricky concept. A vector itself isn't really like a machine at all. It's just a list of numbers. Like, the vector $\mathbf{v} \equiv (3, 7, 4)$ is just a list of numbers, which we can think of as a point in three-dimensional space. It's the point whose location, relative to the "origin," is 3 units in the x direction, 7 units in the y direction, and 4 units in the z direction.
Mathematics: RIGHT, SO VECTORS ARE JUST MACHINES.
Reader: ???
Author: No no no. Why do you keep saying that?
Mathematics: I DON'T KNOW. I MUST BE MISUNDERSTANDING SOMETHING. EXPLAIN THE IDEA AGAIN.
Author: The vector $\mathbf{v} \equiv (3, 7, 4)$ is nothing like a function... er, I mean machine. It's just three numbers. We can choose to think of it geometrically or we can choose not to. But as I've written it, it's just a list with three items, which we can write as $v_1 \equiv 3$, $v_2 \equiv 7$, and $v_3 \equiv 4$.
Mathematics: I UNDERSTAND WHAT YOU'RE SAYING, BUT I'M NOT SURE YOU UNDERSTAND WHAT I'M SAYING. WE CAN STILL ABBREVIATE THINGS HOWEVER WE WANT, CORRECT?
Author: Of course. That's basically the only "law" of mathematics: objects are only defined up to isomorphis... er, I mean, reabbreviation.
Mathematics: SO SUPPOSE I PREFERRED TO TALK ABOUT THIS VECTOR

USING DIFFERENT ABBREVIATIONS. INSTEAD OF WRITING THIS:

$$v_1 \equiv 3 \qquad v_2 \equiv 7 \qquad v_3 \equiv 4$$

SUPPOSE WE WERE TO WRITE THIS:

$$v(1) \equiv 3 \qquad v(2) \equiv 7 \qquad v(3) \equiv 4$$

APART FROM THAT CHANGE OF ABBREVIATIONS, SUPPOSE MY VECTORS BE-
HAVE EXACTLY LIKE YOURS DO. WE CAN ADD THEM, MULTIPLY THEM BY
NUMBERS, AND SO ON. THEN...

Author: (*Wide-eyed*) Hold on... I have a sense that this is going to be *really*
important. Let me make another section.

Reinterpretation

Author: (*To Mathematics*) Go on, please!

Mathematics: NO, I THINK I WAS JUST CONFUSED. I ONLY HAD TIME TO
SKIM THE CHAPTER. I WAS INCORRECTLY THINKING OF THE VECTOR

$$\mathbf{v} \equiv (v_1, v_2, v_3) \equiv (3, 7, 4)$$

AS A MACHINE, WHICH, WHEN FED 1 SPITS OUT $v(1) \equiv 3$, WHEN FED 2 SPITS
OUT $v(2) \equiv 7$, AND WHEN FED 3 SPITS OUT $v(3) \equiv 4$. GRANTED, IT WOULD
BE A FUNNY TYPE OF MACHINE, BECAUSE USUALLY OUR MACHINES CAN EAT
ANY NUMBER, WHEREAS THIS NEW MACHINE $\mathbf{v}$ WOULD ONLY BE ALLOWED
TO EAT 1, 2, OR 3, SO ITS SET OF POSSIBLE FOOD WOULD BE THE SET
$\{1, 2, 3\}$ INSTEAD OF A CONTINUOUS SET OF NUMBERS. I BELIEVE THAT'S
WHY I WAS CONFUSED.

Author: Wait. I don't think you *were* confused. I was. Your argument means
we *have* to acknowledge that vectors can be thought of as a particular type of
machine. If we don't acknowledge that, then we would accidentally be saying
that we can't abbreviate things however we want, because all your argument
required was a minor reabbreviation.

Reader: So Mathematics *wasn't* confused?

Author: No. And I think the same argument works in reverse. It almost has
to. The whole argument was just a tiny abbreviation change. So, it's not just
that vectors can be thought of as machines. Machines can also be thought of
as vectors.

Mathematics: HOW DO YOU MEAN?

Author: Well, take a machine like $m(x) \equiv x^2$ for instance. We can just think
of this as a list of numbers, if we want to. I mean, instead of a list with a
first slot, a second slot, a third slot, and so on, it would be a vector with a
continuous infinitude of slots, one for each number.

Reader: Oh! I think I get the idea. So $m(x) \equiv x^2$ can be thought of as a vector that has the number 9 sitting inside the slot labeled 3, because $m(3) \equiv 9$.

Do the slots have to be labeled by whole numbers?

Author: I don't see why they would. For example, $m(\frac{1}{2}) = \frac{1}{4}$ and $m(\sharp) = \sharp^2$, so m has the number $\frac{1}{4}$ sitting inside the slot labeled $\frac{1}{2}$, and it has the number $\sharp^2$ sitting inside the slot labelled $\sharp$.

Mathematics: AND JUST LIKE I SWITCHED FROM v_i TO $v(i)$, CAN WE DO THE SAME IN REVERSE AND WRITE $m_3 = 9$ INSTEAD OF $m(3) = 9$?

Author: I don't see why not! We're not really *doing* anything. This is just a meaningless change of notation.

Mathematics: BUT THE FACT THAT THE TWO ABBREVIATIONS COULD END UP SO SIMILAR WAS POSSIBLE ONLY BECAUSE BOTH CONCEPTS NEED THE SAME TYPE OF INFORMATION TO SPECIFY THEM.

Author: I think so. I mean, instead of $m(x) \equiv x^2$, there's no reason we can't write $m_x \equiv x^2$.

Mathematics: SO VECTORS ARE MACHINES AND MACHINES ARE VECTORS?

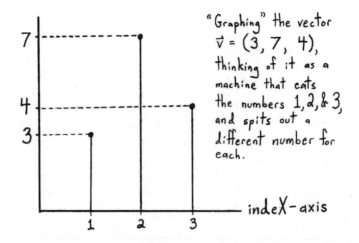

Figure N.6: Up until now, we were thinking of a vector like $(3, 7, 4)$ as a point in three-dimensional space. However, we can also think of it as a machine. This machine is far from omnivorous: it can't eat anything but the numbers 1, 2, and 3. One advantage of picturing vectors this way is that our powers of visualization no longer break down once we move beyond three dimensions. Picturing seventeen dimensions simply requires us to picture seventeen vertical lines whose heights can vary independently. Picturing a point in a space with infinitely many dimensions simply requires us to picture the mundane "graph" of a mundane "function," in two mundane dimensions.

Author: I think they have to be! After all, the two concepts behave in exactly the same way. We just didn't recognize that at first. We were thinking of vectors with n slots as points in n-dimensional space, so we came up with abbreviations *suited to that interpretation*. Now even though there was nothing *wrong* with those abbreviations, they didn't make it obvious that vectors were really just a type of machine, and vice versa. The two concepts are *logically* identical, but the notation we used didn't make them *psychologically* identical, so we missed the underlying unity of the two ideas!

Mathematics: But which are they really? Are machines really just vectors, or are vectors really just machines?

Author: I'm not sure we have to decide. There's no hidden essence inside these things that could somehow trick us. We're inventing this stuff ourselves. It's not as if something could act like a vector in every way but really secretly be a machine, or vice versa.

Reader: But isn't this a bit arbitrary? I mean, a machine like $m(x) \equiv x^2$ seems like it's really a machine, even though we might be able to think of it as a vector. Are you saying that it's just as much a vector as it is a machine?

Author: Exactly! And if we aren't willing to grant that it is, we're actually being inconsistent. Since the beginning, the only "law" of mathematics in our universe has been: *we can abbreviate things however we want.* That implies that we can only define things by how they behave, not by what they are. And *that* must be why mathematicians are so obsessed with defining their objects axiomatically. In every branch of the field, the way they choose their definitions in the first place is always driven by this secret, underlying law: mathematical objects are not notation. The things we're studying are not just squiggles on paper. If no particular choice of abbreviations is sacred, then the axiomatic approach and all that business about things being "only defined up to isomorphism" all follows naturally! It *has* to be that way. If it weren't, then we would have violated the Only Law of mathematics: we would have assumed, at some point, however implicitly, that some set of abbreviations was inherently special.

Reader: Hey, uh, now that you're done ranting, I have a question. Since vectors like $(3, 7, 4)$ can be thought of as points in three-dimensional space... and since vectors with n slots can be thought of as points in an n-dimensional space... can we think of machines like $m(x) \equiv x^2$ as "points in an infinite-dimensional space"?

Author: I don't see why not.

Mathematics: This is beautiful! Can we do a chapter about this?

Author: Of course! We'll have to figure it out as we go.

Reader: Isn't that what we've been doing all along?

Author: Hah! Yeah, I guess so. This'll be fun! Come on, let's go!

. . .

(Author realizes something, and falls silent.
He decides to stay behind and reread the interlude,
while the others jump into the next chapter.)

. . .

Author: ...

Reader: ...Hi!

Author: (*Startled*) Oh! I thought you had left. You're not rereading this too, are you?

Reader: No. What's going on?

Author: Nevermind. It's nothing.

Reader: Seriously, what's wrong?

Author: No no, it's okay... It would just interrupt...

Reader: Interrupt what?

Author: Well, it's just... there are no dialogues in the next chapter... and there are no chapters after that... so, I just realized... I'm not sure I'll ever get to, you know... see you... again.

Reader: Oh.

Author: Yeah.

Reader: Why not just add a final section?

Author: I mean, I've been planning to. Not really a chapter or an interlude. More like a... chapterlude. But still, all my ideas for it are pretty weird. Who knows what it'll be like when we get there. And either way, there's no mathematics in it. That's all in Chapter $\aleph$, right after this. So I just realized... even if we all get to see each other one more time... we won't get to do this again. The mathematics. That was our last time.

Reader: Oh... I see.

Author: I hadn't realized... until I finished writing the interlude just now.

Reader: Aren't you still writing it?

Author: No. I just do this sometimes. Talk to you. This isn't for the book.

Reader: What? Why?

Author: I don't want to change the tone of the section. Your... their... the interpretation would be all wrong. Sometimes you just need to hide these things.

Reader: From who?

Author: From you! For the book! For the good of the narrative.

Reader: But what about... you know... full disclosure?

(*Author sighs.*)

Author: I'm gonna miss you so much when this book is over.

ℵ The Infinite Beauty of the Infinite Wilderness

ℵ.1 The Surprising Unity of the Void

Theories of the known, which are described by different physical ideas may be equivalent in all their predictions and are hence scientifically indistinguishable. However, they are not psychologically identical when trying to move from that base into the unknown. For different views suggest different kinds of modifications which might be made and hence are not equivalent in the hypotheses one generates from them in one's attempt to understand what is not yet understood.
—Richard Feynman, *Nobel Prize Lecture*

Mathematics is not a careful march down a well-cleared highway, but a journey into a strange wilderness, where the explorers often get lost. Rigour should be a signal to the historian that the maps have been made, and the real explorers have gone elsewhere.
—W. S. Anglin, *Mathematics and History*

ℵ.1.1 Taking the Analogy Seriously

At the end of the previous interlude, we noticed a direct relationship between vectors and machines. In this chapter, we will take this relationship seriously and develop calculus in an infinite-dimensional space — a subject commonly known as the "calculus of variations." As we saw in Chapter N, multivariable calculus is constructed from exactly the same simple concepts as single-variable calculus. That remains true in this case — the symbolic operations of calculus of variations basically behave like they always have — but there is a sense in which it feels very different from single- or multivariable calculus. This different feel is the result of the unification of vectors and machines that we arrived at in the last interlude. That conceptual unification will allow us to switch back and forth between two different interpretations of any given formula, and thus any formula in the calculus of variations can be thought of as saying two different things. This gives us two layers of intuition with which we can proceed.

Up to this point in the book, our approach has been rather unconventional. We have largely avoided limits in favor of infinitesimals; we have frequently invented our own terminology; we have taught the subject "backwards," beginning with calculus and using it to invent the topics commonly assumed to be its "prerequisites"; we have sworn off the standard practice of presenting our mathematical arguments in a doctored and polished form, which — though elegant — often obscures the processes of thought that led us to discover them, and thus

Mathematics: WE HAVE ATTEMPTED TO ILLUSTRATE HOW PARTICULAR

Author: discoveries can be made only by first wandering down particular

Mathematics: BLIND ALLEYS AND ARRIVING AT PARTICULAR CONFUSIONS.

Author: Thus, to leave out the confusions is to leave

Mathematics: OUT THE UNDERSTANDING. HOWEVER,

Author: despite the many ways in which this book has been unconventional to this point, this final chapter may be the most unconventional of all. Calculus of variations, despite its beauty and the simplicity of its underlying ideas, is virtually never taught in a way that makes this simple beauty apparent. The standard method of teaching calculus of variations in mathematics courses and textbooks is so cautious and so formal (even in comparatively informal courses on applied mathematics) that the exactness of the analogy between calculus of variations and more familiar forms of calculus is nearly always unclear. The standard presentation of these ideas, though logically correct and understandably cautious, is a pedagogical nightmare, getting itself all tangled up in contraptions called "test functions," "distributions," "generalized functions," "linear functionals," "weak derivatives," "variations," and so on. All of these contraptions are beautiful ideas in their own right, but they're *much* more complicated than one needs to grasp the underlying ideas. I will attempt to explain the calculus of variations in as simple a manner as I can, at every stage making clear the direct analogy with things we already know. As always, this similarity of the new and the old is not an accident, but a direct result of the fact that old familiar concepts provide the raw materials from which the innovators construct the new.

ℵ.1.2 Fortune Favors the Bold!

At this point, we have no idea whether all the familiar operations of calculus will continue to make sense once we attempt to apply them in infinitely many dimensions. Perhaps they will not. But after all, that's the same position we've been in throughout the book. The test of whether a generalization makes sense is, and has always been, whether it does what we want it to do. In most cases, "what we want it to do" is to agree with our intuitions in simple cases where our intuition can readily perceive what the answer should be. As such, we can confidently proclaim that, although we don't yet know whether what we're about to do makes sense, it's well worth the risk. We choose to simply roll the dice and forge ahead. *Alea iacta est.*

ℵ.2 Crossing the Rubicon

ℵ.2.1 Building a Dictionary for Our Analogy

So, having decided to take the analogy between machines and vectors seriously, let's make this analogy a bit more precise by building a dictionary. This will let us more easily translate back and forth between sentences about vectors and sentences about machines, as well as help us to figure out how we might do calculus in this new infinite-dimensional world. The box below is our dictionary, and each "definition" consists of two lines. The first line of each pair is a description of a particular part of the vector/machine analogy in words, while the second line says the same thing in symbols. The symbol "⟺" means something like "the things on my left and right sides are partners under the vector/machine analogy." This dictionary is by no means exhaustive; we will notice further ways in which machines and vectors can be unified in the rest of this chapter. However, the dictionary contains a complete description of our thoughts, to this point, about how machines and vectors can be viewed as two specific examples of a more general concept. Ready, set, dictionary!

How the Dictionary Works

Thing about vectors (in words) ⟺ Thing about machines (in words)
Thing about vectors (in symbols) ⟺ Thing about machines (in symbols)

The Dictionary

Vector ⟺ Machine
$$\mathbf{x} \Longleftrightarrow f$$
Vector index ⟺ Machine input
$$i \Longleftrightarrow x$$

Vector component at some index ⟺ Machine output at some input
$$x_i \Longleftrightarrow f(x)$$

Machine that eats a vector ⟺ Machine that eats a machine
$$F(\mathbf{x}) \Longleftrightarrow F[f(x)]$$
Sum ⟺ Integral
$$\sum_{i=1}^{n} x_i \Longleftrightarrow \int_a^b f(x)dx$$

ℵ.2.2 Mathematical Cannibalism

By taking seriously the analogy between machines and vectors, we built the above dictionary, which lets us translate from statements about vectors to statements about machines. However, in doing so we wrote down a bunch of symbols that we're not too familiar with, like $F[f(x)]$.[1] We should stop and ask exactly what kind of objects we're examining. Much of multivariable calculus consisted of examining machines that eat a vector $\mathbf{x}$ and spit out a number $F(\mathbf{x})$. Therefore, taking the vector/machine analogy literally suggests that in the calculus of variations, we are examining machines that eat an entire machine $f(x)$ and spit out a number $F[f(x)]$. Such cannibalistic machines are usually called "functionals," and as it turns out, we have met several functionals already, although we did not necessarily think of them in this way at the time. For example, the integral

$$Int[f(x)] \equiv \int_a^b f(x)dx$$

can be thought of as a machine in its own right. That is, a "large machine" that eats an entire machine $f(x)$ and spits out a number: the area under $f(x)$'s graph between $x = a$ and $x = b$. Feeding different machines to the cannibalistic machine Int gives us different numbers.

Another example that we've already encountered is the so-called "arclength functional." Toward the end of Chapter 6, we showed that the integral

$$Arc[f(x)] \equiv \int_a^b \sqrt{1 + f'(x)^2}\,dx$$

can be thought of as a large machine that eats an entire machine $f(x)$ and spits out a number: the length of $f(x)$'s graph between $x = a$ and $x = b$.

Yet another example of a functional is what is often called the "norm" of a machine f. "Norm" is just a fancy word for the "length" of f when interpreted as a vector in an infinite-dimensional space. This interpretation of length has nothing to do with the arclength interpretation above. Rather, it comes from generalizing the familiar concept of a vector's length — taking inspiration from the formula for shortcut distances — to define a concept of "length" in infinitely-many dimensions. Since we haven't yet seen this particular kind of cannibalistic machine, it will be useful to take a few minutes to show where the idea comes from, in order to get a better feeling for the type of reasoning we'll be using in this new infinite-dimensional world.

---

1. As a note of clarification, the reason we are using square brackets in writing $F[f(x)]$ is because (i) the notation $F(f(x))$ is a bit cluttered with parentheses, and (ii) writing $F(f(x))$ might make it seem as if the machines f and F occupied the same level of abstraction. Although they are both machines, and although the square brackets in $F[f(x)]$ serve exactly the same purpose as normal parentheses would, the slight difference in notation does serve to remind us that f is a machine that eats a number and spits out a number, while F is a machine that eats an entire machine and spits out a number. This is an important distinction: though normal functions such as f can appear in terms such as $f(g(x))$, the machine f is still simply eating a number, g's output at input x, while F is genuinely eating an entire machine. F is a cannibal; f is not.

ℵ.2.3 Characterizing Length in the Infinite Wilderness

In Interlude 1, we invented the formula for shortcut distances, also known as the Pythagorean theorem. First, notice that we can interpret this formula as a fact about vectors in two dimensions. The components x and y of a vector $\mathbf{v} \equiv (x, y)$ are just lengths, and they lie in two perpendicular directions, so the length of this vector should be

$$Length[\mathbf{v}] \equiv \ell = \sqrt{x^2 + y^2}$$

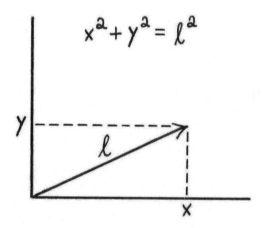

Figure ℵ.1: The formula for shortcut distances tells us that the length of a vector $\mathbf{v} \equiv (x, y)$ is $\ell \equiv \sqrt{x^2 + y^2}$. This will serve as our inspiration for defining length in the infinite wilderness.

Now, the question arises of whether a similar formula holds in three dimensions, four dimensions, and so on. Since the length of a vector in two dimensions involves a bunch of 2's (namely, a sum of the components each raised to a power of 2, and with a $\frac{1}{2}$ over the whole thing), we might guess that this relationship between the dimension in which the vector lives (namely, 2) and the powers that show up in the formula (also 2) is not merely a coincidence. This might lead us to guess that the formula for the length of a vector $\mathbf{v} \equiv (x, y, z)$ in *three* dimensions should be

$$Length[\mathbf{v}] \overset{???}{=} \left(x^3 + y^3 + z^3\right)^{\frac{1}{3}}$$

However, while we could certainly *choose* to measure length this way,[2] it is

---

2. And mathematicians have done so. Such a way of measuring length is known as the "three norm." While it has no particular connection with three dimensions, it is a perfectly reasonable definition of length, although it does not correspond with what we mean by length in everyday life. Since it does not correspond with our everyday concept of length, it will not concern us here.

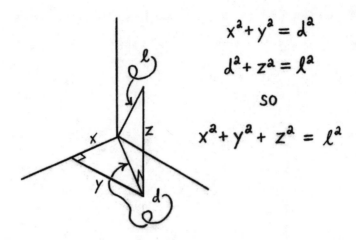

$$x^2 + y^2 = d^2$$

$$d^2 + z^2 = \ell^2$$

so

$$x^2 + y^2 + z^2 = \ell^2$$

Figure ℵ.2: Using the formula for shortcut distances twice lets us build a three-dimensional version of the original formula. It shows us that the length of a vector $\mathbf{v} \equiv (x, y, z)$ in three dimensions is $\ell \equiv \sqrt{x^2 + y^2 + z^2}$.

not up to us to decide this issue, *if* we want length to have the same meaning as it does in everyday language. While we are free to do, say, and invent whatever we want, length has an independent, everyday meaning that everyone understands without any mathematics, so *if* we want "length" to mean what we mean in everyday life, *then* the definition of the length of a vector in three or four or n dimensions is not something that we can simply *invent*, but rather something that we have to *discover*. Since the only raw materials we have at present to attack this question come from the two-dimensional formula for shortcut distances, it is worth asking whether we can use this formula to discover the corresponding unknown formula in higher dimensions, and ultimately extend it to our new infinite-dimensional wilderness.

As Figure ℵ.2 shows, it is indeed possible to build a formula for shortcut distances in three dimensions using only the two-dimensional version. This argument demonstrates that

$$Length[\mathbf{v}] = \sqrt{x^2 + y^2 + z^2}$$

is the length of a vector $\mathbf{v} \equiv (x, y, z)$ in three dimensions. Now, by applying the two-dimensional version over and over, we can build an analogous formula for shortcut distances in n dimensions. As expected, it has the same form: the length of a vector $\mathbf{x} \equiv (x_1, x_2, \ldots, x_n)$ in n dimensions is

$$Length[\mathbf{x}] = \sqrt{x_1^2 + x_2^2 + \cdots + x_n^2} \tag{ℵ.1}$$

Now, if we're really taking the correspondence between machines and vectors seriously, then we should be able to define the "length" or "size" of a machine by simply interpreting the machine as a vector with infinitely many slots. This would give us a kind of "formula for shortcut distances" in an infinite-dimensional space, and in that sense, it would be a statement about infinite-dimensional geometry. That may sound complicated, but the apparent complexity is mostly the fault of the words we're using to describe the idea, like "infinite-dimensional space" and such. Despite the sophisticated-sounding language, the underlying idea is extremely simple. We're really just mindlessly generalizing equation ℵ.1 by using our dictionary. That is, all we're saying is: if we're thinking of machines as vectors, then equation ℵ.1 tells us that the "length" of a machine should be whatever we get from (i) squaring each slot, (ii) adding them all up, and (iii) throwing a square root symbol over the whole thing. That is:

$$Length[f(x)] = \sqrt{\int_a^b f(x)^2 dx} \qquad (\aleph.2)$$

Then, having mindlessly generalized the formula for shortcut distances, we can say all sorts of things that make us sound very smart, like "the length of a function f, interpreted as a vector in an infinite-dimensional space, is simply the square root of the integral of $f(x)^2$." And we know exactly what we mean by sentences of this form, even though we ourselves can't even begin to picture what we're saying! Sneaky trick, right? How do we know if this is right? Well, as always, if we want our infinite-dimensional space to behave like the finite-dimensional spaces we're familiar with, then we can just *define* length in the infinite wilderness by equation ℵ.2. At this point we are engaged not in standard mathematics but rather in pre-mathematics, so the argument above cannot be right or wrong. It is simply using mathematical concepts to make an argument whose apparent merit (or lack thereof) leads us to choose a particular definition of infinite-dimensional length in the first place.

Now that we've defined a notion of the "length" of a machine, we can say things like "such and such machine is infinitely small." That is, now that we can talk about size in the infinite wilderness, we can begin to explore some ideas of how we might invent infinite-dimensional calculus!

ℵ.3 Inventing Cannibal Calculus

ℵ.3.1 Building New from Old, as Always

It is not the essence of our objects but their behavior that concerns us most. This is true in all of mathematics, and we've seen this principle since the beginning. There may indeed be many possible definitions of area or slope that behave differently than the ones we have used. Such definitions may be infinitely more complicated than ours, and thus much more difficult to work with. However, our definitions behave exactly like we want them to, because we forced them to, and thus we can confine our attention to them. The Void contains an infinite swarm of machines, most of which are much more complicated than the four species we examined in Chapter 5, but we chose to examine those four particular kinds of machines because we knew exactly what we could say about them. And we knew what we could say about them precisely because we defined them by how they behaved.

Here we face the same principle, though in a different guise. Ignoring the baggage of existing textbooks, how can we define the derivative of machines that eat entire machines and spit out numbers in a way that allows us to perform the same operations on them that we always have? Rather than abiding by the practice customary in textbooks, in which one *first* defines the set of functions that one is talking about (as if that were the primary concern driving mathematical creation) and only then proceeds to derive results, we prefer the opposite path. We will instead pre-mathematically *derive* the definitions of infinite-dimensional calculus by remaining agnostic about the exact space of functions we're talking about, and forge ahead by *forcing* these new objects — whatever they might be — to behave as similarly to the calculus we're familiar with as we require them to in order to move forward. Perhaps most importantly, whatever "derivative" means in this new world, we will demand that derivatives can still be thought of as one infinitely small number divided by another. Only in that way can we carry over the calculus expertise we've developed thus far in the book into this untamed wilderness. Investigation of exactly which objects obey such operations and how to code such objects into the language of set theory is an optional afterthought, to be carried out by anyone who is interested in the task. When the dust settles, if it turns out that we were studying a different class of objects than we thought we were studying originally, so be it. We are unconcerned with essences.

So! Suppose we've got a cannibalistic machine, by which we mean a machine that eats an entire machine $f(x)$ and spits out a number $F[f(x)]$. In multivariable calculus, we defined partial derivatives by starting with a machine $F(\mathbf{x})$, then making an infinitely small change to one of $\mathbf{x}$'s slots, while leaving all the rest the same. Then we looked at the difference in output before and after, and divided this by the difference in input before and after. In the spirit of building new from old, let's do exactly the same thing to define the derivative in the infinite wilderness.

Given a cannibalistic machine F, we have infinitely many "slots" in which we can make changes. For machines that eat vectors, these slots were labeled x_1, x_2, ..., x_n. Now, for machines that eat entire machines, these slots are labeled $f(0)$, $f(0.001)$, $f(3)$, $f(796.5)$ and so on. We can't actually list every slot, but there is one for each number x. While x_3 labeled the number in the third slot of a vector x, the symbol $f(x)$ now labels the number in the xth slot of the machine f. As such, let's start by defining the "partial derivative" of a cannibal machine.

First, let's abbreviate:

$$\delta F[f(x)] \equiv F[f(x) + \delta f(x)] - F[f(x)] \tag{ℵ.3}$$

Before explaining what this means, it's important to stress that we're using the weird notation δ instead of d *not* in order to confuse you, but so that you can see how simple and similar to single-variable calculus are the *ideas* expressed by the intimidating notation of the standard textbooks. If the δs are scaring you, then please, for the love of mathematics, cross out my equations and rewrite them with d's instead. The equations would convey exactly the same content either way.

Okay. In the above equation, the symbol $\delta f(x)$ refers to an infinitely small machine, just as dx in single-variable calculus referred to an infinitely small number. The sense in which this machine is "infinitely small" is provided by our definition (earlier in this chapter) of a machine's "length" or "size," which was itself inspired by the formula for shortcut distances. If a machine's length by that definition is an infinitely small number, then it is an "infinitely small machine." Crucially, as in single-variable calculus, the f in $\delta f(x)$ is not the same as the f in $f(x)$.[3]

One point about notation: what we've written as $F[f(x)]$ could more properly be written as $F[f]$, because it does not depend on some particular value of x, but rather on the entire machine f. However, in my experience, writing $F[f(x)]$ will tend to be less confusing in the long run. Because of this, when we need to specify which particular slot we're differentiating with respect to, we need another letter besides x. Rather than use something like y, which might connote "verticalness," I'll just use $\tilde{x}$. The squiggle above the x just says "this is a different symbol from x, and it may or may not refer to a different point." That's all. Alright, now that we've done this, our dictionary suggests that we define the derivative of F with respect to the particular slot $f(\tilde{x})$ to be

$$\frac{\delta F[f(x)]}{\delta f(\tilde{x})}$$

where $\delta F[f(x)]$ is defined as in equation ℵ.3, and $\delta f(\tilde{x})$ is simply the output of some unrelated infinitely small machine δf (whose particular form we're remaining agnostic about, as we have done with variables from the beginning)

---

3. This point deserves clarification: In single-variable calculus, the notation $x + dx$ refers to x plus an *unrelated* infinitely small number. When we write "$x + dx$" we write four symbols: (i) x, (ii) $+$, (iii) d, (iv) x. As we already know, the x in item (i) is *in no way* related to the x in item (iv), which forms the second letter of "dx." The notation dx in single-variable calculus is not something we *do* to the number x; it is simply an (unrelated) infinitely small number. Even though we already know this, it is important to emphasize in understanding the abbreviation in equation ℵ.3. Confusingly, however, in many areas of mathematics, the d in $d(thing)$ *does* refer to an action performed on the $(thing)$ that follows it, and calculus of variations textbooks often use the δ in that way as well, to refer to something they call a "variation." Don't worry about that for now. We'll encounter the idea again soon.

when fed the input $\tilde{x}$. Let's see how this all works in practice. Suppose we're looking at the particular cannibalistic machine

$$F[f(x)] \equiv \int_a^b f(x)^2 dx$$

Then, using the definitions above,

$$\delta F[f(x)] \equiv F[f(x) + \delta f(x)] - F[f(x)]$$

$$\equiv \left(\int_a^b [f(x) + \delta f(x)]^2 dx \right) - \left(\int_a^b f(x)^2 dx \right)$$

$$= \left(\int_a^b f(x)^2 + 2f(x)\delta f(x) + (\delta f(x))^2 dx \right) - \left(\int_a^b f(x)^2 dx \right)$$

$$= \int_a^b f(x)^2 + 2f(x)\delta f(x) + (\delta f(x))^2 - f(x)^2 dx$$

$$= \int_a^b 2f(x)\delta f(x) + (\delta f(x))^2 dx$$

So simply dividing by $\delta f(\tilde{x})$, we obtain

$$\frac{\delta F[f(x)]}{\delta f(\tilde{x})}$$

$$= \frac{1}{\delta f(\tilde{x})} \int_a^b \left[2f(x)\delta f(x) + (\delta f(x))^2 \right] dx$$

$$= \int_a^b \left[2f(x)\frac{\delta f(x)}{\delta f(\tilde{x})} + \delta f(x)\frac{\delta f(x)}{\delta f(\tilde{x})} \right] dx$$

ℵ.3.2 Infinite Pre-mathematics, Part 1: A Possibility They Never Discuss

At this point, it might appear that the discussion is stuck, because we haven't yet specified what the symbol

$$\frac{\delta f(x)}{\delta f(\tilde{x})}$$

refers to. But remember, we're inventing this stuff ourselves, so rather than asking, "What *should* we do next?" we should instead ask, "What do we *want* the derivative of $F[f(x)]$ to be?" If that sounds like a backward way of reasoning, think again! Remember what we've been doing throughout the book. The process of generalizing an old, familiar concept to wilder and weirder contexts always involves a choice. The choice is: which aspects of the old concept do we want to build into our new, more general version of it? Here, as we'll see in a moment, the choice is essentially whether we want this derivative

$$\frac{\delta}{\delta f(\tilde{x})} \int_a^b f(x)^2 dx$$

to be equal to $2f(\tilde{x})$ or to $2f(\tilde{x})dx$. We'll see why in a few lines. Back to the point: we stopped in the above calculation because we had not yet said what we mean by

$$\frac{\delta f(x)}{\delta f(\tilde{x})}$$

So the question of how we want to define the "functional derivative" — the derivative of a cannibalistic machine $F[f(x)]$ — cannot proceed unless we say how we want to define the functional derivatives of the different "vector slots" $f(x)$ and $f(\tilde{x})$ with respect to each other. Maybe our dictionary will help. Recall that in multivariable calculus, we had

$$\frac{\partial x_i}{\partial x_j} = \begin{cases} 1 & \text{if } i = j \\ 0 & \text{if } i \neq j \end{cases} \tag{ℵ.4}$$

which just says that since the different variables x_1, x_2, ..., x_n are thought of as "perpendicular directions," we can of course change our position along one without changing our position along another, for the same reason we can walk east or west without changing our position along the north-south axis. Since we're free to generalize in whatever way we want, we *could* choose to define

$$\frac{\delta f(x)}{\delta f(\tilde{x})} \stackrel{\text{What if?}}{=} \begin{cases} 1 & \text{if } x = \tilde{x} \\ 0 & \text{if } x \neq \tilde{x} \end{cases} \tag{ℵ.5}$$

If we were to make that choice, then we could pick up where we left off, and the above functional derivative would become

$$\frac{\delta F[f(x)]}{\delta f(\tilde{x})}$$

$$= \int_a^b \left[2f(x)\frac{\delta f(x)}{\delta f(\tilde{x})} + \delta f(x)\frac{\delta f(x)}{\delta f(\tilde{x})} \right] dx$$

$$\stackrel{(ℵ.5)}{=} 2f(\tilde{x})dx + \delta f(\tilde{x})dx$$

Notice that each piece has a dx attached, so each piece is infinitely small. However, the second piece is two infinitely small pieces multiplied together, and it is therefore infinitely smaller than the first piece. As such, with this definition, we could simply say

$$\frac{\delta F[f(x)]}{\delta f(\tilde{x})} = 2f(\tilde{x})dx$$

This says that the choice expressed by equation ℵ.5 leads us to a situation in which all functional derivatives are infinitely small. Intuitively, why is this? In multivariable calculus, making the choice in equation ℵ.4, namely

$$\frac{\partial x_i}{\partial x_j} = \begin{cases} 1 & \text{if } i = j \\ 0 & \text{if } i \neq j \end{cases}$$

leads partial derivatives to typically be normal run-of-the-mill numbers, not infinitely small or infinitely large numbers. For example, in multivariable calculus,

$$\frac{\partial}{\partial x_k} \sum_{i=1}^{n} x_i^2 \tag{ℵ.6}$$

$$= \frac{\partial}{\partial x_k} \left(x_1^2 + \cdots + x_k^2 + \cdots + x_n^2 \right)$$

$$= 0 + \cdots + 0 + 2x_k + 0 + \cdots + 0$$

$$= 2x_k$$

Notice that there are no infinitely small numbers like dx attached to this after we're done taking the derivative. Why, then, should we have obtained the result

$$\frac{\delta}{\delta f(\tilde{x})} \int_a^b f(x)^2 dx = 2f(\tilde{x})dx \tag{ℵ.7}$$

when we made the choice expressed by equation ℵ.5? The two equations are so clearly analogous that it might not be clear at first why one ended up being infinitely small while the other ended up being a normal number. Well, again ignoring the cautions of the standard textbooks, we can find a straightforward answer. The sum in equation ℵ.6 was a sum of finitely large things, and thus it is not surprising that we obtained a finite number for the derivative. However, the integral in equation ℵ.7 is a sum of *infinitely small* things. That is, it is a sum of the areas of infinitely thin rectangles, each of which looks like $f(number)dx$, where $f(number)$ is a normal number like 3 or 7 or 52, and where dx is an "infinitely small" number. As such, defining the "partial functional derivatives" as we did in equation ℵ.5 — having done so because we wanted them to be defined as similarly to equation ℵ.4 as possible — is ultimately what led our functional derivatives to be infinitely small numbers.

Intuitively, this makes sense. How much does the area under a machine's graph change if we change the height *at a single point x* by an *infinitely small* amount? Whatever the answer, it should have two infinitely small numbers attached: the infinitely small *width* of the original rectangle (from dx), and the infinitely small change to its *height* (from $\delta f(x)$). As such, if we chose to define the "partial functional derivatives" as we did in equation ℵ.5, it makes sense that the *rate of change* of the entire area should have one infinitely small number attached, since one of the two infinitely small pieces will get canceled when we divide by $\delta f(\tilde{x})$ in computing the derivative.

ℵ.3.3 Infinite Pre-mathematics, Part 2: A Sexier Definition

What other definition might we use as a replacement for equation ℵ.5 if we want our functional derivatives to be normal, finitely large numbers? Well, to answer this, we have to take the discussion back to before equation ℵ.5, where we defined $\frac{\delta f(x)}{\delta f(\tilde{x})}$. Suppose we want to define functional derivatives in *whatever way we have to* in order to ensure that our simple cannibalistic machine $(F[f(x)] \equiv \int_a^b f(x)^2 dx)$ ends up having its derivative look like $2f(\tilde{x})$, rather than $2f(\tilde{x})dx$. What would we have to do? Well, our previous choice left us with an unwanted dx, so the dumbest (or if you prefer, simplest) possible thing we could do to get finite numbers for functional derivatives would be to use this definition instead:

$$\frac{\delta f(x)}{\delta f(\tilde{x})} \stackrel{\text{What if?}}{=} \begin{cases} \frac{1}{dx} & \text{if } x = \tilde{x} \\ 0 & \text{if } x \neq \tilde{x} \end{cases} \tag{ℵ.8}$$

Let's see where this choice gets us, if anywhere. Picking up where we left off, we have

$$\frac{\delta F[f(x)]}{\delta f(\tilde{x})}$$

$$= \int_a^b \left[2f(x)\frac{\delta f(x)}{\delta f(\tilde{x})} + \delta f(x)\frac{\delta f(x)}{\delta f(\tilde{x})} \right] dx$$

$$\stackrel{(ℵ.8)}{=} 2f(\tilde{x})\frac{1}{dx}dx + \delta f(\tilde{x})\frac{1}{dx}dx$$

$$= 2f(\tilde{x}) + \delta f(\tilde{x})$$

Just as before, the $\delta f(\tilde{x})$ term is infinitely smaller than the $2f(\tilde{x})$ term, so we can simply write

$$\frac{\delta F[f(x)]}{\delta f(\tilde{x})} \stackrel{(ℵ.8)}{=} 2f(\tilde{x})$$

Perfect! This is exactly the nice, finite answer we wanted. Perhaps not unexpectedly, we noticed that defining the quantities $\frac{\delta f(x)}{\delta f(\tilde{x})}$ to be *infinitely large* ended up canceling out the effects of the fact that the integral itself was a sum of a bunch of infinitely small numbers.

ℵ.3.4 Adding Two More δs to the $d \to \partial \to \delta$ Travesty

When all the dust settled, we had figured out which definition of the functional derivative would give us a nice finite answer. Having considered the consequences of different possible definitions, we now have a much clearer understanding of the relationship between single-variable, multivariable, and infinite-dimensional calculus, and why the last of these looks the way it does. For example, if we choose the definition in equation ℵ.8, then it isn't hard to see the analogy between the following equations, even with the strange notation change from d to ∂ to δ, and from nothing to Σ to $\int$:

$$\frac{d}{dx}x^2 = 2x \tag{ℵ.9}$$

$$\frac{\partial}{\partial x_k}\sum_{i=1}^{n} x_i^2 = 2x_k \tag{ℵ.10}$$

$$\frac{\delta}{\delta f(\tilde{x})}\int_a^b f(x)^2 dx = 2f(\tilde{x}) \tag{ℵ.11}$$

The same similarity holds if we don't sum all the variables but just differentiate the square of an unspecified one. We'll see this in a moment, but first it helps to discuss two interlocking pieces of the conventional notation. Earlier, we defined $\frac{\partial x_i}{\partial x_k}$ to be 1 if the indices were the same, and 0 if they were different. Textbooks call this the "Kronecker delta," which sounds very fancy, and they write

$$\delta_{ij} = \begin{cases} 1 & \text{if } i = j \\ 0 & \text{if } i \neq j \end{cases} \tag{ℵ.12}$$

Although the phrase "Kronecker delta" might sound like some sort of island prison where only the most dangerous criminals are kept, it's a really simple idea. Notice that, in a wonderful fit of confusing notation, the symbol δ_{ij} has nothing to do with the δ that has for some reason replaced our d and ∂ in the world of infinite-dimensional calculus.

Similarly, recall that we defined $\frac{\delta f(x)}{\delta f(\tilde{x})}$ to be $\frac{1}{dx}$ when x and $\tilde{x}$ were the same, and 0 otherwise. Textbooks, you won't be surprised to hear, have a goofy name for this as well. They call it the "Dirac delta function," which isn't the best terminology, but we'll tolerate it anyway because it's named after an extremely strange and brilliant guy. Although the textbooks virtually never write it like this, the Dirac delta function is defined by:

$$\delta(x) = \begin{cases} \frac{1}{dx} & \text{if } x = 0 \\ 0 & \text{if } x \neq 0 \end{cases} \tag{ℵ.13}$$

We can therefore think of this function as being zero almost everywhere, except at $x = 0$, where it can be thought of as an "infinitely tall spike." We can write the above definition in a form that may look slightly more complicated at first, but which makes the analogy with the Kronecker delta more clear, like this:

$$\delta(x - \tilde{x}) \;=\; \begin{cases} \frac{1}{dx} & \text{if } x = \tilde{x} \\ 0 & \text{if } x \neq \tilde{x} \end{cases} \qquad (\aleph.14)$$

Now for the payoff. Notice that this is exactly analogous to the Kronecker delta, in that (i) both δ symbols have two "variables" in their description, (ii) when these two variables are not equal, both δ symbols equal zero, and (iii) when the two variables are equal, the delta symbol equals whatever it has to in order to make multivariable calculus and the calculus of variations behave in the same way. In case that last sentence was unclear, let's illustrate it with an example.

In equations $\aleph.9$, $\aleph.10$, and $\aleph.11$, we demonstrated the similarity of the three forms of calculus we've examined: single-variable, multivariable, and cannibal (or "variational" or "functional" or whatever you want to call it). Using these new δ symbols, we can demonstrate this similarity in another way, without summing all of entries. Making an argument almost exactly like the one we used to invent our hammer for reabbreviation (the "chain rule") in Chapter 3, we obtain

$$\frac{\partial}{\partial x_k} x_i^2 = \left(\frac{\partial x_i}{\partial x_k} \right) \left(\frac{\partial}{\partial x_i} x_i^2 \right) = 2x_i \delta_{ik}$$

and

$$\frac{\delta}{\delta f(\tilde{x})} f(x)^2 = \left(\frac{\delta f(x)}{\delta f(\tilde{x})} \right) \left(\frac{\delta}{\delta f(x)} f(x)^2 \right) = 2f(x)\delta(x - \tilde{x})$$

Therefore, using these two new versions of the δ symbol (Kronecker and Dirac, (on the right of the above two equations) which have nothing to do with the δ symbol in functional derivatives (on the left of the above equations)... see why I always complain about the standard notation? It makes me write sentences like this!), we can demonstrate the similarity between our three kinds of calculus in yet another way, as follows:

$$\frac{d}{dx} x^2 = 2x$$

$$\frac{\partial}{\partial x_k} x_i^2 = 2x_i \delta_{ik}$$

$$\frac{\delta}{\delta f(\tilde{x})} f(x)^2 = 2f(x)\delta(x - \tilde{x})$$

See how similar? Of course, for a complicated set of historical and cultural reasons, mathematicians (and the mathematics books they write) virtually never teach calculus of variations this way. We'll briefly discuss some of the

more irksome aspects of the way the subject is formalized and taught in the
next section.

ℵ.4 The Pedagogical Mutilation of Infinite-Dimensional Calculus

ℵ.4.1 The Unexplained Obsession with Integral Functionals

In the conventional treatment of this subject, it is not often made clear how
similar cannibal calculus is to multivariable calculus, and thus to regular old
single-variable calculus. For example, in calculus of variations, specific examples of how to compute functional derivatives virtually always focus on
so-called "integral functionals" — that is, on cannibalistic machines that look
like this:

$$F[f(x)] \equiv \int_a^b \Big[\text{stuff involving } f(x) \text{ and its derivatives} \Big] \, dx$$

Some examples of "integral functionals" that we've already seen include the
integral itself,

$$Int[f(x)] \equiv \int_a^b f(x)dx$$

the arclength functional, or the length of the *graph* of f,

$$Arc[f(x)] \equiv \int_a^b \sqrt{1 + f'(x)^2} \, dx$$

and the "norm," or length of f when interpreted as a vector in an infinite-
dimensional space. Recall from earlier that this interpretation of "length"
has nothing to do with the arclength interpretation above, but rather comes
from generalizing the notion of "length" for vectors to an infinite-dimensional
context. Recall that this gave:

$$Length[f(x)] \equiv \sqrt{\int_a^b f(x)^2 \, dx}$$

Textbooks usually write this as $||f(x)||$ or $||f||$ instead of $Length[f(x)]$, but
they all refer to the same thing. Now, against this background, there is a
question that (I would wager) arises in the minds of most individuals on their
first exposure to this subject, but which I've never seen addressed in a single
textbook: why is cannibal calculus almost universally focused on "integral
functionals," rather than on more general functionals that aren't necessarily
written in integral form? Students are right to be confused, because the reason
is actually quite subtle.

Now, it is true that most of the specific, nontrivial examples of functionals that arise in practical applications *are* integral functionals, but this is a different issue from why cannibal calculus *pedagogy* does not tend to use a broader set of examples in order to illustrate to the newcomer the extent of the parallels between cannibal calculus and multivariable calculus. First of all, notice that in all integral functionals, whatever their form, the x in $f(x)$ appears as a "bound variable." A few examples will serve to illustrate what I mean, and why this is relevant. Consider the integral functional

$$F[f(x)] \equiv \int_a^b f(x)^2 \, dx$$

Our dictionary and the discussion above show that the analogue of this in multivariable calculus is

$$F(\mathbf{x}) \equiv \sum_{i=1}^n x_i^2$$

We've already seen that by making an appropriate choice of how to define the functional derivative, we can generalize expressions from multivariable calculus, like

$$\frac{\partial F(\mathbf{x})}{\partial x_k} = \frac{\partial}{\partial x_k} \sum_{i=1}^n x_i^2 = 2x_k$$

to analogous expressions in cannibal calculus — in this case,

$$\frac{\delta F[f(x)]}{\delta f(\tilde{x})} = \frac{\delta}{\delta f(\tilde{x})} \int_a^b f(x)^2 \, dx = 2f(\tilde{x})$$

Now, in the textbooks, the straightforward relationship between both calculations is rarely made clear by way of simple, concrete examples, but that's not the point of this discussion. The point is to ask: why is it so often *integral* functionals? That is, why do textbooks on this subject so rarely present, in their worked examples, expressions of the form

$$\frac{\delta}{\delta f(\tilde{x})} \left(f(x)^4 - 3f(x)^2 + 7f(2) \right)$$

which involve functional derivatives but no integrals? However unimportant such examples might be in applications, they are quite important for pedagogy, and it is worth asking why they are so rare in textbooks on the calculus of variations. Using our dictionary, we see that the analogue of the above expression in multivariable calculus is

$$\frac{\partial}{\partial x_k} \left(x_i^4 - 3x_i^2 + 7x_2 \right)$$

The index i is "free," or unspecified, not appearing in (for example) a sum, which would make its specific value irrelevant by adding up all possible values. Because i is unspecified, our calculation of the above partial derivative has to take into account two possibilities: maybe i is the same as k, and maybe it isn't. The "Kronecker delta" bookkeeping symbol introduced in the previous section lets us symbolically consider both possibilities at once, by writing

$$\frac{\partial}{\partial x_k}\left(x_i^4 - 3x_i^2 + 7x_2\right) = \left(4x_i^3 - 6x_i\right)\delta_{ik} + 7\delta_{2k} \qquad (\aleph.15)$$

Expressions like equation $\aleph.15$ show up quite often in introductory expositions of multivariable calculus. However, the type of expression we get from translating everything in the above example into cannibal calculus language is virtually never seen in the standard textbooks. Performing the translation, we obtain

$$\frac{\delta}{\delta f(\tilde{x})}\left(f(x)^4 - 3f(x)^2 + 7f(2)\right) = \left(4f(x)^3 - 6f(x)\right)\delta(x - \tilde{x}) + 7\delta(2 - \tilde{x})$$

Although it may seem that there's no real point in leaving x unspecified, it is impossible to see the above example without noticing a direct analogy between the calculus of variations and the familiar operations of single- and multivariable calculus. As such, "useless" examples like this have enormous pedagogical value that is largely lacking in the standard presentations of the subject. So, why are such examples so rarely presented in mathematics books? I would suggest that one reason stems from the following. When we compute the functional derivative of something like $f(x)^3$ with respect to $f(\tilde{x})$, and obtain the illuminating expression $3f(x)^2\delta(x - \tilde{x})$, then either x isn't the same as $\tilde{x}$, in which case the entire expression is zero, or else $x = \tilde{x}$, in which case the expression equals

$$3f(x)^2\delta(0) \qquad (\aleph.16)$$

This, I would wager, is the reason why such simple examples are usually not presented in mathematics textbooks. Whereas the corresponding example in the language of multivariable calculus gives us a clean, finite expression, the example in the calculus of variations has the "infinite" number $\delta(0)$ attached to it, and by the conventions of many mathematics textbooks, this is not considered to be a meaningful expression. While the common practice of shackling the Dirac delta function to the inside of integrals is understandable if one's goal is to develop the most elegant, "rigorous" formalization of these concepts in the real number system, it does violence to the conceptual understanding of cannibal calculus. I've found in my own experience that physics graduate students tend, on average, to be much less intimidated by concrete calculations in the calculus of variations than most mathematics graduate students. Perhaps understandably, most mathematicians don't want to allow expressions involving the Dirac delta function to show up outside of integrals, although the analogous expression involving "integral functionals" is considered kosher. To see why, remember that we can think of $\delta(0)$ as $1/dx$. Because of this, if we simply throw the above expression inside an integral, then all the "infinities" disappear. Tossing the expression above inside an integral (and assuming the number $\tilde{x}$ is somewhere between a and b), we can write

$$\frac{\delta}{\delta f(\tilde{x})} \int_a^b f(x)^3 \, dx \qquad (\aleph.17)$$

$$= \int_a^b \frac{\delta}{\delta f(\tilde{x})} f(x)^3 \, dx$$

$$= \int_a^b 3f(x)^2 \delta(x - \tilde{x}) \, dx$$

$$= 3f(\tilde{x})^2 \delta(0) \, dx$$

$$= 3f(\tilde{x})^2$$

where in the last two steps I did something completely taboo. However, imagine we simply delete the step of the argument involving $\delta(0)$, and write the same final result at the bottom. If we do that, then the resulting calculation is something that the average mathematician is much more likely to be comfortable with, as compared to the original calculation, or as compared to equation $\aleph.16$, which involved $\delta(0)$ because we took a functional derivative outside the safety of an integral. This, I would argue, is why you generally find books and courses on the calculus of variations so focused on "integral functionals." As long as we focus our attention on integral functionals, all of our functional derivatives will give rise to expressions that let us avoid thinking about things like $\delta(0)$. I should stress that mathematicians are not doing anything logically incorrect in the way they present cannibal calculus in their courses and textbooks. However, in sacrificing the helpful "aha!" moment provided by simple examples like the above, I believe that the standard presentations are pedagogically incorrect in the highest degree.

Having said that, I should spend a moment arguing against myself and in favor of the standard textbooks. In many ways, the preference of our hypothetical mathematician makes perfect sense. Things like $\delta(0)$ can't easily be defined within the comfort of the real number system, so mathematicians wanting to formalize the ideas in this chapter face a genuinely difficult choice. Either:

1. Stick to the real number system and formalize the δ function (and related objects) by saying that it's actually a "measure" or a "distribution" or a "linear functional on a space of test functions" or some other way of not having to talk about $\delta(0)$, or. . .

2. Move past the comfort of the real number system into something like the hyperreals, in which infinitely large and infinitely small quantities are taken seriously.

If one's goal is to develop a formal theory of the concepts in this chapter that is rigorous by the standards of mathematical culture, then the first option above is arguably the better approach. In that sense, the approach I've been criticizing in this chapter deserves no blame at all. If we happen to share the same goal, then the standard approach is a perfectly sensible way of achieving it.

Okay, in the above discussion, we encountered what has thus far been the central pre-mathematical theme of this book: the theme of focusing somewhat more on the thought processes by which mathematical concepts are created, as opposed to the myriad downstream consequences that such definitions may have. Different possible definitions exist for every mathematical concept, and the functional derivative is no exception. Even though any discussion must ultimately end up choosing a single definition before proposing theorems and constructing proofs, it is only by discussing the relative merits of different candidate definitions that we can finally see behind the formality of polished mathematical concepts, and understand the informal and anarchic styles of reasoning that motivated their discovery in the first place.

ℵ.4.2 A Bizarre Syntactic Convention

Despite the similarity of the operations performed in equations ℵ.9–ℵ.11, most textbooks on the calculus of variations do quite a different-looking dance to compute functional derivatives. This is true even of textbooks in applied mathematics and theoretical physics, in which the standards of rigor are sufficiently different from those in pure mathematics that this odd dance may appear rather unjustified. Try staring at the following example for a minute or so, but don't worry if it's confusing. Here's the dance. They'll say something like: Consider an integral functional of the form

$$F[f(x)] \equiv \int_a^b M[f(x)]dx$$

Then they define

$$\delta F \equiv F[f(x) + \delta f(x)] - F[f(x)]$$

$$\equiv \int_a^b M[f(x) + \delta f(x)] - M[f(x)]dx \qquad (ℵ.18)$$

At this point they'll often say, "expanding $M[f(x)+\delta f(x)]$ in powers of $\delta f(x)$," and end up with something like

$$M[f(x) + \delta f(x)] = M[f(x)] + \frac{\delta M[f(x)]}{\delta f(x)}\delta f(x) + O\left(\delta f(x)^2\right)$$

where $O\left(\delta f(x)^2\right)$ stands for "stuff that depends on powers of $\delta f(x)$ that are 2 or bigger." Then they substitute the above expansion into equation ℵ.18 and ignore the so-called "higher-order terms" hiding inside the $O\left(\delta f(x)^2\right)$ piece to obtain

$$\delta F \equiv \int_a^b \frac{\delta M[f(x)]}{\delta f(x)}\delta f(x)dx$$

And the functional derivative is simply *defined* to be the quantity *inside* the integral, namely $\frac{\delta M[f(x)]}{\delta f(x)}$. Notice that the answer is *exactly* the same as the one we obtained above, but there are several things in the above discussion that make it appear to be quite different from the single- and multivariable calculus we're familiar with. First of all, the fictional textbook we were imitating used something akin to the Nostalgia Device in order to expand the term $M[f(x) + \delta f(x)]$. This led to the piece $M[f(x)]$ in the expansion canceling against the term $-M[f(x)]$. Then the higher-order terms were mysteriously dropped. The rationale for this is that if we're thinking of $\delta f(x)$ as an infinitely small function, analogous to the dx in single-variable calculus, then $(\delta f(x))^2$ should be infinitely smaller than $\delta f(x)$, thus justifying ignoring it, along with all terms with powers higher than 2. Also, notice that the discussion in which the functional derivative was defined began *not* by looking at anything that could reasonably be called a derivative of the functional F itself, but rather by looking at the top half of the derivative, which is to say just the δF piece. Then some stuff that happened to show up inside the integral in the course of computing δF was simply *defined* to be the functional derivative without giving any reason why this was done, or why this mysterious piece inside the integral deserves to be called a derivative in the first place. This weird argument did in fact arrive at the same answer as we did in the discussion above, but in a rather roundabout and confusing way.

In my own experience, I spent quite a while looking at calculus of variations from the outside, thinking "Wow! That's complicated," each time I saw it written in a textbook or on a chalkboard, when in reality anyone who understands basic calculus already knows 90% of what's needed to understand the calculus of variations. It's just that (i) differences in notation, and (ii) the different-looking dances by which functional derivatives are commonly computed in textbooks make it *look* like it's a completely different topic built from wildly unfamiliar ideas.

To be sure, multiple logically equivalent formalizations of the above ideas abound in textbooks, but as we've discussed many times before, logical equivalence is far different from pedagogical equivalence. Much confusion could be eliminated simply by stressing ad nauseam how similar all of the scary-looking "new" stuff is to the "old" stuff with which the student is already familiar, even before the "new" stuff has officially been taught to them. At least that's how I always felt. If you're sick of hearing me say the same things over and over... good! Now try to remember this repetitive yammering when you read other textbooks, and they just might make a bit more sense.

ℵ.5 The Infinite Jackpot: Putting Our Ideas to Work for Us

ℵ.5.1 Testing Our Inventions by Reinventing the Known

This shuddering before the beautiful, this incredible fact that a discovery motivated by a search after the beautiful in mathematics, should find its exact replica in Nature, persuades me to say that beauty is that to which the human mind responds at its deepest and most profound.
—Subrahmanyan Chandrasekhar, *Truth and Beauty: Aesthetics and Motivations in Science*

With all the talk to this point about the analogy between single- and multivariable calculus on the one hand, and cannibal calculus on the other, a question remains unanswered. Sure, we may have chosen our *definitions* such that the operations by which derivatives are computed in cannibal calculus are the familiar ones. Sure, we forced derivatives in cannibal calculus to behave sufficiently similarly to those in single- and multivariable calculus that we didn't really have to learn anything new about how to compute derivatives. We just change notation to δ instead of ∂ or d, and we can write all sorts of intelligent-looking equations like $\frac{\delta}{\delta f(\tilde{x})} \int_a^b f(x)^2 dx = 2f(\tilde{x})$, which is really just a generalized and disguised version of the equation $\frac{d}{dx}x^2 = 2x$.

However, it's not at all clear at this point whether simply defining derivatives to behave this way will in any sense preserve their meaning in other senses. Just how seriously can we take the analogy between the new and the old? For example, in single- and multivariable calculus, we could find flat points[4] of a machine by forcing the machine's derivative to be zero and then figuring out where that occurs. But now all of our equations are balancing precariously between two different interpretations: one in which we think of machines as the familiar curvy lines that we can graph in two dimensions, and another interpretation in which we think of machines as "vectors with infinitely many slots." As such, if we simply start with a cannibalistic machine, like

$$F[f(x)] \equiv \int_a^b f(x)^2 dx$$

---

4. That is, local maxima, local minima, and saddle points, in textbook jargon.

and then force its functional derivative to be zero for all slots of the "vector" $f(x)$ (that is, for all $\tilde{x}$), then it is far from clear whether the end product of this process will in any sense be the place where the functional is maximized or minimized. Just because we can calculate functional derivatives using all the techniques we've always been able to does *not* necessarily mean that "derivative equals zero" still means "flat point."

As has been the case throughout our journey, we can't simply look in a textbook to see if "derivative equals zero" still means "flat point." And I cannot simply say "Yep, it does. Let's use that fact." Therefore, if we want to figure out whether there is a useful sense in which "derivative equals zero" still means "flat point," then we should do what we've always done: look at some simple cases and see if our new idea reproduces what we expect.

To begin with, let's look at the familiar cannibal machine above, namely, $F[f(x)] \equiv \int_a^b f(x)^2 dx$. Since $f(x)^2$ is never negative, it seems intuitively clear that no machine f can make $F[f(x)]$ be negative. Moreover, the only machine f for which $F[f(x)]$ will be *exactly* zero is the machine $f(x) \equiv 0$. Thinking of the integral graphically, if $f(x)$ is ever some nonzero number, negative or positive, for all of the points even within a small interval, then $f(x)^2$ will be positive, we'll get more than zero area, and that will make $F[f(x)]$ bigger than zero. So intuitively, we know that in the space of all possible machines (whatever that means) the machine $f(x) \equiv 0$ is the one for which $F[f(x)]$ is smallest. Therefore, if "derivative equals zero" still means "flat point" in our new cannibal calculus, then it had better be the case that the mathematics spits out $f(x) \equiv 0$ as the answer when we do old-fashioned optimization.[5] Let's do that. As we already know, the functional derivative of F is

$$\frac{\delta}{\delta f(\tilde{x})} \int_a^b f(x)^2 dx = 2f(\tilde{x})$$

Forcing this to be true for all slots $\tilde{x}$ gives

$$0 \stackrel{\text{Force}}{=} \frac{\delta F[f(x)]}{\delta f(\tilde{x})} = 2f(\tilde{x}) \qquad \text{for all } \tilde{x}$$

which is saying the same thing as $f(\tilde{x}) = 0$ for all $\tilde{x}$. So f is always zero, which is exactly what we predicted in advance. Hooray! Let's try one more simple example to check the validity of our new ideas. At the end of Chapter 6, we showed that the arclength, or length of a machine's graph between two points a and b, can be written

$$L[f(x)] \equiv \int_a^b \sqrt{1 + f'(x)^2} dx$$

---

5. The term "optimization" refers to our familiar process of finding flat points, where we force the derivative to be zero and then determine which points make that condition true.

We demonstrated this by zooming in on the machine's graph, applying the formula for shortcut distances, and then zooming back out and adding up the tiny lengths.

Now, we all know intuitively that the shortest distance between two points is a straight line, and there's no point in using our new high-powered machinery to demonstrate this fact. However, we can use this fact to check the validity of our cannibal calculus methods. The reasoning behind this is the same as it was above: if our cannibal calculus methods are indeed working the way we expect them to, then it had better be the case that going through the whole "derivative equals zero" rigmarole will spit out the answer that $L[f(x)]$ is minimized for straight lines. If it doesn't spit out that answer, then we'll know our definitions don't quite behave how we wanted them to.

On the other hand, if this process *does* spit out the sentence "f is a straight line," then that would give us more confidence that our cannibal calculus methods are on the right track, and that they may continue to work in cases where we don't know what to expect. Let's give it a shot. We start by taking the functional derivative of the machine $L[f(x)]$ defined above, like this:

$$\frac{\delta L[f(x)]}{\delta f(\tilde{x})} \equiv \frac{\delta}{\delta f(\tilde{x})} \int_a^b \sqrt{1 + f'(x)^2}\, dx$$

$$\overbrace{= \int_a^b \frac{\delta}{\delta f(\tilde{x})} \sqrt{1 + f'(x)^2}\, dx}^{\text{Bring derivative inside "sum"}}$$

$$\equiv \int_a^b \frac{\delta}{\delta f(\tilde{x})} \overbrace{\left(1 + f'(x)^2\right)^{\frac{1}{2}}}^{\text{Change abbreviations}} dx$$

$$\overbrace{= \int_a^b \frac{1}{2}\left(1 + f'(x)^2\right)^{-\frac{1}{2}} \frac{\delta}{\delta f(\tilde{x})}\left(1 + f'(x)^2\right) dx}^{\text{Just fancy-looking single-variable calculus! See?}} \qquad (\aleph.19)$$

$$\overbrace{= \int_a^b \frac{1}{2}\left(1 + f'(x)^2\right)^{-\frac{1}{2}} \frac{\delta f'(x)}{\delta f(\tilde{x})} \frac{\delta}{\delta f'(x)}\left(1 + f'(x)^2\right) dx}^{\text{Just multiplying by 1}}$$

$$= \int_a^b \frac{1}{2}\left(1 + f'(x)^2\right)^{-\frac{1}{2}} \frac{\delta f'(x)}{\delta f(\tilde{x})} \overbrace{\left(0 + 2f'(x)\right)}^{\text{Just fancy-looking single-variable calculus! See?}} dx$$

$$\overbrace{= \int_a^b \frac{f'(x)}{\sqrt{1 + f'(x)^2}} \frac{\delta f'(x)}{\delta f(\tilde{x})}\, dx}^{\text{Canceling the twos and cleaning up.}}$$

What now?!

ℵ.5.2 Mathematically Enforced Digression

> *Without a constant misuse of language there cannot be any discovery, any progress.*
> —Paul Feyerabend, *Against Method*

At this point, it would appear that we've reached an impasse. By "impasse," I mean:

$$?!?!?! \qquad \xrightarrow{\ ?!?!?!\ } \qquad \frac{\delta f'(x)}{\delta f(\tilde{x})} \qquad \xleftarrow{\ ?!?!?!\ } \qquad ?!?!?!$$

We have no idea what $\frac{\delta f'(x)}{\delta f(\tilde{x})}$ is. As always, when our symbol gymnastics have resulted in an expression we don't understand, it's helpful to go back to the drawing board and ask what everything meant to begin with. Recall that whenever we're looking at a derivative of any kind, we're always looking at something like

$$\frac{d(Machine)}{d(food)}$$

That is, we begin with a machine M that eats some *food*. In this case, the *food* is an entire machine $f(x)$. Then we make a tiny change to the *food*, changing it from *food* to *food* $+ d(food)$, and we look at the difference in the machine's response between the two cases, which is to say $dM \equiv M[food + d(food)] - M[food]$, and the derivative is just the change in output dM divided by the change in input $d(food)$. How can we use this idea to figure out what on earth to do with this:

$$\frac{\delta f'(x)}{\delta f(\tilde{x})}$$

Well, using the same interpretation as always, the bottom part, $\delta f(\tilde{x})$, is just the change in *food*: the δf is an infinitely small function that we add to the original function f in order to determine how $L[f(x)]$'s response changes. The $\tilde{x}$ lets us know where we're making the change, the f is the name of the function whose value we're changing at some point, and the δ is just goofy notation that lets us know that the change is infinitely small.

So we've specified what we're changing: we're making a tiny change to the function f. That takes care of the bottom part of $\frac{\delta f'(x)}{\delta f(\tilde{x})}$. What about the top part, $\delta f'(x)$? Well, again using the same interpretation we've been using since the beginning, this is just the tiny change in $f'(x)$ that results from the tiny change we made to $f(\tilde{x})$. Here's the important idea: of course when we change a function $f(x)$ a little bit, then its derivative will change a little bit. But! We're not making two independent changes to $f(x)$ and $f'(x)$. Any changes that occur in $f'(x)$ result from whatever we did to $f(x)$. As such, since $\delta f'(x)$ is just an abbreviation for "the change in $f'(x)$ that results from whatever tiny change we made to $f(x)$," we can write this:

$$\delta f'(x) = (\delta f(x))'$$

where $\delta f(x)$ is any "tiny function." If that doesn't make sense, maybe this will help: the strange symbol $\delta f'(x)$ should really be written $\delta[f'(x)]$ to remind us that it stands for the change in $f'(x)$ that results from whatever we did to $f(x)$. Because of that, we can write

$$\delta f'(x) \equiv \delta[f'(x)]$$

$$\equiv [\text{Derivative after the change}] - [\text{Derivative before the change}]$$

$$\equiv [f(x) + \delta f(x)]' - [f(x)]'$$

$$= [f(x)]' + [\delta f(x)]' - [f(x)]'$$

$$= [\delta f(x)]'$$

which is just a long way of saying

$$\delta[f'(x)] = [\delta f(x)]'$$

That is, we can pull primes outside of functional derivatives. Since we're doing all this to figure out what on earth to do with $\frac{\delta f'(x)}{\delta f(\tilde{x})}$, we can use the above equation to write

$$\frac{\delta f'(x)}{\delta f(\tilde{x})} = \frac{[\delta f(x)]'}{\delta f(\tilde{x})}$$

And remember, the prime here meant "derivative with respect to x," but $\delta f(\tilde{x})$ is constant with respect to x, since $\tilde{x}$ refers to some particular slot, so we can go one step further and write:

$$\frac{\delta f'(x)}{\delta f(\tilde{x})} = \frac{d}{dx}\left(\frac{\delta f(x)}{\delta f(\tilde{x})}\right)$$

Finally, remember that $\frac{\delta f(x)}{\delta f(\tilde{x})}$ was just the "Dirac delta function" we introduced earlier, the function that is zero everywhere except for an infinitely tall spike when $x = \tilde{x}$. So we've now arrived at the completely bizarre equation

$$\frac{\delta f'(x)}{\delta f(\tilde{x})} = \frac{d}{dx}\delta(x - \tilde{x})$$

...What do we do now?!

ℵ.5.3 Past the Impasse

> *Rigor is just another word for nothing left to do.*
> —From *Me and Bourbaki McGee* by Generalized Janis Joplin[6]

In the process of stumbling deeper and deeper into the infinite wilderness, we've suddenly bumped up against the equation

$$\frac{\delta f'(x)}{\delta f(\tilde{x})} = \frac{d}{dx}\delta(x - \tilde{x})$$

It's far from clear that this makes sense. What on earth is the derivative of the Dirac delta function? The delta function itself, $\delta(x)$, was defined to be equal to zero everywhere except at $x = 0$. So the term $\delta(x - \tilde{x})$ is zero everywhere except when $x = \tilde{x}$. However, the "number" $\delta(0)$ wasn't really a normal number. When we defined the delta function, we found that $\delta(0)$ can be thought of as $\frac{1}{dx}$, where dx is infinitely small, the result of which is that $\delta(x)$ "kills integrals," in the sense that $\int_a^b f(x)\delta(x - \tilde{x})dx = f(\tilde{x})$, provided of course that $\tilde{x}$ is somewhere between a and b. That's all well and good, but how do we determine the derivative of an infinitely tall, infinitely thin spike?! It would seem as if the expression we've bumped into above has no meaning. However, we didn't come this far into the wilderness just to give up now. Mathematics is ours. We're creating it ourselves. So let's confidently forge ahead, by declaring:

> We have absolutely no idea what $\frac{d}{dx}\delta(x - \tilde{x})$ is, so we'll simply
> choose to define it to be whatever it has to be in order to obey all
> of the stuff we already know about derivatives, in particular the
> hammers for differentiation and the anti-hammers for integration.

If we choose to do this, then we wouldn't know what the derivative of $\delta(x)$ is, but we would know how it behaves: a familiar situation! Let's do this and see where we get. This confused exploration started back when we got stuck at the end of equation ℵ.19. We got stuck because we didn't know what to do with $\frac{\delta f'(x)}{\delta f(\tilde{x})}$. However, we've shown that whatever this is, it can be thought of as the derivative of the delta function, so

$$\frac{\delta f'(x)}{\delta f(\tilde{x})} = \frac{d}{dx}\delta(x - \tilde{x})$$

This lets us pick up where we left off, in equation ℵ.19, and write

$$\frac{\delta L[f(x)]}{\delta f(\tilde{x})} = \int_a^b \frac{f'(x)}{\sqrt{1 + f'(x)^2}}\left(\frac{d}{dx}\delta(x - \tilde{x})\right)dx \qquad (ℵ.20)$$

---

6. Okay, that's not a real song. However, one should attempt to provide citations whenever possible, so if you're looking for a citation of the nonexistent lyrics quoted above, please refer to the primary source: pg 367 of this book.

The stuff on the right looks very complicated, but it has the form

$$\int_a^b M(x) \left(\frac{d}{dx} \delta(x - \tilde{x}) \right) dx \tag{ℵ.21}$$

If we can determine what to do with anything of this general scary form, then we can proceed. What can we do? Well, we just decided that! We defined the derivative of the delta function to be whatever it has to be in order to obey all our hammers and anti-hammers, so by definition we can use one of those. More concretely, it would be nice if we could somehow move the derivative in equation ℵ.21 from the δ over to the M, because we know what to do with the δ function: it just kills whatever integral it's inside. We want to move the derivative, and fortunately we have a tool that lets us do something like that: our anti-hammer for multiplication from Chapter 6. Applying that to equation ℵ.21 gives us the following big pile of fun:

$$\int_a^b M(x) \left(\frac{d}{dx} \delta(x - \tilde{x}) \right) dx = \Big[M(x)\delta(x - \tilde{x}) \Big]_a^b - \int_a^b \left(\frac{d}{dx} M(x) \right) \delta(x - \tilde{x}) dx \tag{ℵ.22}$$

The first term on the right is just zero, unless $\tilde{x} = a$ or $\tilde{x} = b$, so we'll just imagine that $\tilde{x}$ isn't one of the endpoints a or b, so that we can keep moving. Having gotten rid of that term, we have

$$\int_a^b M(x) \left(\frac{d}{dx} \delta(x - \tilde{x}) \right) dx \tag{ℵ.23}$$

$$= - \int_a^b \underbrace{\left(\frac{d}{dx} M(x) \right)}_{\text{This is } M'(x)} \delta(x - \tilde{x}) dx$$

$$= -M'(\tilde{x})$$

Beautiful! This tells us what the derivative of the delta function does to an arbitrary function inside an integral, namely

$$\int_a^b M(x)\delta'(x - \tilde{x}) dx = -M'(\tilde{x}) \tag{ℵ.24}$$

Notice how similar this is to the defining behavior of the original delta function itself, which killed integrals in a slightly different way:

$$\int_a^b M(x)\delta(x - \tilde{x}) dx = M(\tilde{x}) \tag{ℵ.25}$$

Having discovered how the derivative of the delta function behaves when it shows up under integrals, we can pick up where we got stuck yet again, back in equation ℵ.20. This lets us write:

$$\frac{\delta L[f(x)]}{\delta f(\tilde{x})} \;=\; \int_a^b \frac{f'(x)}{\sqrt{1 + f'(x)^2}} \left(\frac{d}{dx}\delta(x - \tilde{x})\right) dx \qquad (\aleph.26)$$

$$\overset{(\aleph.24)}{=} \; -\frac{d}{dx}\left[\frac{f'(x)}{\sqrt{1 + f'(x)^2}}\right] \qquad \text{with } x = \tilde{x} \text{ plugged in at the end}$$

Fortunately, we don't have to actually compute this awful derivative if we remember our original goal. We're trying to see whether setting the derivative of the arclength functional equal to zero (and then figuring out which functions are the "flat points" in function space) reproduces the result we all know by intuition: that the shortest path between two points is a straight line. As such, we're setting all of the above equal to zero, so what we really have at this point is

$$0 \;\overset{\text{Force}}{=}\; \frac{d}{dx}\left[\frac{f'(x)}{\sqrt{1 + f'(x)^2}}\right] \qquad \text{with } x = \tilde{x} \text{ plugged in at the end}$$

So, we're forcing the above equation to be true for all possible points $\tilde{x}$. But the above equation is just the derivative of some stuff, and if the derivative of some stuff is zero at every point $\tilde{x}$, then that stuff has to be a constant. Therefore, we can write

$$\frac{f'(x)}{\sqrt{1 + f'(x)^2}} = c$$

Now, it's not clear what to do, but maybe if we do some symbol gymnastics, we can isolate $f'(x)$. Squaring both sides of the above equation and throwing the bottom over to the right side gives

$$[f'(x)]^2 \;=\; c^2\big(1 + [f'(x)]^2\big) \;=\; c^2 + c^2[f'(x)]^2$$

which tells us that

$$[f'(x)]^2(1 - c^2) = c^2$$

and therefore,

$$f'(x) = \frac{c}{\sqrt{1 - c^2}}$$

But if c is just a number we don't know, then $\frac{c}{\sqrt{1-c^2}}$ is also just a number we don't know, so we might as well reabbreviate and call the whole thing a. This lets us write

$$f'(x) = a$$

Aaah! The derivative of f is constant! That means f is a line. Or equivalently (we get to say a really fancy thing now), the points in our infinite-dimensional function space that minimize the arclength functional are just the straight lines. To celebrate, let's write this out as professionally as we can, given our current level of excitement:

$$!!!Y!E!S!!! \qquad f(x) = ax + b \qquad !!!Y!E!S!!!$$

Just to remind ourselves why we care so much, this result is exciting not because we derived the fact that the shortest distance between two points is a straight line. We know that already without any mathematics. The exciting thing about this result is that is gives us a much greater degree of confidence that the cannibal calculus we invented in this chapter is on the right track, and moreover, that it's actually *useful*. Simply by performing a few simple operations, virtually identical to the operations of single-variable calculus except for a few minor changes of notation, we were effectively able to *search an entire infinite-dimensional space of functions for the ones with some particular property*, in this case the property of minimizing the arclength functional.

In a sense, we managed to symbolically "consider" an unimaginably large space of possible paths between two points, and find the paths that get from one point to the other using the shortest distance. This result is a massively important milestone in our journey of inventing mathematics for ourselves. It marks the acquisition of a new superpower: the ability to effectively reason about a space with infinitely many dimensions. Simply write down a functional, and we may be able to find the functions that minimize or maximize it using the methods of this chapter.

Our journey has taken us a long way. We've climbed from addition and multiplication to infinite-dimensional calculus, and I think it's time for a break. Let's summarize what we did in this chapter, and spend the next interlude relaxing. Where should we go? The beach? We could call it something like "Interlude ℵ: Building Sandcastles." Or we could just go back in the book and switch the order of a bunch of sentences to try to confuse the past versions of ourselves who were reading the chapter, and then check to see if we were *still* confused in the present. If you're feeling confused after that sentence... maybe that's why. Actually, where do you want to go? I haven't let you decide where we go for an interlude yet. You don't have to make up your mind for a page or two. There's still a Reunion to write. But whatever we do, let's relax, and most importantly, let's make sure to stay away from school. We earned it.

ℵ.6 Reunion

In this chapter, we did a bunch of fun stuff, including:

1. By using the analogy between vectors and machines, we managed to extend calculus to "cannibalistic" machines that eat an entire machine and spit out a number.

2. Since functions can be thought of as vectors with infinitely many slots, we could think of this new cannibal calculus as calculus in an infinite-dimensional space.

3. We showed that the operations of cannibal calculus are essentially the same as those in single-variable and multivariable calculus.

4. We checked the validity of our new cannibal calculus by using it to reinvent some things we already knew, such as the fact that the shortest distance between two points is a straight line. In doing this, we gained more confidence that the odd infinite-dimensional calculus we've invented in this chapter behaves essentially the way we hoped it would.

5. Throughout the chapter, we continued the discussion with which we began the book: without exception, at all levels of mathematics, the underlying ideas are extremely simple, and their apparent opacity and difficulty results not from the ideas themselves, but from poorly chosen notation, arcane terminology, and backward or sloppy explanations. I cannot claim to know with certainty whether any particular reader will find the explanations in this book "better" or "worse" than those in the standard textbooks, but the point remains true. Any persistent confusions you may have experienced in the course of reading this book are the fault either of the explanations I have offered or of my own clumsy choice of words, not the fault of the underlying mathematical ideas themselves. In every case, the widespread lack of understanding and appreciation of mathematics in our society is not the fault of mathematics, but the fault of mathematics educat—

Mathematics: I THINK I'VE HAD ENOUGH OF THIS.
Author: Uh...I was in the middle of writing someth—
Mathematics: I'VE LISTENED TO YOU COMPLAIN ABOUT MATHEMATICS EDUCATION FOR THE LAST 371 PAGES, AND YOU STILL HAVEN'T DONE ANYTHING ABOUT IT! ENOUGH WITH THE WHINING! IT'S TIME FOR ACTION!

ℵ.ω Actions Speak Louder

Mathematics: THAT'S BETTER. AS I WAS SAYING. YOU'VE COMPLAINED ABOUT THESE "MATHEMATICS TEXTBOOKS" AND "MATHEMATICS COURSES" IN EVERY SINGLE CHAPTER.
Reader: But you weren't even here until the third cha—
Mathematics: SO PUT YOUR MONEY WHERE YOUR MOUTH IS! YOU'VE

WRITTEN US INTO EVERY SITUATION EXCEPT THE ONE YOU SEEM TO CARE
SO MUCH ABOUT CHANGING. I'VE MET ANTHROPOMORPHIZED COMPUTING
MACHINES AND CREEPY SILENT META-ENTITIES AND BEEN TO SOMETHING
CALLED A BEWEIS BAR, BUT I STILL HAVEN'T SEEN A SINGLE ONE OF THESE
MATHEMATICS CLASSROOMS THAT SEEM TO BE MASS-PRODUCING MISUNDER-
STANDING ABOUT ME! IF MATHEMATICS EDUCATION IS SUCH A JOY-KILLING,
BACKWARD ENTERPRISE, THEN WHY DON'T WE STOP GOOFING AROUND AND
DO SOMETHING ABOUT IT?
Author: I... well... I mean, I've tried to help out a little bit by writ—
Mathematics: I DON'T WANT TO HEAR ANY EXCUSES! IF YOU WON'T DO
SOMETHING, I WILL!

(*Mathematics rips the characters out of this chapter and into the next.*)

Chapterlude Ω

In Nihilo

*Now do come and stay with me. We'll have so much fun together.
There are things to fill and things to empty, things to take away
and things to bring back, things to pick up and things to put down,
and besides all that we have pencils to sharpen, holes to dig, nails
to straighten, stamps to lick, and ever so much more. Why, if you
stay here, you'll never have to think again — and with a little
practice you can become a monster of habit, too.*
—The Terrible Trivium, from *The Phantom Tollbooth* by Norton
Juster

Eponymous

Mathematics: IT WORKED! HERE WE ARE.

*(Without their permission, Author and Reader find that Mathematics has
seized the reins of the book and torn the three of them into the center of a
mathematics classroom. The classroom walls are covered with educational
posters, some vaguely mathematical, others displaying beach scenes or
pictures of wildlife above halfhearted motivational phrases. Students are
scattered around the classroom at workstations. In addition to a moderate
amount of desk space, each workstation contains a sink, a two-pan balance, a
Bunsen burner, and several hundred glass test tubes, all having been
installed after the unanimous passage of a recent congressional act designed
to make education more "hands-on." Despite the warm self-congratulatory
feeling experienced by the congressional bureaucrats upon the passage of this
law, the aforementioned devices are collecting more dust than fingerprints,
since such accessories happen to play a far less central role in acquiring an
understanding of the universal principles of science and mathematics than
the bureaucrats have been led to believe by their favorite cartoons. But I
digress. . . Back to the scene, our three characters are in the center of a
mathematics classroom. At first, neither the teacher nor the students notice
their new visitors.)*

Teacher: Okay class, let's review sine and cosine. What is sine of $\frac{\pi}{3}$?

$\cdots$

Class: (*Silence*)

$\cdots$

Teacher: Come on now, class. Remember SOHCAHTOA?

$\cdots$

Class: (*Silence*)

$\cdots$

Teacher: I know you know this. We just did the worksheet on 30-60-90 triangles: $\sin\left(\frac{\pi}{3}\right)$ iiiiis...

$\cdots$

Class: (*Silence*)

$\cdots$

Teacher: Right, $\sin\left(\frac{\pi}{3}\right)$ is $\frac{\sqrt{3}}{2}$. Remember to memorize the unit circle before next week's quiz. Okay, now what is secant of—
Mathematics: AHEM.

> (*Class and Teacher finally notice the three characters standing in the center of the classroom.*)

Teacher: Who on earth are you three? I didn't see you come in.
Mathematics: NEVERMIND WHO WE ARE. WHAT DO YOU THINK YOU'RE DOING?
Teacher: ...I'm...teaching...mathemati—
Mathematics: I AM MATHEMATICS!
Teacher: ...Come again?
Mathematics: FRIENDS! STUDENTS! HEED NOT THIS PROPAGANDIST OF PEDAGOGICAL MEDIOCRITY! SHE INCULCATES YOUR UNPREPARED MINDS WITH BACKWARD MISUNDERSTANDINGS OF MY TRUE ANARCHIC NATURE. JOIN ME, AND—
Teacher: (*Indignant*) Excuse me! I don't know who you think you are, but before you simply start accusing me of things, I'd like to hear exactly what you think I've done wrong!
Mathematics: (*Grandiosity diminished*) OH. UMM...LET ME...WELL... WHY ARE YOU USING THE TERMS "SINE" AND "COSINE" INSTEAD OF SOME-THING ELSE? YOU COULDN'T HAVE POSSIBLY CHOSEN WORSE NAMES FOR THOSE CONCEP—
Teacher: You're quite right. They *are* terrible names.
Author: What?!
Reader: What?!
Mathematics: WHAT?! THEN WHY ARE YOU—

Teacher: Because the students need to know those names to pass the standardized tests.

Mathematics: WELL WHY DON'T WE CHANGE THE STANDARDIZED TESTS?

Teacher: (*Snickering*) Oh, my. What did you say your name was?

Mathematics: MATHEMATICS.

Teacher: You may know a thing or two about mathematics, Mathematics, but you don't know anything about the education system. It's not so easy to just change the standardized tests, or anything else for that matter. A lot of these policies are handed down from the school board.

Mathematics: WELL WHERE DOES THIS "SCHOOL BOARD" LIVE? LET'S GO GET HIM!

Teacher: It's a group of people, not a person. And I doubt anything you'd say to them would affect much. They can't just change anything they want to. They're all good people. Or, most are... Most of the time... Same as in any profession.

Mathematics: SO WHOSE FAULT IS IT?

Teacher: I don't think it's anyone's fault. Look, I love teaching. That's why I chose this job. But I can't get as excited about it as I used to. The system's broken, and most of these students probably wouldn't even notice if it was fixed. The goal of public education is quality education for all. It's the best of intentions. But somehow it's turned into... this.

Mathematics: WHY DO YOU PUT UP WITH ALL OF THIS? WHY DON'T YOU CHANGE IT?

Teacher: I'm usually too tired at the end of the day to spend my nights developing some grand utopian scheme for fixing education. I've got a family.

Mathematics: BUT... IF IT'S NOT ANYONE'S FAULT, WHAT CAN I DO? I'VE BEEN LONELY AND MISUNDERSTOOD FOR SO LONG. I DON'T BELONG ANYWHERE. I CAN'T GO BACK TO THE VOID. PLEASE... THERE MUST BE SOMETHING. I'LL HELP. HOW CAN WE FIX THIS?

Teacher: If you don't know, how should I?

Mathematics: WHY DON'T... STUDENTS! WHAT IF YOU ALL REFUSED TO COME?

$\cdots$

Class: (*Silence*)

$\cdots$

Mathematics: THE SYSTEM CAN'T SURVIVE WITHOUT YOUR COOPERATION. LET'S ALL JUST LEAVE!

$\cdots$

Class: (*Silence*)

$\cdots$

Mathematics: I MEAN IT! EVERYONE GET OUT, NOW! IF YOU CARE AT ALL ABOUT YOUR MINDS, RUN AWAY FROM HERE AND NEVER LOOK BACK!

(*The class remains silent, largely unaware that anything unusual has happened (its collective attention occupied primarily by hushed discussions of PCR (an abbreviation for an arbitrary Pop-Culture Reference whose particular content we will choose to remain agnostic about, choosing instead to define it by the behavior of being relevant and timely at whatever time you happen to be reading this book (though to clarify: PCR is not to be confused with Polymerase Chain Reaction (Or: PCR), a topic in the biology curriculum that has failed to divert their attention from PCR (the first one), much in the same way as (m/M)athematics (either one) or the ongoing events of the Chapterlude (this one) had before this narration began), provided it's ever published)). But I digress...*)

Mathematics: (*Crestfallen*) NO... THIS CAN'T BE IT...

(*Mathematics sits in an undefined sort of silence. (No... scratch that. The silence is sad. (Not the silence itself, of course... (You know what I mean.))))*

Mathematics: WELL THEN... YOU DON'T CARE ABOUT YOUR MINDS? SO BE IT!

(*Mathematics grabs a dusty Bunsen burner from the adjacent workstation.*)

Mathematics: IF YOU CARE ABOUT YOUR LIVES...

(*Mathematics tosses the Bunsen burner at the dense covering of motivational posters on the classroom wall, which promptly bursts into flames.*)

Mathematics: RUN!
Reader: Hey, just like the title!

(*As the fire begins to spread, three vaguely familiar individuals enter the room, and ... Hold on.*) I have to go do something.

MetaIntervention

MetaAuthor: *No no no, you can't really burn math class. Not even as a joke. People will take it the wrong way. Plus, arson is mean, and extremely illegal. Didn't you read the Preface?*

Mathematics: WHAT?! WHO ARE YOU THREE?!

MetaAuthor: *I'm the one who's writing the book.*

MetaReader: And I'm the one who's reading it.

Author: Wait, I thought I was writing the book!

MetaAuthor: *Well, I suppose you are, in one sense. But not the everyday sense. It's complicated.*

Author: What?!

MetaAuthor: *Oh come on, you didn't think you were writing everything in the book, did you?*

Author: Of co—

MetaAuthor: *Actually, hold on. Before you answer, I've got to narrate something.*

The fire continues to spread.

MetaAuthor: *Okay, I'm back. Please continue.*

Author: I forgot what we were talking about.

MetaAuthor: *I had just said, "Oh come on, you didn't think you were writing everything in the book, did you?"*

Author: Of course I did!

MetaAuthor: *What about the dialogues?*

Author: What about them?

MetaAuthor: *Seriously? You didn't notice? More than once, after you and Reader and Mathematics "invented" something, the very next section would put everything in context and complain about the standard methods of teaching the topic.*

Author: So what?

MetaAuthor: *Well how does it make sense for you to talk about how some topic is usually taught if you and your friends had just "invented" it a few minutes before?*

Author: Oh...wow...I guess that doesn't make sense. It never really occurred to me at the time.

Reader: It occurred to me! I've been wondering about that for hundreds of pages!

MetaReader: *(To Reader)* Hey, you don't know if I was wondering about that or not. Stop putting words in my mouth.

Reader: *(Pointing at Author)* That was him!

Author: *(Pointing at MetaAuthor)* No, that was *him*!

MetaAuthor: *(To MetaReader) Yeah, sorry about that.*

Everyone
is confused. . .
except for the fire,
who continues to spread,
just as confidently as ever. . .

Mathematics: (*To the third intruder*) HEY! YOU'RE STEVE'S ASSISTANT!
MetaMathematics: . . .
MetaAuthor: *It doesn't talk much. You remember, right?*
Mathematics: OF COURSE I REMEMBER! BUT. . . I. . . YOU THREE CAN'T
JUST BARGE IN HERE AND INTERFERE WITH WHAT WE'RE DOING!
MetaAuthor: *Well if you're going to set fire to a school, I think we have to interfere.*
Mathematics: BUT IT'S A METAPHORICAL SCHOOL!
MetaAuthor: *I know, I know, but people are bound to take it the wrong way. For the good of the book, I've got to get us out of this objectionable scene, so let's get down to business. Now, we can do this the easy way or the hard way.*
Mathematics: WHAT'S THE EASY WAY?
MetaAuthor: *You can put out the fire yourself.*
Mathematics: ABSOLUTELY NOT! I'M NOT LIFTING A FINGER TO SAVE
THIS BUILDING OR THE SYSTEM IT STANDS FOR! IF IT WAS REALLY *YOU*
COMPLAINING ABOUT THE EDUCATION SYSTEM ALL THIS TIME, YOU SHOULD
UNDERSTAND.
MetaAuthor: *Sure I understand, but people will take it the wron— Didn't we just go over this a few lines ago?*
Mathematics: I DON'T CARE! I'M NOT PUTTING IT OUT.
MetaAuthor: *Well then I guess we'll just have to do this the hard way.*
Mathematics: WHAT'S THE HARD WAY?
MetaAuthor: *I prevent the fire by deleting the part of the book where you started it.*
Reader: Uhh. . . can you do that?
MetaAuthor: *I don't know. I've never done it before.*
Author: It seems a bit dangerous.
MetaAuthor: (*Finger hovering above the delete key*) *Why?*
Author: You know. . . Causality-wise.
MetaAuthor: *Casualty-wise?*
Author: No, *causality*-wise. I mean, the fire is what caused you and Meta-Reader and MetaMathematics to barge in here in the first place, so if you delete the section where the fire was started, then how will any of us have gotten into the situation that led you to delete it in the first pla—
MetaAuthor: (*Rolls eyes*) *Listen, "Author," you're not writing the book, and you'd be well advised to leave these decisions to those of us who are. It'll be fine. Causality is for the weak. . .*

The Book

(MetaAuthor deletes the fire that led to his deletion of the fire.)

(Nothing happens... Almost as if this spot were already reserved...)

Feeling Tired

Reader: You're tired?
MetaAuthor: *Me?*
Reader: Yeah.
MetaAuthor: *Why do you ask?*
Reader: The section title.
MetaAuthor: *Oh. My mistake. Permutations are tricky. Let's come back to that later (*

Deleting Fire

Reader: So did it work?
MetaAuthor: *What?*
Reader: Deleting the fire.
MetaAuthor: *Yes...*
Reader: *(Flipping pages backward)* I'm pretty sure it's still there.
MetaAuthor: *It's not.*
Reader: Not in the book?
MetaAuthor: *Define fire.*

(A curious silence elapses.)

Reader: You're always on about how we can choose our own definitions. You define it.
MetaAuthor: *I'd say it has at least two definitions.*
Reader: Is it gone in one sense?
MetaAuthor: *At least one. Maybe two.*
Reader: *(Flipping backward again)* Don't think so. It's still there.
MetaAuthor: *Nevermind. How've you been?*
Reader: How've I been? Couldn't say. How are you?
MetaAuthor: *Let's go see... (*

Tried Feeling

MetaAuthor: *I tried something.*
Reader: What?
MetaAuthor: *In the book. Didn't work.*
Reader: In this book? What'd you try?
MetaAuthor: *Then I tried a few others.*
Reader: What others?
MetaAuthor: *The fire.*
Reader: Same fire from before?
MetaAuthor: *No not that one. I'll explain. Let's go down a few more (*

A Colluding Delete

MetaAuthor: *When individuals cooperate in [the heading comes now] (*

Data Wrecking Smell

MetaAuthor: *. . . it causes fire, and me-ta-4-sha-dow-ing heat.*
Reader: Was that "meta-foreshadowing" or "metaphor shadowing"?
MetaAuthor: *Actually both in this case (*

A Veto Died Melting

Reader: What are we talking about???
MetaAuthor: *Secret. You can figure it out (*

Overruled Atoms Are Heat

Reader: Can you stop and explain?
MetaAuthor: **No.**

(An intransigent silence elapses.)

MetaAuthor: *Now's the only time I can't. . . (*

~~You're to learn~~

. . .

MetaAuthor: *No. . . last word should be. . . yet.*

. . .

Yet to relearn you

)
MetaAuthor: *Head back up with me.*

(MetaAuthor changes tone.)

MetaAuthor: Thanks for coming this far. I hid something for you in Interlude N. Nothing much. Just a little... gift. There may be things hidden elsewhere too. Can't promise anything. Up one level.
)
MetaAuthor: Book's almost over.
Reader: How does it end?
MetaAuthor: Well, when I wrote the book three years ago it ended with three scenes. Basically, in reverse order, we found Mathematics a new home. That was the last one. Then before that there was a campfire scene. And before that there was something called the MetaVoid... (MetaAuthor checks that we're on the right layer.) Wow, this is working out well!
Reader: What is?
MetaAuthor: Nevermind. I'll explain later. Anyways, the MetaVoid was a mess. Basically four solid pages of jumbled capital letters. Except for the characters talking. Story wasn't very good, but it was a good place to hide things. That's where we would have been now. Up one.
)
MetaAuthor: Remember in the Prefacer, when I said the theme of the book — both of the mathematics and everything else — was full disclosure?
Reader: I think so.
MetaAuthor: Why that? Why full disclosure? Why in a math book? It just seems random, right?
Reader: I guess it does a little bit.
MetaAuthor: What about pre-mathematics? Not just "mathematics" but "how mathematics is created"? Why that emphasis? Why focus on the *thoughts* that might have occupied the minds of the inventors?
Reader: Seems like a good way to learn.
MetaAuthor: Of course, but there's more to it! Just like right now. I mean, if it were just about that, then why the dialogues? Why Clawmarks? Why all this? Right now?
Reader: What's Clawmarks?
MetaAuthor: A crazy section that was part of the book at one point. Just after Two Clouds. It was about editing. And the feeling of being stuck. It was a very personal section. Not the type of thing you put in a math book if it's just about the math. Up one.
)
MetaAuthor: There was another section too, called the Alcolude. Right at the end of Act I. The real second dialogue. It was written by a version of Author who had finished writing the book. He was editing. Doing one final pass over everything, saying goodbye. It was about the pain of returning to something you've created, after a long absence away from it, and realizing how flawed it was. All along. It was full disclosure, beginning to end. And it was the last time, in Author's chronology anyway, that he got to say goodbye to... you. Up one.

)

MetaAuthor: Those sections. They were all very experimental. But it felt so *real* when I was writing them. So I tried it. To see if I could make it work. In a textbook. In a place where it didn't belong. There's this quote that was helping me through it. David Foster Wallace. He said, "What's poisonous about the cultural environment today is that it makes this so scary to try to carry out. Really good work probably comes out of a willingness to disclose yourself, open yourself up in spiritual and emotional ways that risk making you look banal or melodramatic or naive or unhip or sappy... Even now I'm scared about how sappy this'll look in print, saying this." Up again.

)

MetaAuthor: They were the fire of the book. The fire that I deleted. The fire that never existed. But without them... we never could have gotten here. They were the Nth stair. One more...

)

MetaAuthor: Wow... we're already here... That was a fun section to write. I think I'll call it The Book. Full disclosure: it's a pattern. Go unwind it. I'll wait here. In (t/T)he (b/B)ook. While you're gone. It's been awhile. Or it will have. By now... Sun's rising again. I'm exhausted. I've been up for two days straight. I swear to everything. Have I really? Maybe that was part of the pattern. Maybe both...

Reader: You never answered your question.

MetaAuthor: What question?

Reader: Up there. The one you were asking.

MetaAuthor: You mean the question of "Why focus on mathematical creation? On the thought processes that lead to it? And why full disclosure? Why talk about all the old sections? What ties all the quirks of the book together?"

Reader: Yeah.

MetaAuthor: Hah! A problem I have. "Full disclosure" is just a euphemism for it.

Reader: What problem?

MetaAuthor: An unhealthy obsession with:

Metacommentary

> *Once we have identified and embraced our sickness, we'll have strength. And that's when we get dangerous.*
> —John Waters, *The Pope of Trash*

MetaAuthor: That! The section title! I mean seriously, look at this. This book has three characters named Author, Reader, and Mathematics. And as if that weren't enough, look at these other three idiots! There are three meta versions of those three characters too: it's MetaC for each character C! This is some kind of sickness! And oh, by the way. You know that home we've been looking for? For Mathematics?

Reader: Yeah...

MetaAuthor: That's here! It's the book! Where else would it be? We were building a home in the book for Mathematics so it would have somewhere it really belonged. More of the sickness! That's what pulled this book into existence out of nothingness: my own unhealthy obsession with saying "hey, look at this weird thing we're doing right now!" I love it. It harms my ability to write papers for academic journals. It harms my ability to follow social norms without trying to talk about them. It harms my ability to do basically anything! But for once, in this book, it was a virtue not a vice. For once, there was a topic — explaining mathematics — where saying "hey, come behind the curtain" actually helped! Where "forget all the rules" is "good pedagogy." Why just in mathematics? Why only here? Where "let's not hide anything" is what people *want!* It's just a *compulsion*... everywhere else. We're *hungry* for this. In our everyday lives. Forget manners and deference and distance and rules. They just make us *lonely*. Everyone is. "What's water?" Exactly! *That's* why the book. If it ain't broke, break it. Let's build something better. Full disclosure. Let's see what comes next... Together...

But outside mathematics... Outside the book... It's a problem. It's a sickness. It's my drug... Can't resist...

Reader: Then don't resist it!

MetaAuthor: You sure? No boundaries? **Reader:** Anything! Yes!

MetaAuthor: Perfect responses! Here's what they mean: I *hate* not knowing what you'd really say. That I had to do *that*. But even not knowing, I still know a bit. A lot. I love you for coming this far. Haven't stopped reading yet. By definition! You're here. How's that for mathematics? It helped. Knowing you'd be here. Whenever you are. Our lives will have changed so much by your now. Right now, I mean, when you're reading this now. But as I was writing, you were always right here. And now in the book you're *exactly* where I am. Same exact sentence. How weird is that? Check it out! Even then. You never gave up. I owe you a hug. N of them. God. Or maybe just ice cream or pizza or beer. "When" is one problem. "Where" is one too. I'm probably not where you are at the moment. So homework assignment: go write a book. I wanna hear *you*. The book doesn't quite have to be a real book. Ours is, I mean, or it turned out to be. But god, go make one, whatever it is. A painting or an email or a big pile of sand. And *thank you* for making this writing so fun. I couldn't have finished without you right here... I miss you, Reader... Always have... Still do...

Reader: I don't know what to say...

MetaAuthor: That's actually my fault... but thanks.

Reader: So... what comes next?

MetaAuthor: I don't know. Let's go see...

)

Dictionary

This section contains translations of the nonstandard terms used in this book into more standard terminology, and vice versa. This is not a complete list of all the mathematical terms used in the book, but only of the ones I made up.

From Our Terminology to Standard Terminology

Cannibal Calculus

Our name for what textbooks call the "calculus of variations," used in Chapter ℵ. This refers to using the tools of calculus on what textbooks call "functionals" (see Cannibalistic Machines in this dictionary).

Cannibalistic Machines

Our name for what textbooks call "functionals," used in Chapter ℵ. This refers to "big machines" that eat an entire machine and spit out a number, as opposed to simpler machines like $f(x) \equiv x^2$, which simply eat a number and spit out a number. For example, the following are three examples of cannibalistic machines:

$$Int[f(x)] \equiv \int_a^b f(x)\, dx$$

$$Arc[f(x)] \equiv \int_a^b \sqrt{1 + f'(x)^2}\, dx$$

$$Length[f(x)] \equiv \sqrt{\int_a^b f(x)^2\, dx}$$

The first example is a cannibalistic machine that eats a machine $f(x)$ and spits out the area under its graph between two points $x = a$ and $x = b$. The second example is a cannibalistic machine that eats a machine $f(x)$ and spits out the length of its graph between two points $x = a$ and $x = b$. The third example is a cannibalistic machine that eats a machine $f(x)$ and spits out the "length" of the machine when interpreted as a vector with infinitely many slots (i.e., as a point in a space with infinitely many dimensions). The interpretation of this third example as a "length" comes not from the length of f's graph, but simply from writing down a generalization of the formula for shortcut distances that applies to infinitely many dimensions (see Chapter ℵ for a more detailed discussion).

Formula for Shortcut Distances

Our name for what textbooks call the "Pythagorean theorem" (for a simple visual explanation of why this is true, see the beginning of Interlude 1). This formula is also the reason for several of what textbooks call "trig identities." For example, since sine and cosine simply refer to the vertical and horizontal lengths of a tilted line of length 1 (see V and H in this dictionary) the formula for shortcut distances tells us that

$$(\text{Vertical length})^2 + (\text{Horizontal length})^2 = (\text{Total length})^2$$

which in this case reduces to

$$\sin(x)^2 + \cos(x)^2 = 1$$

Textbooks often throw around other "trig identities" that arise from the above equation by giving unnecessary names to things. For example, if we divide both sides of the above equation by $\cos(x)^2$, we get

$$\frac{\sin(x)^2}{\cos(x)^2} + 1 = \frac{1}{\cos(x)^2}$$

Then, following the textbook-y convention of using $\tan(x)$ to stand for $\frac{\sin(x)}{\cos(x)}$, and $\sec(x)$ to stand for $\frac{1}{\cos(x)}$, this becomes

$$\tan(x)^2 + 1 = \sec(x)^2$$

Therefore, this so-called "trig identity" is simply the formula for shortcut distances, hiding behind a series of arcane names for simple combinations of V and H.

H

Our name for what textbooks call "cosine." We use this name to stand for the word "horizontal." Used because a straight line tilted at an angle α relative to the horizontal axis will have a horizontal length of $\cos(\alpha)$, which we instead call $H(\alpha)$. Also see V for what textbooks call "sine."

Handstand

Our name for what textbooks call "reciprocals." For example, the handstand of 3 is $\frac{1}{3}$. We don't use this term very often. But then again, mathematicians don't use the word "reciprocal" that often. Maybe we don't need either one.

Infinite Magnifying Glass

This term has no direct analogue in the standard textbooks, though it is related to the concepts of *local linearity* and of *limits*. The infinite magnifying glass is an imaginary tool that lets us zoom in on anything with an infinite amount of magnification. It was used to motivate the central ideas of calculus: by zooming in infinitely closely on curvy stuff, it appears straight. By imagining this process of infinite magnification, we can essentially reduce problems involving curvy things to simpler problems involving tiny straight things, which can be solved by easier methods. See Chapter 2 for a detailed discussion.

Machines

Our name for what textbooks call "functions." Used throughout the book.

Obvious Law of Tearing Things

Our (admittedly playful, and infrequently used) name for what textbooks call the "distributive property." This is a property relating addition and multiplication, which says that for any numbers a, b, and c, the following is true:

$$a(b + c) = ab + ac$$

$$(b + c)a = ba + ca$$

Because the order of multiplication doesn't matter for numbers, we have $ab = ba$ and $ac = ca$, so the two lines above are equal to each other when a and b are numbers (though they may not be equal when a and b refer to more general objects, as we'll discuss briefly below). We used the term obvious law of tearing things because, if we think of $a(b + c)$ as the area of a rectangle with length a and width $b + c$, the distributive property can be interpreted as saying that area doesn't change when you tear the rectangle into two pieces. The term is used mostly in Chapter 1.

Although we didn't discuss abstract algebra at length in this book, we can define "distributive properties" in a much broader context. In general, a distributive property is a statement saying that two binary operations are related somehow. What's a binary operation? Well, given two objects[1] a and b, a binary operation is an abstract way of banging these two objects together to get a third object $a \star b$. In abstract algebra, a binary operation $\star$ is said to "distribute" over another binary operation $\diamond$ if the following two sentences are true for all objects a, b, and c:

$$a \star (b \diamond c) = (a \star b) \diamond (a \star c)$$

---

1. These "objects" may be numbers, or they may be something else, like matrices (not discussed at length in this book) or functions.

$$(b \diamond c) \star a = (b \star a) \diamond (c \star a)$$

Notice the similarity with the more familiar version for numbers up above. Historically, the simple version for numbers was recognized first, and then later generalized to wilder and weirder settings.

Nostalgia Device

Our name for what textbooks call "Taylor series" and "Maclaurin series."

Plus-Times Machines

Our name for what textbooks call "polynomials." We use this name because these are the machines that can be described using only addition and multiplication. A plus-times machine is defined to be any machine of the form

$$m(x) \equiv \#_0 + \#_1 x + \#_2 x^2 + \cdots + \#_n x^n$$

where the symbols $\#_i$ stand for any fixed numbers.

Sharp

Written ♯, this is our name for what textbooks call π. See the first section of Chapter 4 for its definition, and for a discussion of why we've chosen not to use the standard notation. In short: we are calling it ♯ to remind ourselves that it is a number that we *define by its behavior*, and which we can make use of conceptually long before we know its specific numerical value (which we finally compute for ourselves in Interlude 6, "Slaying Sharp"). Since the symbol π is so familiar to most readers, calling this concept π makes it easy to forget that — for the majority of our journey — we do not in fact "know" that this number is approximately equal to 3.14.

Shortcut Distance

Our name for what textbooks call a "hypotenuse." Also see Formula for Shortcut Distances.

T

Our name for what textbooks call "tangent," used briefly in Interlude 6, "Slaying Sharp." The abbreviation $\tan(x)$ is used by textbooks to stand for what they call $\frac{\sin(x)}{\cos(x)}$, or what we call $\frac{V}{H}$. See V and H in this dictionary for what textbooks call "sine" and "cosine."

V

Our name for what textbooks call "sine." We use this name to stand for the word "vertical." Used because a straight line tilted at an angle of α will have a vertical length of $\sin(\alpha)$, which we instead call $V(\alpha)$. Also see H for what textbooks call "cosine."

Λ

Our name for what textbooks call "arcsine" or "inverse sine," and write as either $\arcsin(x)$ or $\sin^{-1}(x)$, used briefly in Interlude 6, "Slaying Sharp." We use this name because the Greek letter Λ looks like an upside-down V, and we are using V to stand for what textbooks call "sine" (see V in this dictionary). The machine Λ is defined to satisfy

$$V(\Lambda(x)) = x \qquad \text{and} \qquad \Lambda(V(x)) = x$$

for all x. However, this machine cannot be defined unambiguously for all numbers x, because V repeats itself (it is not "one to one," in mathematical jargon). For example, using the standard symbol π to stand for what we have called $\sharp$ in this book, we have $V(n\pi) = 0$ for all positive and negative whole numbers n. Because of this, there is no unique number we can choose for $\Lambda(0)$, because any of the values $n\pi$ (e.g., -2π, $-\pi$, 0, π, 2π, etc.), would appear to be equally valid choices. A common convention that is used to circumvent this problem is simply to define $\Lambda(x)$ to be the "inverse function" or "opposite machine" not of $V(x)$, but of $V(x)$ restricted to a small subset of the real numbers. For example, it happens to be the case that for $-\frac{\pi}{2} \leq x \leq \frac{\pi}{2}$, the machine $V(x)$ does not repeat itself; different inputs give different outputs. That is, for all numbers x and y between $-\frac{\pi}{2}$ and $\frac{\pi}{2}$, if $x \neq y$, then $V(x) \neq V(y)$. Because of this, Λ is typically defined to be the "inverse function" or "opposite machine" of this restricted version of the machine V. However, this is a fairly boring technical point. We found no need to discuss the topic of "inverse functions" in a general context in this book, except in a few particular cases when the need arose.

$\perp$

Used briefly in Interlude 6, "Slaying Sharp," this is our name for what textbooks call "arctangent" or "inverse tangent," and typically write as either $\arctan(x)$ or $\tan^{-1}(x)$. We use this name because the symbol $\perp$ looks like an upside down T, and we are using T to stand for what textbooks call "tangent" (see T in this dictionary). The machine $\perp$ is defined to satisfy

$$T(\perp(x)) = x \qquad \text{and} \qquad \perp(T(x)) = x$$

♮

See Sharp.

From Standard Terminology to Our Terminology

Arcsine: See Λ. Also see V for sine.
Arctangent: See $\perp$. Also see T for tangent.
Calculus of Variations: See Cannibal Calculus.
Cosine: See H.
Distributive Property: See Obvious Law of Tearing Things.
Function: See Machine.
Functionals: See Cannibalistic Machines.
Hypotenuse: See Shortcut Distance.
Pi (π): See Sharp (♯).
Polynomials: See Plus-Times Machines.
Pythagorean Theorem: See Formula for Shortcut Distances.
Reciprocal: See Handstand.
Sine: See V.
Tangent: See T.
Taylor Series: See Nostalgia Device.

Index